Natural Fibre Composites from Agricultural Crops

Natural Fibre Composites from Agricultural Crops

Edited by

Mohd Sapuan Salit,
Teuku Rihayat,
Januar Parlaungan Siregar
and Agung Efriyo Hadi

Cambridge Scholars Publishing

Natural Fibre Composites from Agricultural Crops

Edited by Mohd Sapuan Salit, Teuku Rihayat, Januar Parlaungan Siregar and Agung Efriyo Hadi

This book first published 2025

Cambridge Scholars Publishing

Lady Stephenson Library, Newcastle upon Tyne, NE6 2PA, UK

British Library Cataloguing in Publication Data
A catalogue record for this book is available from the British Library

ISBN: 978-1-0364-4865-3
ISBN (Ebook): 978-1-0364-4866-0

TABLE OF CONTENTS

CHAPTER I

HYBRID NATURAL FIBER COMPOSITES IN CHEMICAL ENGINEERING APPLICATIONS

AIDA SAFITRI, SURYANI, TEUKU RIHAYAT, WASLIATUL HASANAH

Abstract

Incorporating natural fibers into composite materials offers a harmonious synergy between renewable resources and engineering innovation. Hybridization, involving the combination of various natural fibers, allows for tailoring properties to meet specific performance requirements. This abstract delves into the pivotal role of hybridization, highlighting the enhanced mechanical, thermal, and sustainability attributes it imparts to composite materials. These materials work well in a wide range of applications, from structural reinforcement to filtration and separation. We look at case studies that demonstrate the effectiveness of hybrid natural fiber composites in improving equipment durability, process efficiency, and environmental impact reduction in chemical engineering applications. The interfacial compatibility of different natural fibers and polymer matrices is also highlighted in this abstract. Strategies to optimize interfacial adhesion, such as surface modification techniques, are discussed in the context of chemical engineering applications. This exploration highlights the importance of achieving effective bonding between the diverse components within hybrid composites.

1. Introduction

One of the most notable applications of natural fiber composites in chemical engineering lies in enhancing the sustainability and durability of infrastructure within chemical plants. Utilizing natural fibers such as jute, flax, and sisal as reinforcements in polymer matrices has shown remarkable promise in fabricating corrosion-resistant pipes, tanks, and vessels. Fiber-reinforced composites have been strengthened for many years by the research community. Several researchers proposed various methods for strengthening and minimizing the brittleness of these materials [1]. Fiber-reinforced composite matrix toughness, which significantly affects matrix-dominated composite properties, can be increased through various methods. However, the correlation between structural properties and failure resilience has grown to uncover a new toughness mechanism for biological composites [2-3]. The two kinds of fibers are synthetic and natural. In the past twenty years, synthetic fibers like glass and carbon have become more and more popular [4]. Composites made of synthetic fibers have a high strength-to-weight ratio and are frequently used in automotive and aerospace applications. When it comes to waterproofing, stretching, and stain resistance, synthetic fibers are more convenient. On the other hand, natural fibers are becoming more and more well-liked in research because of their resource sustainability, biodegradability, affordability, and light weight [5–6].

In many industrial sectors, natural fibers, particularly plant fibers, are replacing synthetic glass fibers as reinforcement or composite materials, in order to reduce environmental effects and thereby support long-term growth and a circular economy [7-9]. The production of these eco-friendly materials, however, faces technical and technological difficulties.

Natural fibers are gradually taking the place of synthetic fibrers in engineering applications. By reducing the use of synthetic fibers in terms of weight reduction, affordable material costs, and renewability, this replacement helps to preserve the environment. Despite their poor mechanical properties, natural fibers perform well in impact tests. As a result, high strength with higher mechanical properties is required for advanced applications. Hybridization may be a viable option for improving mechanical properties in such cases [10-12]. There are some restrictions in nonstructural applications, though, like a high moisture content and weaker strength and stiffness. As a solution, fiber hybridization with chemical modifications has been recommended. Another type of hydrophobic fiber can be added to hydrophilic fiber composites as an effective method for improving moisture resistance and reducing the deterioration of natural fiber polymer composite properties [13–14].

Figure 1. Natural fibers are widely used as fillers

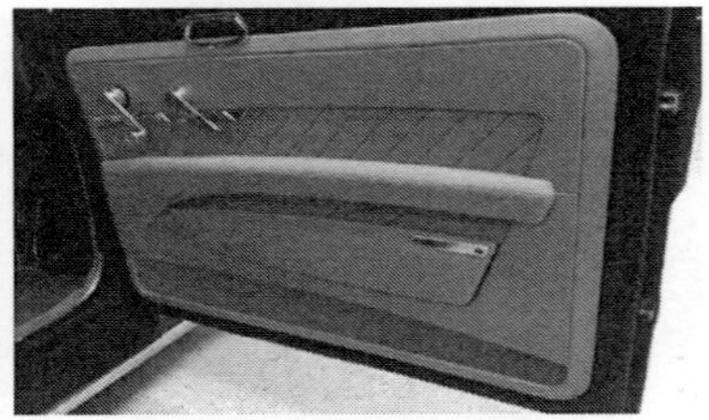

Figure 2. The use of natural fiber in the production of daily needs

The most important parameter influencing the machinability of synthetic composites was discovered to be the direction of fiber alignment in the composite [15]. The cutting force is usually affected by the fiber's orientation in relation to the cutting direction. The chip is formed in front of the cutting tool edge by cutting the matrix along the fiber/matrix interface, depending on the fiber orientation [16]. Fiber alignment, referring to the orientation of reinforcing fibers within a composite material, has been found to exert a profound influence on the mechanical and thermal properties of the material. This alignment, typically categorized as unidirectional, woven, or randomly oriented, significantly affects the anisotropic behavior of composites, resulting in varied responses to external forces. Importantly, the same fiber alignment that impacts mechanical properties also extends its influence to machinability, dictating the ease with which the composite can be machined without causing undue damage to the cutting tool, surface quality, and overall structural integrity [17]. The machinability of a material is governed by a complex interplay of factors, including cutting forces, heat generation, chip formation, and tool wear. In synthetic composites, the direction of fiber alignment dictates the interactions between the cutting tool and the reinforcing fibers. When machining along the direction of fiber alignment (parallel to fibers), the cutting tool encounters a continuous sequence of fibers, leading to higher cutting forces and increased tool wear. Conversely, machining across the direction of fiber alignment (perpendicular to fibers) involves intermittent fiber cutting, resulting in reduced cutting forces and enhanced machinability [18].

One of the primary concerns when machining composite materials is the potential for delamination – the separation of layers within the material. The direction of fiber alignment plays a pivotal role in mitigating or exacerbating delamination. Machining parallel to fiber alignment often leads to higher delamination propensity due to the continuous cutting of fibers, potentially compromising the structural integrity of the component. In contrast, machining perpendicular to fiber alignment minimizes delamination risks by promoting fiber deflection and minimizing fiber pullout [19-20].

Researchers have directed their attention towards examining the influence and resilience of composite structures in the past few decades, uncovering numerous design imperfections. These complexities arise due to the fact that outcomes are influenced not solely by material parameters, but also by the configuration of tests and samples. Moreover, damage manifests in diverse manners, encompassing phenomena like the separation of fibers from the matrix, surface micro buckling, matrix fissuring, and fiber rupture. Grasping the response to impact, the energy absorbed upon impact, and the strength exhibited during impact prior to material failure holds paramount importance in formulating a pragmatic design for structures. The impact test serves as a technique for deciphering the notch sensitivity and fracture toughness of a composite material. An essential determinant influencing the behavior of an impact within a composite resides in the fracture toughness of the resin system. Understanding the impact of fiber alignment direction on machinability is of paramount importance for industries relying on composite materials. Manufacturers can leverage this knowledge to optimize machining strategies by selecting the most suitable cutting direction based on the intended application and performance requirements. For instance, components subjected to predominantly axial loading may benefit from machining perpendicular to fiber alignment to enhance machinability while preserving structural integrity.

2. Overview of Natural Fiber-reinforced Hybrid Polymer Composite

Hybrid composites are developed through the integration of multiple reinforcing materials in the same polymer matrix in order to enhance the composite's properties. When several strengthening agents are combined, they can have a synergistic or antagonistic effect. Furthermore, properties can be adjusted by balancing the advantages of one material against the disadvantages of another. The incorporation of multiple reinforcements in a matrix provides a broader range of properties that single fiber-reinforced composites cannot accomplish. As a result, scientists and researchers have given close attention to the hybridization of reinforcement materials of natural and synthetic origin.

Since synthetic fiber composites have been demonstrated to be of a higher quality than natural fibers, their use in the manufacturing and production sectors is unavoidable. Synthetic fibres are used to create high-performance polymer matrix composite products like building panels, fiber-reinforced plastic tanks, and aircraft and vehicle parts. Some of the most popular synthetic fiberres in the composites industry include glass, carbon, and aramid.

Due to their outstanding mechanical performance, synthetic fiber-reinforced polymer composites with glass, carbon, aramid, and boron fibers have been used extensively and frequently in engineering applications [21]. Due to their poor non-biodegradable and recyclable qualities, synthetic

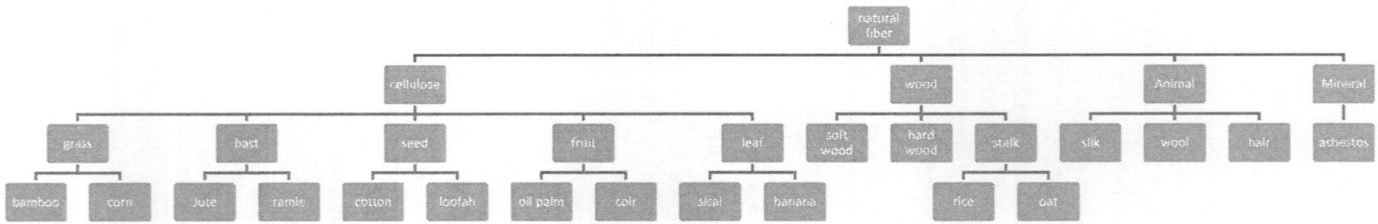

Figure 3. Natural fiber classification

fiber-reinforced composites have raised environmental concerns. Because of their advantages for the environment, such as their use of natural resources, low pollutant emissions, improved energy recovery, biodegradability, and high specific stiffness and strength, natural fibers are increasingly being used as an alternative to synthetic fibers in composites. Natural fiber composites' recyclability is an additional benefit that raises the sustainability aspects of these materials.

Cellulose is the primary constituent of natural fibers and is composed of long chains of glucose molecules. These chains are organized into microfibrils, which form the fundamental structural framework of the fiber. The arrangement and orientation of cellulose microfibrils significantly influence the fiber's strength, stiffness, and overall mechanical properties [22]. Alongside cellulose, hemicellulose is another polysaccharide present in natural fibers. Hemicellulose molecules surround the cellulose microfibrils, acting as a matrix that holds the microfibrils together. This matrix contributes to the fiber's flexibility, water absorption characteristics, and interactions with other materials.

Lignin is a complex and amorphous polymer that fills the spaces between cellulose microfibrils, providing rigidity and structural support to the fiber. However, it can also make fibers less accessible to chemical treatments. Lignin content can vary between different natural fibers, influencing their strength, stiffness, and resistance to degradation. Pectin, another complex carbohydrate, is found in the middle lamella that holds individual plant cells together. Proteins are present in minor quantities and contribute to the fiber's mechanical properties. These components can impact fiber-matrix interactions in composite materials.

Natural fibers possess a lumen, which is the central open space within the fiber, and a cell wall surrounding it. The cell wall consists of the primary cell wall and the secondary cell wall. The secondary cell wall, which is more rigid and dense, develops after the primary cell wall and significantly contributes to the fiber's mechanical properties. Within the macroscopic structure of natural fibers, microstructures such as pits, microfibril angles, and crystalline regions exist. These microstructures influence properties like fiber flexibility, mechanical strength, and the degree of water absorption.

Bark-derived fiber has one of the highest cellulose contents (30–76% by weight) and lowest microfibril angles (2–80) of the different natural fiber sources. Other natural fiber components like pectin and hemicellulose regulate additional characteristics like water absorption, wet strength, swelling, and fiber bundle integration. To achieve the ideal reinforcement in natural fiber-reinforced composites, complete characterization of natural fibers thus becomes critical.

Within composites, the matrix assumes a pivotal role as the primary component safeguarding fibers or fillers against abrasion and challenging environmental conditions. Additionally, it serves to bind these fibers or fillers together, endowing the composite with heightened strength by absorbing energy amidst stress-induced deformation. The transmission of a load onto the composite material becomes facilitated through the matrix when subjected to external forces. To this end, factors such as type, attributes, and processing methodologies demand careful consideration. Consequently, the selection of a matrix capable of enduring greater fracture limits than the reinforced fibers becomes imperative. Two primary classifications of polymer resins function as matrices: thermosets and thermoplastics. Thermosets undergo curing at elevated temperatures, acquiring an irreversible solidified state. This transformation precludes their reversion to uncured forms. Thanks to intricate cross-linking catalyzed reactions, thermosets exhibit robustness and stability even under adverse circumstances. On the contrary, thermoplastics remain pliable, lacking intermolecular cross-linkages and exhibiting a propensity to melt at elevated temperatures. The viscosity of thermoplastics and their requirement for high processing temperatures restrict their practical applicability relative to thermosets. Commonly encountered thermosets encompass epoxy, vinyl ester, phenolic resin, polyester, and urethane.

Certain natural fibers like wood, jute, kenaf, rice husk, and hemp find application in bolstering the strength of thermosetting polymers such as epoxy, urethane, vinyl ester, phenolic, polyester, polymide, and polyurethane (PU), as well as thermoplastics including polyethylene (PE), polypropylene (PP), nylon, polycarbonate (PC), polyvinyl chloride (PVC), polyether-ether ketone (PEK), and acrylonitrile-butadiene-styrene (ABS) polymers and elastomers. In contrast to thermosets and elastomers with intricate cross-linked structures, thermoplastics offer enhanced design adaptability and simplified processing methodologies. Nevertheless, limited research has been conducted regarding the incorporation of agricultural residues as reinforcements in these polymer matrices, encompassing materials like soybean husks, pineapple fibers, and banana leaf fibers.

Table 1. The hybrid effect was investigated for a number of different reinforcing materials and matrices.

Matriks	Reinforcement	Tensile strength (Mpa)	Referencee
Epoxy	Flax/e-glass	164.23	24
Epoxy	Powder eggshell	73.83	25
HDPE	Palm oil shell	1.17	26
Epoxy	Jute/ carbon nano tube	36.00	27
Epoxy	Jute/tea leaf fiber	75.60	28
Corn starch	cornhusk fiber/sugar palm fiber	19.09	29
Epoxy	coir /woven-carbon fiber	354.51	30
polyester	Coconut coir/sugarcane leaf sheath	16.88	31
Cardanol resin	bagasse/coconut shell	37.28	32
Epoxy	banana bract/palm fiber	40000	33

Matrices necessitating high processing temperatures prove unsuitable for producing composites reinforced with natural fibers due to the inherent instability of most natural fibers at temperatures exceeding 200°C.

Numerous different matrixes and reinforcing materials were examined in order to determine the hybrid effect. The highest tensile strength values were produced by the mechanical properties of the E-glass/flax mixture. Glass fibers have an amorphous atomic structure connected by ionic bonds. The tensile strength of composite materials is greatly affected by the interfacial bond quality. Flax fiber's hydrophilic properties, which result in a weak interfacial area and reduce stress transfer between the fiber and the epoxy, have an impact on its tensile strength. This is because more fiber interactions result in ten times higher tensile strength for the flax/E-glass hybrid composite.

Studies by Alfatah et al. [26] demonstrated an undesirable connection at the interface between the filler and matrix when employing a solitary filler, leading to a reduction in tensile strength. This effect transpired due to the absence of a coupling agent that would have otherwise modified the surface particles of palm shells. Consequently, the lack of this agent resulted in diminished tensile strength, flexibility, and water absorption. Furthermore, natural fibers display a heightened sensitivity to elevated processing temperatures compared to glass fibers. Through meticulous process optimization and parameter adjustment, it becomes feasible to compound polyamides at temperatures exceeding 200°C. Additionally, cellulose fibers are lighter and exhibit lower tensile values than glass fibers due to their lower density.

The inherent mechanical properties and other properties of natural fibers, which are listed in Table 1, determine the processing and properties of natural fiber hybrid composites. The tensile properties of fibers vary greatly. Many of these NFRC properties, such as the fiber-matrix bond, interaction, thermal stability, moisture sensitivity, and so on, are directly influenced by some of these traits. Natural fiber and matrix hydrophilicity and hydrophobicity, for example, threatens fiber-matrix compatibility and can reduce the mechanical performance of these composites. As a result, a plethora of techniques for chemically and physically altering fibers, as well as adding bridging materials during processing, have emerged.

3. Detailed explanation of each manufacturing technique's steps, requirements, and outcomes

3.1 Alkali Treatment

Alkali treatment involves the immersion of natural fibers, such as jute, flax, or hemp, in alkali solutions, typically sodium hydroxide (NaOH) or potassium hydroxide (KOH). This treatment induces a series of chemical reactions, including deacetylation, mercerization, and increased surface roughness. Deacetylation removes acetyl groups from the fiber surface, exposing hydroxyl groups that enhance fiber-polymer interactions. Mercerization, on the other hand, leads to fiber swelling and alignment of cellulose molecules, contributing to improved fiber crystallinity and mechanical properties.

The chemical modification brought about by alkali treatment imparts several improvements to natural fibers. Enhanced hydrophilicity, increased surface area, and improved fiber-matrix adhesion result from the removal of impurities and the exposure of active functional groups. These changes lead to enhanced wettability and interfacial adhesion, critical factors in optimizing composite properties such as tensile strength, modulus, and impact resistance. Additionally, the increased roughness and swelling influence moisture absorption behavior, influencing dimensional stability [34]. This treatment removes some of the lignin, wax, and oil from the fiber cell wall's outer surface, depolymerizes the cellulose, and exposes short-sized crystallites [35]. The addition of sodium hydroxide (NaOH) solution to natural fibers increases hydroxyl group ionisation into alkoxides. The inclusion of sodium oxide (NaOH) in the fiber with surface modification facilitates the ionization of alcohol bodies by hydroxyl, which can be expressed chemically as follows:

$$\text{Fiber} - \text{OH} + \text{NaOH} \rightarrow \text{Fiber} - \text{O} - \text{Na} + \text{H}_2\text{O}$$

Successful alkali treatment necessitates careful control of treatment parameters such as concentration, treatment duration, and temperature. The extent of fiber modification depends on these parameters, with higher concentrations and longer durations generally leading to increased modifications. However, a balance must be struck, as excessive treatment can degrade fibers, resulting in reduced mechanical properties. Optimal conditions vary based on fiber type and intended application. The impact of alkali treatment on natural fibers extends to their applications in composite materials. Chemically-modified fibers offer improved compatibility with polymer matrices, resulting in composite materials with enhanced mechanical, thermal, and barrier properties. In sectors ranging from automotive to construction, the incorporation of alkali-treated natural fibers has led to lightweight, environmentally-friendly composites that find applications as structural components, packaging materials, and even reinforcement in concrete.

3.2 Silane treatment

Silane treatment involves the chemical bonding of organosilane compounds to the surface of natural fibers. The functional groups of silane molecules form covalent bonds with hydroxyl groups on the fiber surface, thereby creating a strong and durable interface. The resulting fiber-matrix adhesion significantly improves composite properties by reducing interfacial stress, enhancing load transfer, and mitigating moisture-induced degradation [36]. The primary advantage of silane treatment lies in its capability to enhance fiber-matrix adhesion. By effectively bridging the hydrophilic fiber surface with the hydrophobic polymer matrix, silane-treated natural fibers exhibit improved compatibility and wettability. This not only bolsters mechanical properties such as tensile strength and modulus but also leads to a reduction in moisture absorption and increased resistance to environmental degradation.

Silane treatment offers a versatile platform for tailoring composite properties based on the choice of organosilane compound. By selecting specific functional groups, researchers can tailor the surface chemistry of natural fibers to match the characteristics of diverse polymer matrices. This tailoring allows for optimization of composite properties, making silane treatment a versatile tool for achieving desired outcomes in various applications [37-38]. It is well known that silk fibers have a central hollow region, the lumen, which allows for water to penetrate via capillarity, especially when the composite contains a high fiber content. However, the silan-hydroxyl group linkages are also susceptible to hydrolysis, and the breakdown under certain conditions must be investigated.

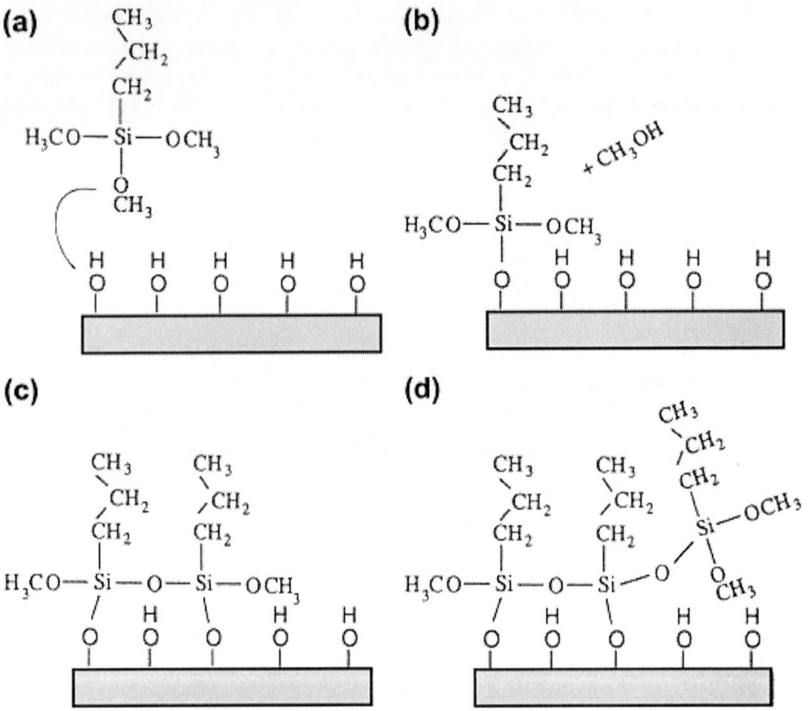

Figure 4. Silane surface modification process: (a) The hydroxylated surface is immersed within a solution containing n-propyl trimethoxysilane (nPTMS); (b) One of the methoxy groups within nPTMS forms a pair with a hydroxyl group, resulting in the release of methanol; (c) Two methoxy groups on a different nPTMS molecule undergo reaction, with one attaching to the hydroxyl group and the other binding to the methoxy group of the first nPTMS molecule; (d) The third nPTMS molecule solely reacts with methoxy groups. These molecules establish a connection not directly with the surface but rather with the silane film network.

Considering this knowledge, an optional strategy is activating the alkoxysilane by hydrolyzing the alkoxy groups and forming the more reactive silanol groups. To form a macromolecular interaction, silanol may interact with the hydroxyl groups of fibers or condense on the surfaces of fibers and/or in cell walls. Under heating conditions, the interfacial adhesion of treated fibers and polymer matrices becomes better, as are the properties of the resulting composites, by blocking the hydroxyl groups, which is reversible to hydrolysis, and by the formation of the macromolecular network, which is permanent. Water must be present during the hydrolysis of the alkoxy groups.

3.3 Acetylation

The reaction of organic compounds with the acetyl ion CH3COO- has been described by the surface modification of acetylation. The esterification method evolved from the plasticization of cellulose fibers. This process is additionally referred to as fiber acetylation. Ethanoic anhydride reacts in the hydroxyl groups of cellulose in the reaction of ethanoic anhydride and lingo-cellulosic materials [39].

Fiber – OH – CH3COOCOCH3 → Fiber – OCOCH3 – CH3COOH

The polar hydroxyl groups in the fiber's cell wall are encouraged to be replaced by the less polar acetyl groups as a result of this chemical change [40]. As a result, the hydrophilic properties of the fibers are reduced, which enhances their compatibility with nonpolar matrices. With acetylation treatment, natural fiber moisture absorption can be reduced, and the composite's mechanical properties, such as interfacial compression and tensile and flexural strength, can be enhanced. Sometimes an alkaline pretreatment is followed by acetylation for better results [41–42].

3.4 Treatment with Benzoyl chloride

In the benzoylation process, benzoyl chloride is used to decrease the hydrophobicity of the fibers and increase the adhesion of the fibers to the matrix, increasing the strength of the composite. In the benzoylation procedure, the fibers' hydroxyl groups were first activated by an alkaline pre-treatment before being immersed in a benzoyl chloride solution for 15 minutes [43]. Benzoyl chloride that had adhered to the surface of the fiber was removed with an ethanol solution, then the material was washed with water and dried in the oven. Sisal fiber composites that had undergone alkali pre-treatment demonstrated greater thermal stability after being treated with benzoyl chloride.

In the synthesis of organic compounds, benzoylation is a crucial transformation. The fiber treatments have been where benzoyl chloride is most frequently used. A good example of this is benzoyl (C6H5C=O), which is linked to a decrease in the hydrophilic nature of the treated fiber and an increase in interaction with the hydrophobic matrix [44]. The treated composite was found to have greater thermal stability than the untreated fiber composite [45].

$$\text{Fibre-OH} + \text{NaOH} \longrightarrow \text{Fibre-O}^-\text{Na}^+ + H_2O$$

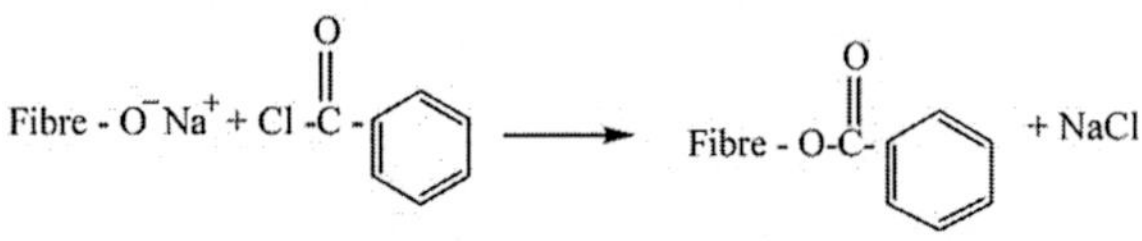

The interfacial adhesion of the hemp fiber and the PE matrix was found to be improved by this treatment. The cellulose and lignin hydroxyl groups in the fiber are first activated by an alkaline treatment, and then the fiber is suspended in a 10% NaOH solution and benzoyl chloride for 15 minutes [46]. The isolated fibers were then washed with water and dried in an oven at 80 °C for 24 hours after being soaked in ethanol for an hour to remove the benzoyl chloride. The treatment was responsible for the treated fiber's reduced hydrophilicity.

4. Cotton fibers composite

The most significant natural fiber used in textile production is cotton, and this can be discovered in other products like fine paper [47]. Cotton has undergone extensive study due to its value. However, there is still a lot to learn about the intricate details of cotton fiber structure as of the time of writing. Understanding the relation between the structure and the performance characteristics of the fibers requires knowledge of these structural specifics. It's necessary for knowledge-based development.

Figure 5. Cotton plants that grow in Indonesia

The cotton plant is cultivated for fiber in more than 90 countries. The Republic of Indonesia's Ministry of Agriculture said national cotton production was 127 tons with a planting area of 703 ha in 2020. This number decreased by 54.6% from the previous year which was 280 tons with a planting area of 1,620 ha. East Nusa Tenggara is the largest cotton-producing province with production reaching 70 tons with a planting area of 147 ha in 2020. Then following that, Central Java province produces 21 tons of cotton with a planting area of 131 ha and East Java 15 tons.

Each cotton fiber is the result of the growth of epidermal cells on the surface of the cottonseed, inside the fruit of the cotton plant [48]. Cotton fiber processing includes unit operations from harvesting the boll in the field to separating the fiber from the seeds at the milking site and finally turning it into yarn or other engineered products. Throughout this processing stage, the different parts of the machine pull, drag, catch, and rub the fibers against another surface or against each other until they are aligned into the desired shape.

Mature cotton fibers contain average (absolute dry matter percentage) cellulose, 94.0; waxy substance, 0.6; pectin (calculated as pectic acid), 0.9; organic acids, 0.8; nitrogenous substance (calculated as protein), 1.3; ash, 1.2; non-cellulosic polysaccharides, 0.3; and, unknown substance, 0.9 (total, 100%) [49]. Cotton fiber, a highly elongated and thickened single cell of the seed epidermis, is the most important natural textile fiber in the world [50]. Cotton fiber begins with swelling above the surface of the ovule near the day of the flower opening (or flower opening).

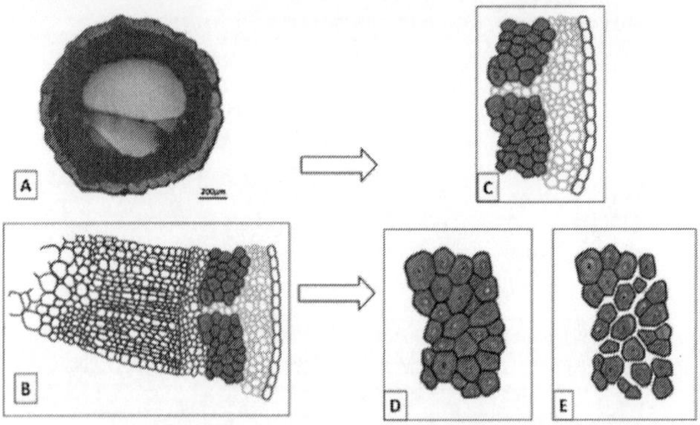

Figure 6. Fibers' location in stem

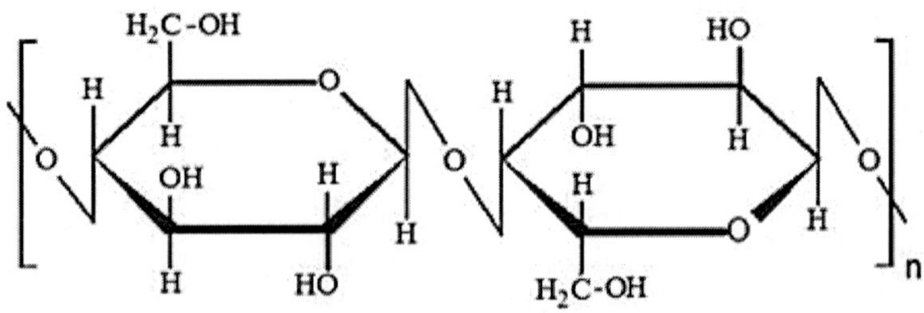

Figure 7. Chemical structure of cotton fiber

Since there is increased turgor pressure on some epidermal cells, the fibers' initial enlargement is diametrical, which likely involves loosening of the cell walls. The cotton fibers go through the following processes over the next 50 days: (a) extreme elongation (up to >2.5 cm) caused by primary wall synthesis; (b) transition wall expanding and primary wall remodeling; and (c) secondary wall expanding because of accumulation of nearly pure cellulose. Before the boll opens to reveal the fine fibers inside, a mysterious process of "ripening" and cell death takes place. The cell wall of cotton fiber determines parameters of industrial importance for fiber quality in addition to morphogenesis. For spinning yarn, the fiber length created by primary wall synthesis is crucial. Layers of thin "winding" cell walls and cellulose secondary walls, which also enable the fibers to dye deeply and absorb water, contribute to the fiber's strength. The grade of intermediary needed is determined by the ratio of cellulose secondary wall to fiber perimeter. However, cotton fiber has a number of shortcomings that require an effort to change its properties depending on the intended use. Certain limitations are associated with cotton fibers:

- Considerable water absorption, contingent on the fiber type, results in expansion, thereby influencing the product's performance and reliability.
- Inadequate bonding at the fiber-matrix interface and reduced mechanical properties often arise due to incongruous water absorption (where fibers are hydrophilic and matrices hydrophobic).
- Incompatibility exists between several polymer matrices and specific fibers.
- The substantial variability in cotton fiber characteristics and properties stems from the extensive diversity in fiber attributes.
- Susceptibility to fungal infestations and insect attacks.
- Flammability is a concern, and the degradation temperature remains relatively low. Natural fiber quality and attributes can substantially fluctuate based on factors such as weather conditions and agricultural practices (including harvest season, soil quality, climate, fertilization, location, etc.).
- Thermal conductivity in Natural Fiber-reinforced Composites (NFRCs) is comparatively lower in contrast to glass fiber-reinforced composites.
- The supply chain is intricate, marked by geographical constraints on availability.
- High water absorption (depending on the type of fiber) leads to swelling, which affects the performance and dependability of the product.

It represents a large proportion of natural fibrous materials based on cellulose (the chemical structure representing repeating units of cellulose fibers), and is highly flammable. A typical LOI value of cellulosic materials is centered around 20, the actual value is highly dependent on the morphology and other structural features of the constituent fibers as well as on the final shape and density of the fabric itself. Cotton is widely used to produce textiles suitable for domestic applications such as clothing, bedding, furniture, curtains, etc. However, cotton is also known as one of the most flammable textile materials with an LOI value of up to 18.4. It is very important for public safety to make this material non-flammable. Here are some treatments that can be done to modify the fiber properties, including:

Cotton fibers have been employed as reinforcement in diverse organic matrices like polyester, vinyl ester, and epoxy matrices. However, despite the benefits they offer, research exploring the utilization of natural fibers for enhancing inorganic matrices is currently limited. Although the authors of a prior study investigated the mechanical properties of geopolymer composites reinforced with short cotton fibers, the available literature is sparse in this domain. The findings from this study revealed that a further increase in short cotton fiber content beyond 0.5 wt% resulted in diminished mechanical properties. This decrease was attributed to challenges in workability, leading to the formation of voids and clumping of fibers. The impact of various parameters related to cotton fiber, such as length, weight, volume fraction, and other influencing factors, on the mechanical attributes of different thermosetting polymer composites, is a topic of significant concern. In general, sisal fiber assumes diverse configurations, including unidirectional short fibers, random orientations, varied weaves (plain, twill, and woven patterns), and distinct ply fiber orientations such as 0, 90, and 45 degrees.

Figure 8. Household needs product using cotton as a raw material

Due to increasing interest in environmental aspects, eco-friendly fibers with high specific strength and low cost are widely used to reinforce the matrix, which in turn is used to transfer loads to the reinforcement and to protect reinforcement from environmental degradation. The characterization technique of natural fiber-reinforced thermoset composites makes it possible to determine the mechanical performance of the resulting composites and it is primordial to assess their quality and select the appropriate application for using them. For example, the degree of vacancies in a composite can affect the quality of the biocomposite material and its mechanical properties.

5. Exploring Environmental Advantages of Natural Fiber Composites in Chemical Engineering Applications

The use of natural fiber composites can potentially lead to reduced energy consumption during manufacturing. These composites typically require less energy-intensive processing compared to their synthetic counterparts, translating to lowered greenhouse gas emissions. By examining empirical evidence, we unveil the correlation between energy-efficient production processes and the application of natural fiber composites in chemical engineering. Compared to many synthetic fibers, natural fibers have the advantage of lower melting or degradation temperatures. This property allows for processing at lower temperatures during composite fabrication. Traditional manufacturing methods for synthetic composites often require higher temperatures, which in turn demand more energy for heating and cooling. Natural fiber composites can be produced at temperatures that are not only energy-efficient but also mitigate the risk of thermal degradation.

The manufacturing process for natural fiber composites is often simpler compared to their synthetic counterparts. This streamlined approach can result in fewer processing steps, reducing the overall energy input required. Additionally, some natural fibers possess inherent self-binding properties, eliminating the need for additional binding agents or energy-intensive chemical treatments. Natural fibers tend to have lower stiffness and strength compared to synthetic alternatives, which can be advantageous during processing. The reduced stiffness often necessitates less mechanical energy during cutting, shaping, and forming operations. This translates to decreased energy consumption when transforming raw fibers into composite components.

Chemical engineering principles are employed in the formulation and processing of the matrix material. This involves mixing, melting, or curing the polymer to achieve the desired properties such as viscosity, reactivity, and thermal stability. Incorporation of natural fibers into the polymer matrix marks a crucial step. Various techniques are employed, including hand lay-up, compression molding, injection molding, and even advanced methods like resin-transfer molding or pultrusion. These methods involve arranging the fibers in a desired orientation within the matrix and subsequently applying heat and pressure to consolidate the materials into a composite structure.

The trend towards green and sustainable manufacturing practices aligns well with the use of natural fiber composites. Many manufacturing techniques for natural fibers, such as pulping, carding, and weaving, are inherently less energy-intensive than those required for synthetic fibers. Integrating these processes with composite production further contributes to energy reduction. Natural fibers are often sourced locally, reducing the need for extensive transportation. In contrast, synthetic fibers may need to be transported from distant manufacturing facilities, increasing energy consumption due to transportation logistics. The localized sourcing of natural fibers minimizes transportation-related energy expenditure. The lightweight nature of natural fiber composites also has implications for energy consumption beyond manufacturing. The use of these composites in various applications, such as automotive and aerospace, leads to reduced energy consumption during the operational phase due to the lower weight of the end products. This effect contributes to overall energy efficiency throughout the product lifecycle.

The manufactured composite undergoes rigorous quality control and testing procedures. These tests may include mechanical testing (tensile, compression, flexural), thermal analysis, microscopy, and more. These evaluations ensure that the composite meets required performance standards and addresses any potential defects or inconsistencies. The manufacturing process for natural fiber composites in the chemical engineering sector embodies a synergy between materials science and engineering principles. It encompasses the selection of materials, surface modification, polymer processing, composite fabrication, curing, and quality assurance. The resultant natural fiber composites exhibit a unique balance of properties that make them suitable for a wide range of applications across various industries.

6. Future outlook of natural fiber composite

The cultivation of hemp, rice husks, bamboo, coconut, plantain, pineapple, sisal, and wheat husks is widespread across the globe. However, the by-products generated from these sources often go untapped, failing to be recycled into derivative products that possess greater value and utility. This could be attributed to the labor-intensive nature of natural fiber manufacturing processes, while the production of synthetic fibers demands substantial financial resources. As a result, entrepreneurs are actively evaluating the potential of engaging in such ventures. Conversely, when considering the forthcoming five years, the prospects for the natural fiber market appear promising. As per the compound annual growth rate (CAGR), the natural fiber market is projected to achieve a valuation of USD million by 2022 and further elevate to USD million by 2028. This study offers an extensive and comprehensive analysis of the global natural fiber market, achieving this by systematically delineating the current state and trends of the market, conducting a thorough examination of the competitive landscape involving key players, and providing an intricate breakdown of market segments categorized by type, application, and region.

Regenerate. This report's research period runs from 2016 to 2027 in several aspect such as: most popular fiber type (Figure 9), future product design (Figure 10), resin type (Figure 11), and profits earned as well as profits expected in the future (Figure 12).

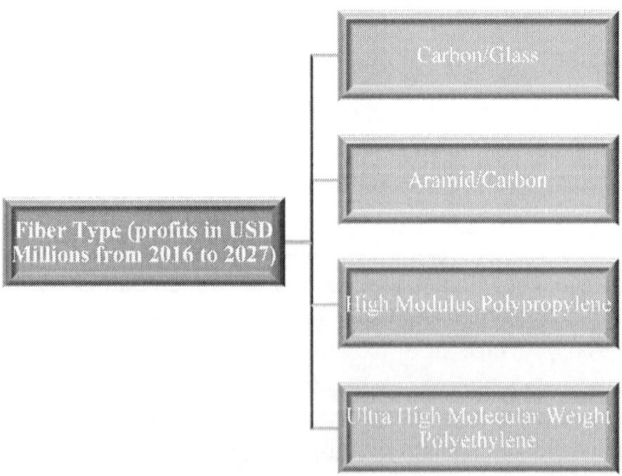

Figure 9. Most popular synthetic fiber and matriks

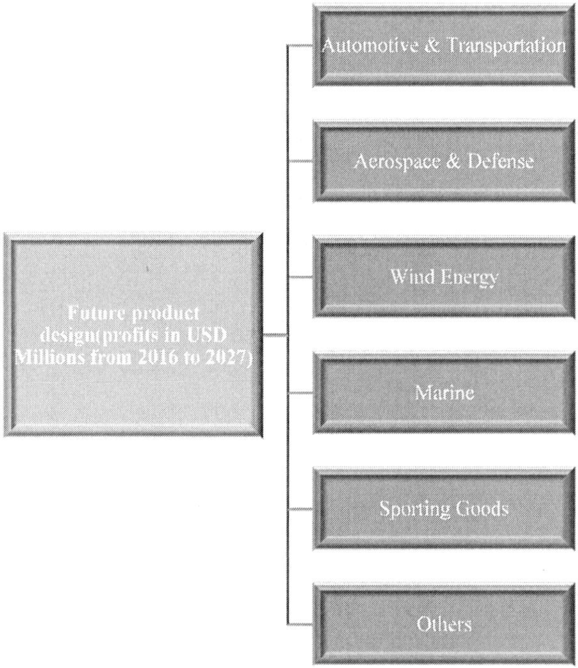

Figure 9. End-use of fiber market

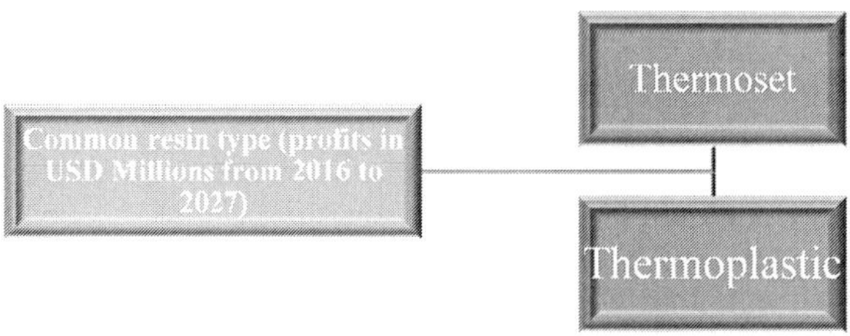

Figure 10. Resin type

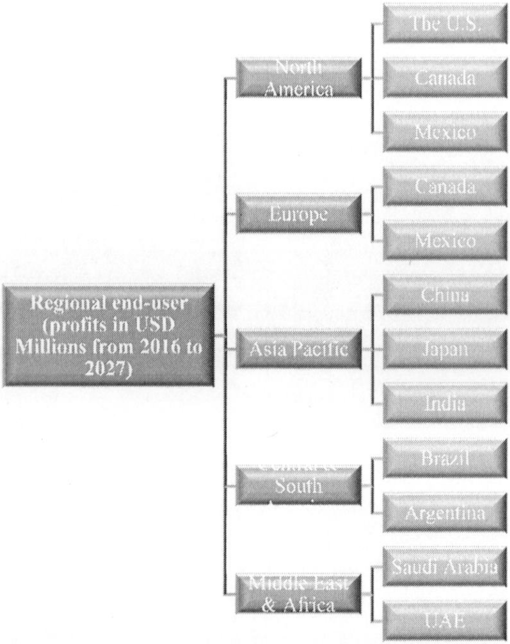

Figure 11. Global Natural Fiber Composites Market Size

The Global Natural Fiber Composites Market's total revenue is expected to be dominated by North America in 2021. This is a result of consumers' growing acceptance of the region's production of fiberboard, particleboard, and composite panels. Due to the abundance of better raw materials at much lower prices in the Asia Pacific region, a major manufacturer of bio-composite materials is moving its operations there. As a result, Asia Pacific nations are attracting more foreign direct investment (FDI) from Western nations due to government initiatives that support the establishment of start-ups for environmentally-friendly materials. To promote the use of environmentally-friendly materials in the building sector and other industries, the governments of many nations, including Japan, Germany, and the United States, issued regulations. Governments around the world are working to use more recyclable materials in response to the growing demand for bio-based products.

7. Conclusion

In the journey through the realm of hybrid natural fiber composites in chemical engineering applications, we've uncovered a dynamic landscape rich with possibilities. This chapter's exploration of these innovative materials has illuminated their capacity to transcend conventional boundaries, delivering heightened performance, enhanced sustainability, and a gateway to a greener future. The pursuit of hybridization, fusing diverse natural fibers within composite matrices, emerges as a pivotal strategy for optimizing properties. The resulting synergy has the potential to revolutionize structural integrity, mechanical strength, and thermal stability. The amalgamation of fibers, each endowed with unique attributes, affords tailored solutions across an array of applications in the chemical engineering domain. Our investigation has traversed varied sectors within chemical engineering. From equipment durability in harsh environments to advanced filtration and separation processes, hybrid natural fiber composites have exhibited their prowess. The interplay of fibers and matrices has led to new horizons, elevating efficiency, efficacy, and environmental consciousness. Interfacials compatibility, a linchpin of composite performance, has been explored with scrutiny. The strategic manipulation of fiber-matrix interactions, through surface treatments and innovative bonding techniques, has been shown to unlock the full potential of these hybrids. This, in turn, has laid the groundwork for more reliable and resilient chemical engineering processes. Sustainability remains the cornerstone of our exploration. As the world seeks solutions that tread lightly on our planet, hybrid natural fiber composites emerge as a beacon. Through reduced reliance on non-renewable resources, lower carbon footprints, and minimization of waste streams, these composites beckon toward a more environmentally responsible future. On this chapter's tapestry, we have woven the threads of innovation, sustainability, and adaptability. The prospects presented by hybrid natural fiber composites in chemical engineering applications are not only promising but transformative. As we draw this chapter to a close, we stand on the precipice of a new era where hybridization serves as the bridge to unlocking unparalleled possibilities. This journey is far from over; it extends into uncharted territories where science, engineering, and nature converge to shape a brighter, more resilient future.

References

1. Hasan, K. F., Horváth, P. G., & Alpár, T. (2020). Potential natural fiber polymeric nanobiocomposites: A review. *Polymers*, *12*(5), 1072.
2. Chandgude, S., & Salunkhe, S. (2021). In state of art: Mechanical behavior of natural fiber-based hybrid polymeric composites for application of automobile components. *Polymer Composites*, *42*(6), 2678-2703.
3. Pickering, K. L., Efendy, M. A., & Le, T. M. (2016). A review of recent developments in natural fibre composites and their mechanical performance. *Composites Part A: Applied Science and Manufacturing*, *83*, 98-112.
4. Wambua, P., Ivens, J., & Verpoest, I. (2003). Natural fibres: can they replace glass in fibre reinforced plastics?. *Composites science and technology*, *63*(9), 1259-1264.
5. Sanjay, M. R., Arpitha, G. R., & Yogesha, B. (2015). Study on mechanical properties of natural-glass fibre reinforced polymer hybrid composites: A review. *Materials today: proceedings*, *2*(4-5), 2959-2967.
6. Mahir, F. I., Keya, K. N., Sarker, B., Nahiun, K. M., & Khan, R. A. (2019). A brief review on natural fiber used as a replacement of synthetic fiber in polymer composites. *Materials Engineering Research*, *1*(2), 86-97.
7. Aida, H. J., Nadlene, R., Mastura, M. T., Yusriah, L., Sivakumar, D., & Ilyas, R. A. (2021). Natural fibre filament for Fused Deposition Modelling (FDM): A review. *International Journal of Sustainable Engineering*, *14*(6), 1988-2008.
8. Prabhu, L., Krishnaraj, V., Sathish, S., Gokulkumar, S., Karthi, N., Rajeshkumar, L., ... & Elango, K. S. (2021). A review on natural fiber-reinforced hybrid composites: Chemical treatments, manufacturing methods and potential applications. *Materials Today: Proceedings*, *45*, 8080-8085.
9. Kumar, R., Ul Haq, M. I., Raina, A., & Anand, A. (2019). Industrial applications of natural fibre-reinforced polymer composites–challenges and opportunities. *International Journal of Sustainable Engineering*, *12*(3), 212-220.
10. Sanjeevi, S., Shanmugam, V., Kumar, S., Ganesan, V., Sas, G., Johnson, D. J., ... & Das, O. (2021). Effects of water absorption on the mechanical properties of hybrid natural fibre/phenol formaldehyde composites. *Scientific Reports*, *11*(1), 13385.

11. Mustafa, B. S., Jamal, G. M., & Abdullah, O. G. (2023). Enhancement of the hardness and impact strength of epoxy-based metal oxide nanocomposites through hybridization. *Modern Physics Letters B*, 2350151.
12. Raju, P., Raja, K., Lingadurai, K., Maridurai, T., & Prasanna, S. C. (2021). Glass/Caryota urens hybridized fibre-reinforced nanoclay/SiC toughened epoxy hybrid composite: mechanical, drop load impact, hydrophobicity and fatigue behaviour. *Biomass Conversion and Biorefinery*, 1-10.
13. Azammi, A. N., Ilyas, R. A., Sapuan, S. M., Ibrahim, R., Atikah, M. S. N., Asrofi, M., & Atiqah, A. (2020). Characterization studies of biopolymeric matrix and cellulose fibres based composites related to functionalized fibre-matrix interface. In *Interfaces in particle and fiber-reinforced composites* (pp. 29-93). Woodhead Publishing.
14. Nurazzi, N. M., Harussani, M. M., Aisyah, H. A., Ilyas, R. A., Norrrahim, M. N. F., Khalina, A., & Abdullah, N. (2021). Treatments of natural fiber as reinforcement in polymer composites—a short review. *Functional Composites and Structures*, *3*(2), 024002.
15. Slamani, M., & Chatelain, J. F. (2023). A review on the machining of polymer composites reinforced with carbon (CFRP), glass (GFRP), and natural fibers (NFRP). *Discover Mechanical Engineering*, *2*(1), 4.
16. Chegdani, F., Takabi, B., El Mansori, M., Tai, B. L., & Bukkapatnam, S. T. (2020). Effect of flax fiber orientation on machining behavior and surface finish of Natural Fiber-Reinforced Polymer composites. *Journal of Manufacturing Processes*, *54*, 337-346.
17. Xue, F., Zheng, K., Liao, W., Shu, J., & Dong, S. (2021). Investigation on fiber fracture mechanism of c/sic composites by rotary ultrasonic milling. *International Journal of Mechanical Sciences*, *191*, 106054.
18. Adda, B., Belaadi, A., Boumaaza, M., & Bourchak, M. (2021). Experimental investigation and optimization of delamination factors in the drilling of jute fiber–reinforced polymer biocomposites with multiple estimators. *The International Journal of Advanced Manufacturing Technology*, *116*, 2885-2907.
19. Pereszlai, C., Geier, N., Poór, D. I., Balázs, B. Z., & Póka, G. (2021). Drilling fibre reinforced polymer composites (CFRP and GFRP): An analysis of the cutting force of the tilted helical milling process. *Composite Structures*, *262*, 113646.
20. Vigneshwaran, S., John, K. M., Deepak Joel Johnson, R., Uthayakumar, M., Arumugaprabu, V., & Kumaran, S. T. (2021). Conventional and unconventional machining performance of natural

fibre-reinforced polymer composites: A review. *Journal of Reinforced Plastics and Composites*, *40*(15-16), 553-567.

21. Ashokavel, S., & Boopathi, M. (2022). Numerical investigation of mechanical properties of hybrid fiber-reinforced polymer composites. *Materials Today: Proceedings*, *52*, 2255-2263.
22. Kim, H. J., & Lee, W. J. (2023). The Influence of Cellulose Nanocrystal Characteristics on Regenerative Silk Composite Fiber Properties. *Materials*, *16*(6), 2323.
23. Koppula, S. B., Karachi, S., Kumar, V., Borra, N. D., Phaneendra, Y., Neigapula, V. S. N., & Hemalatha, S. (2023). Investigation into the mechanical characteristics of natural fiber-reinforced polymer composites: Effects of flax and e-glass reinforcement and stacking configuration. *Materials Today: Proceedings*.
24. Kumar, B. A., Saminathan, R., Tharwan, M., Vigneshwaran, M., Babu, P. S., Ram, S., & Kumar, P. M. (2022). Study on the mechanical properties of a hybrid polymer composite using egg shell powder based bio-filler. *Materials Today: Proceedings*, *69*, 679-683.
25. Alfatah, T., Mistar, E. M., Syabriyana, M., & Supardan, M. D. (2022). Advances in oil palm shell fibre reinforced thermoplastic and thermoset polymer composites. *Alexandria Engineering Journal*, *61*(6), 4945-4962.
26. Saiteja, J., Jayakumar, V., & Bharathiraja, G. (2020). Evaluation of mechanical properties of jute fiber/carbon nano tube filler reinforced hybrid polymer composite. *Materials Today: Proceedings*, *22*, 756-758.
27. Prabhu, L., Krishnaraj, V., Sathish, S., GokulKumar, S., & Karthi, N. (2020). Study of mechanical and morphological properties of jute-tea leaf fiber-reinforced hybrid composites: Effect of glass fiber hybridization. *Materials Today: Proceedings*, *27*, 2372-2375.
28. Ibrahim, M. I. J., Sapuan, S. M., Zainudin, E. S., & Zuhri, M. Y. M. (2020). Preparation and characterization of cornhusk/sugar palm fiber-reinforced Cornstarch-based hybrid composites. *Journal of Materials Research and Technology*, *9*(1), 200-211.
29. Singh, Y., Singh, J., Sharma, S., Lam, T. D., & Nguyen, D. N. (2020). Fabrication and characterization of coir/carbon-fiber-reinforced epoxy-based hybrid composite for helmet shells and sports-good applications: Influence of fiber surface modifications on the mechanical, thermal and morphological properties. *Journal of Materials Research and Technology*, *9*(6), 15593-15603.
30. Arul, M., Sasikumar, K. S. K., Sambathkumar, M., Gukendran, R., & Saravanan, N. (2020). Mechanical and fracture study of hybrid natural

fiber-reinforced composite–Coir and sugarcane leaf sheath. *Materials Today: Proceedings*, *33*, 2795-2797.

31. Balaji, A., Udhayasankar, R., Karthikeyan, B., Swaminathan, J., & Purushothaman, R. (2020). Mechanical and thermal characterization of bagasse fiber/coconut shell particle hybrid biocomposites reinforced with cardanol resin. *Results in Chemistry*, *2*, 100056.
32. Dinesh, S., Elanchezhian, C., Vijayaramnath, B., Sathiyanarayanan, K., & Adinarayanan, A. (2020). Experimental investigation of banana bract fiber and palm fiber-reinforced with epoxy hybrid composites. *Materials Today: Proceedings*, *22*, 335-341.
33. Vinod, A., Vijay, R., Singaravelu, D. L., Sanjay, M. R., Siengchin, S., & Moure, M. M. (2019). Characterization of untreated and alkali treated natural fibers extracted from the stem of Catharanthus roseus. *Materials Research Express*, *6*(8), 085406.
34. Mohana Krishnudu, D., Sreeramulu, D., & Reddy, P. V. (2020). Alkali treatment effect: Mechanical, thermal, morphological, and spectroscopy studies on abutilon indicum fiber-reinforced composites. *Journal of Natural Fibers*, *17*(12), 1775-1784.
35. Jagadeesh, P., Puttegowda, M., Mavinkere Rangappa, S., & Siengchin, S. (2021). A review on extraction, chemical treatment, characterization of natural fibers and its composites for potential applications. *Polymer Composites*, *42*(12), 6239-6264.
36. Sumrith, N., Techawinyutham, L., Sanjay, M. R., Dangtungee, R., & Siengchin, S. (2020). Characterization of alkaline and silane treated fibers of 'water hyacinth plants' and reinforcement of 'water hyacinth fibers' with bioepoxy to develop fully biobased sustainable ecofriendly composites. *Journal of Polymers and the Environment*, *28*, 2749-2760.
37. Fathi, B., Foruzanmehr, M., Elkoun, S., & Robert, M. (2019). Novel approach for silane treatment of flax fiber to improve the interfacial adhesion in flax/bio epoxy composites. *Journal of Composite Materials*, *53*(16), 2229-2238.
38. Sabarinathan, P., Rajkumar, K., Annamalai, V. E., & Vishal, K. (2020). Characterization on chemical and mechanical properties of silane treated fish tail palm fibres. *International Journal of Biological Macromolecules*, *163*, 2457-2464.
39. Zaman, H. U., & Khan, R. A. (2021). Acetylation used for natural fiber/polymer composites. *Journal of Thermoplastic Composite Materials*, *34*(1), 3-23.
40. Koohestani, B. A. B. A. K., Darban, A. K., Mokhtari, P., Yilmaz, E. R. O. L., & Darezereshki, E. S. M. A. E. E. L. (2019). Comparison of

different natural fiber treatments: a literature review. *International Journal of Environmental Science and Technology*, *16*, 629-642.
41. Gholampour, A., & Ozbakkaloglu, T. (2020). A review of natural fiber composites: Properties, modification and processing techniques, characterization, applications. *Journal of Materials Science*, *55*(3), 829-892.
42. Matykiewicz, D., Barczewski, M., Mysiukiewicz, O., & Skórczewska, K. (2021). Comparison of various chemical treatments efficiency in relation to the properties of flax, hemp fibers and cotton trichomes. *Journal of Natural Fibers*, *18*(5), 735-751.
43. Godara, S. S. (2019). Effect of chemical modification of fiber surface on natural fiber composites: A review. *Materials Today: Proceedings*, *18*, 3428-3434.
44. Ganesh, S., Lakshmanan Saraswathy, J., Raghunathan, V., & Sivalingam, C. (2022). Extraction and characterization chemical treated and untreated lycium ferocissimum fiber for epoxy composites. *Journal of Natural Fibers*, *19*(13), 6509-6520.
45. Alhijazi, M., Safaei, B., Zeeshan, Q., Asmael, M., Eyvazian, A., & Qin, Z. (2020). Recent developments in luffa natural fiber composites. *Sustainability*, *12*(18), 7683.
46. Shahroze, R. M., Chandrasekar, M., Senthilkumar, K., Senthilmuthukumar, T., Ishak, M. R., & Asyraf, M. R. M. (2019, April). A review on the various fibre treatment techniques used for the fibre surface modification of the sugar palm fibres. In *Proceedings of the Seminar Enau Kebangsaan* (pp. 48-52).
47. Ezaddin, N. A., & Salih, R. F. (2023). Cultivated Natural Fiber Crops in Iraq and their Uses: A Review. *Passer Journal of Basic and Applied Sciences*, *5*(1), 191-201.
48. Tian, Y., & Zhang, T. (2021). MIXTAs and phytohormones orchestrate cotton fiber development. *Current Opinion in Plant Biology*, *59*, 101975.
49. Han, G., Li, Y., Yang, Z., Wang, C., Zhang, Y., & Wang, B. (2022). Molecular mechanisms of plant trichome development. *Frontiers in Plant Science*, *13*, 910228.
50. Maryum, Z., Luqman, T., Nadeem, S., Khan, S. M. U. D., Wang, B., Ditta, A., & Khan, M. K. R. (2022). An overview of salinity stress, mechanism of salinity tolerance and strategies for its management in cotton. *Frontiers in Plant Science*, *13*, 907937.

CHAPTER II

NATURAL FIBER COMPOSITE PROCESSING, FABRICATION AND APPLICATION

EEN SETIAWATI, ALFIAN PUTRA, SATRIA NANDA, TEUKU RIHAYAT, NELLY SAFITRI

Abstract

The utilization of bamboo fiber as a reinforcing agent in composite materials has witnessed a noteworthy surge, marking a significant juncture in high-tech innovation. This upswing is a direct response to the burgeoning demand for the development of biodegradable, sustainable, and recyclable materials. The amalgamation of a natural matrix with bamboo fiber engenders a composite material that harnesses the optimal attributes of each constituent. Presently, the matrices employed exhibit remarkable softness and flexibility when juxtaposed with their natural fiber counterparts. This synergy culminates in the creation of composites characterized by exceptional strength-to-weight ratios. It is imperative to acknowledge that the rapid evolution of industrial product manufacturing technology has empowered consumers to make informed choices, aligning their preferences with their desired outcomes. In this context, the recent strides in the realm of natural fiber composites have been particularly noteworthy, attributed to their commendable performance and durability. The ascendancy of fiber composites has been precipitated by their increasingly diverse array of applications. Natural fibers, by their intrinsic nature, possess qualities such as biodegradability, lightweight attributes, and renewability. It is pertinent to categorize natural fibers based on their origin, which includes animal, plant, and mineral fibers. Animal fibers, often protein-based, comprise those derived from animal hair or secretions. Pertinently, anatomical distinctions among fibers of the same species exert considerable influence over their density and mechanical properties. Other contributory factors, including environmental conditions, methods of transportation, storage duration, and extraction conditions, exert a defining influence over the size and quality of natural fibers. Amongst these natural fibers, bamboo fiber emerges as a preeminent contender, renowned for its remarkable strength and reliability in the production of high-quality, sustainable industrial products. The composition of bamboo fiber predominantly encompasses cellulose, hemicellulose, and lignin. The morphological examination of bamboo fiber surfaces reveals a distinct absence of overt damage. The surface roughness, a notable feature, fosters adhesive bonding within the fiber matrix, thus serving as the primary utilization of bamboo fiber. The utilization of bamboo composites in product development endeavors surmounts several challenges, most notably in the domains of dimensional stability, longevity, weather resistance, high impact resilience, low maintenance requisites, non-toxicity, and minimal susceptibility to flame propagation.

2.1 Natural Fiber and Composites

2.1.1 Various Kinds of Chemical Fiber and Their Benefits

a. Wool Fiber

The attributes of wool fibers are contingent upon both their length and diameter. Typically, these fibers exhibit a range of lengths, spanning from 3.6 centimeters to 35 centimeters, while their fineness varies within the bracket of 10 to 70 microns. In terms of shape, wool fibers generally manifest as circular to elliptical, featuring distinctive scales or undulations [1], as illustrated in Figure 1. Wool fiber, a venerable and historically significant animal fiber, is sourced from the hair of diverse mammalian species, encompassing goats, sheep, and alpacas. It is pertinent to underscore that the characteristics of wool fibers are subject to inherent variations, contingent upon factors such as the quality of the sheep, the anatomical location from which the fibers are procured, and the presence of extraneous impurities. Notably, wool sourced from the shoulder region of the sheep is lauded for its superior and finer quality, while wool originating from the legs is regarded as of a comparatively inferior grade [2].

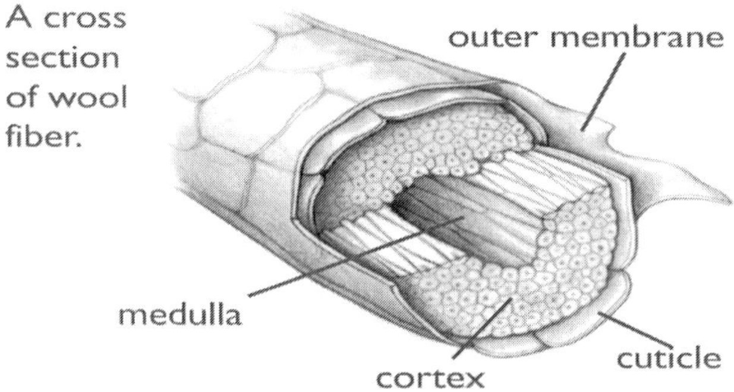

Figure 1. Cross-section view of wool fiber

b. Silk Fiber

Silk fiber is produced from the secretions of certain insects during the formation of their cocoons or webs. This fiber is also one of the ancient fibers known to man. Silk fiber-producing cocoons generally consist of two binders, namely fibroin and sericin. Fibrion in silk fiber is a natural protein that can provide strength to the fiber, while sericin can help hold the fiber structure [3]. The silk fibers produced are usually produced in the form of continuous filaments by silkworms (Bombyx mori). Fabrics made from silk fibers are generally very hygienic because silk fibers are dust-resistant and do not promote bacterial growth. The shape of the silk fiber is triangular and non-uniform (Figure 2). Silk fiber has various application fields such as clothing, coatings, food, cosmetics, pharmaceuticals, and construction [4].

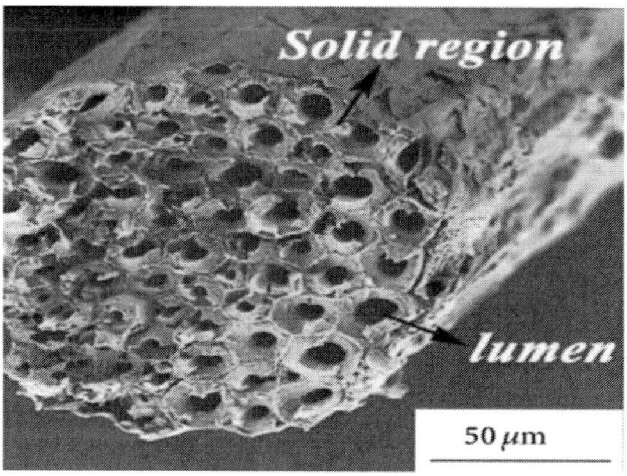

Figure 2. Cross-section of Silk Fibers (Habib Awais et al., 2021)

c. Cotton Fiber

Cotton stands out as an exceedingly popular seed-derived fiber, widely employed in the production of textiles, home furnishings, and a spectrum of industrial applications. Notably, cotton fiber is acknowledged for its status as a preeminent source of pure cellulose, boasting an impressive cellulose content of approximately 90% [5].

Figure 3. Shape of Cotton Fiber

Distinguished by their remarkable capacity for moisture absorption, cotton fibers have been demonstrated to retain up to 25% of their weight in water. In terms of length, cotton fibers typically span the range of 20 to 32 mm. In the present era, organic cotton cultivation has gained considerable momentum, characterized by the absence of chemical inputs or fertilizers. This approach is undertaken with the dual objectives of preserving soil fertility and safeguarding the environment [6]. Intriguingly, the cross-sectional morphology of cotton fibers assumes a bean-like configuration, which undergoes a transformation to a more circular shape when water is absorbed, as depicted in Figure 3. Cotton belongs to the Gossypium hirsutum L. species which is cultivated in more than 50 countries around the world, among which most of the growing areas are in developing countries [7]. Taxonomically, cotton finds its place within the Gossypium hirsutum L. species and is cultivated across a span of more than 50 nations worldwide. Notably, a substantial portion of cotton production is concentrated within developing countries.

d. Jute Fiber

Jute fiber is a type of bast fiber, with the highest production volume compared to other bast fibers. The length and height of the burlap plants range between 2.5–5 m and 1-3.5 m. The water retting process is generally used to obtain jute fiber from the stems of the jute plant [8]. Single jute fibers are generally polygonal in shape with rounded corners (Figure 4).

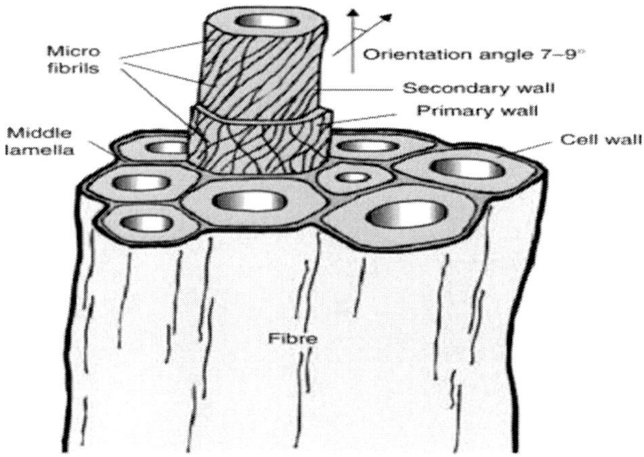

Figure 4. Morphology of Jute Fiber (Habib Awais et al., 2021)

e. Hemp Fiber

Hemp, characterized by its coarser texture and deeper hue, presents a unique challenge in the realm of bleaching when juxtaposed with cotton. Derived from the Cannabis sativa plant species, hemp fiber serves primarily industrial purposes.

In terms of morphology, hemp fibers exhibit a cylindrical shape, as illustrated in Figure 5, featuring discernible joints and surface irregularities [9]. Jute, on the other hand, boasts a robust, long-lasting, rigid, and lightweight natural bark fiber, rendering it an enticing choice not only for woolen composites and insulation materials but also as a polishing agent [10]. The versatile applications of hemp fiber span diverse industries, including textiles, construction, paper production, composites, and animal bedding [11]. Notably, hemp plants typically attain heights exceeding 3 meters, yielding hemp fibers that can extend up to 2 meters in length [11].

Figure 5. Hemp Fiber

a. Bamboo Fiber

Bamboo classified as a fibrous tree bark material exhibits remarkable growth rates, capable of attaining a staggering velocity of up to 3 centimeters per hour. Bamboo fiber, owing to its unique attributes, emerges as an exemplary choice for deployment as a reinforcement component within composite materials. The inherent characteristics of bamboo fiber are contingent upon factors encompassing its ecological niche, dimensions, soil moisture levels, and prevailing temperature conditions [12]. Bamboo fiber's outstanding specific strength positions it akin to glass fiber, rendering it a compelling alternative to conventional steel in various applications. While certain bamboo species can reach maturity within a mere three years, the majority necessitates a more protracted growth period, typically spanning six to seven years, before the fibers attain full maturation [13].

b. Kenaf fiber

Kenaf is classified as a wild dicotyledonous plant due to its distinctive three-layered stem structure, facilitating its ability to thrive in diverse environmental conditions. The growth cycle of kenaf can be successfully completed within a span of 150 to 189 days when provided with attentive care, even in conditions with limited water, pesticide, or fertilizer inputs. Notably, the kenaf plant exhibits impressive vertical growth, attaining heights ranging from 1.5 to 4.5 meters over a span of approximately four to five months. Its fibers possess a diameter of 67.6 micrometers, enabling the production of up to 10,000 kilograms of kenaf fiber per hectare [14]. The cultivation of kenaf has garnered global interest, primarily due to its exceptional ability to absorb phosphorus and nitrogen from the soil, along with its capacity to accumulate substantial levels of carbon dioxide. These characteristics have contributed to the widespread adoption of kenaf cultivation practices worldwide.

c. Sisal

Sisal fiber is a leaf-derived fiber characterized by its coarse texture and categorization as a rigid natural fiber. Typically, sisal leaves exhibit an average weight of approximately 600 grams and contain an estimated 1000 individual fibers [15]. The extraction process for sisal fiber from leaves involves various methods, including retting, scraping, or mechanical decortication. Sisal fiber finds versatile applications, commonly employed in the manufacturing of products such as balers, binding threads, filter paper, and automotive components, among others. In terms of length, sisal fibers typically span a range of 1 to 1.5 meters [16].

Figure 6. Sisal Fiber Cross-Section

d. Coir Fiber

Coir is a fruit fiber obtained from the outer skin or coconut coir. Coconut coir fiber generally has two colors, namely brown or white, depending on the ripeness of the fruit. The use of coir fiber should be limited to products that are softer or smoother and substantial because coir fiber is coarser than other leaf and bark fibers. Coir fiber can inhibit water penetration because it has a cavity surface filled with air (Figure 7) so that it can help to be applied to rope carpets, mats, building materials, and FRP composites [17-18].

Figure 8. Coir Fiber

e. Asbestos Fiber

Figure 9. Morphology of Asbestos Fibers (Habib Awais et al., 2021)

Asbestos constitutes a mineral-based fiber sourced from rocks, with the capacity for rock formations to yield asbestos fibers known for their exceptional heat resistance and non-combustible properties. Commercially available asbestos fibers are typically found in lengths ranging from 12 to 300 meters and possess a fine texture akin to that of glass fibers [19]. The cross-sectional configuration of asbestos strands assumes a circular or polygonal shape. Furthermore, asbestos exhibits corrosion resistance and stands impervious to the corrosive effects of alkalis and acids, rendering it valuable in a multitude of applications, including conveyor belts, industrial packaging materials, electric reels, and theater curtains. It is vital to note, however, that prolonged and uninterrupted exposure to pure asbestos is associated with the risk of developing cancer [20].

2.1.2 Physical and Chemical Properties of Natural Fibers

In table 2 it can be seen that the increase in fiber diameter occurs due to a decrease in density. Table 2 provides insights into the tensile strength values of various natural fibers, juxtaposed with their respective densities. Among these fibers, abaca stands out as a natural fiber with the highest specific strength and the lightest weight. This distinct attribute positions natural fibers as promising candidates for reinforcement materials in composite applications, as evidenced by their impressive tensile strength and specific stiffness values [21]. Natural fibers like sisal and hemp offer the advantage of easy accessibility and cost-effectiveness, rendering them viable substitutes for traditional materials such as glass and carbon fibers. Additionally, a subset of natural fibers, including hemp, sisal, kenaf, coir, and others, exhibit low densities coupled with high tensile strength values [22].

2.2 Bamboo Teaching Studies

2.2.1 Bamboo and Bamboo-Based Materials

The characteristics inherent to bamboo encompass cellulose, hemicellulose, and lignin, collectively constituting a natural composite material characterized by an intricately optimized structure, both on the macro and micro scales, culminating in remarkable mechanical properties. At a macro scale, the bamboo's distinctive hollow stems are segmented, connected by nodes that span along its length [23]. Delving into the microstructure of bamboo, it reveals two distinctive cell types. Firstly, the fiber cells exhibit an elevated aspect ratio and longitudinal alignment within the vascular tissue. Meanwhile, the parenchyma cells, forming brick-shaped structures in the ground tissue, serve a different role. Thick-walled fiber cells envelop the vascular bundles, functioning as primary transport elements and creating a robust supportive shell around them. Notably, the distribution of fibers across the cross-section is uneven, accumulating predominantly in the outermost regions. Within the bamboo fiber cell wall, a highly multi-lamellar structure is evident, with up to ten layers discernible. In contrast, parenchyma cells within the ground parenchyma tissue display a distinct

Table 1. Effect of chemical composition on the mechanical properties of natural fibers (azizatul karimah et al., 2021)

Chemical component of natural fibers	Parameters of mechanical properties				Parameters of physical properties		
	Tensile Strength	Spesific Youngs Modulus	Failure Starin	Microfibril Angel (MFA)	Diameter	Density	Moisture Gain
Cellulose	xxx	Xx	-	-	x	xxx	-
Hemicellulose	-	xxx	xx	xx	x	-	xx
Lignin	-	-	xxx	xxx	-	-	xx
Pectin	-	-	xx	xx	-	xxx	-
Wax	-	Xx	-	-	-	-	x

Note: Symbol (x): positive correlation, (-): negative correlation composition, featuring a reduced proportion of crystalline cellulose microfibrils and a heightened concentration of amorphous polymers [24].

The average fiber volume fraction (FVF) fluctuates within the range of 30-45%. In cross-sectional analysis, the FVF exhibits a gradient, increasing from approximately 15–20% on the interior to around 60–65% on the exterior of the bamboo culm. This distribution gives rise to a robust outer layer characterized by elevated strength. Consequently, the stem's proximity to the root manifests lower tensile strength but heightened stiffness, affording it superior flexural properties capable of withstanding the demands of wind and snow loads. This nuanced distribution classifies bamboo as a naturally adept material [25]. On a macroscopic scale, the inherent heterogeneity of bamboo culms exerts significant influence on their mechanical properties. The strategic entwinement of fibers in the internodes area serves to enhance isotropic properties, though it results in a substantial reduction in tensile strength in these regions as compared to the internodes themselves. Consequently, the knots emerge as the principal points of vulnerability within the bamboo stem under tensile forces [26].

2.2.2 Densified Bamboo

The significant variation in mechanical characteristics arises from the intricate and heterogeneous composition of bamboo. The process of bamboo densification leads to substantial enhancements in mechanical performance. The most straightforward and cost-effective approach to achieve densification involves applying compressive forces in the presence of elevated temperature and/or humidity conditions. The development of advanced bamboo connectors designed to augment mechanical attributes translates into notable improvements in tensile modulus and strength. Through the utilization of the TM compression method, one can attain an impressive tensile modulus ranging from 31 to 35 GPa and strength values reaching up to 350 MPa within a density range of 0.83 to 1.22 g cm^{-3}[27].

An exemplary high-density engineered bamboo product, known as bamboo scrimber, undergoes a meticulous process. This involves coating bamboo sections with resin and subsequently compressing them in parallel alignment. This process yields solid blocks with significantly enhanced mechanical and physical properties [28].

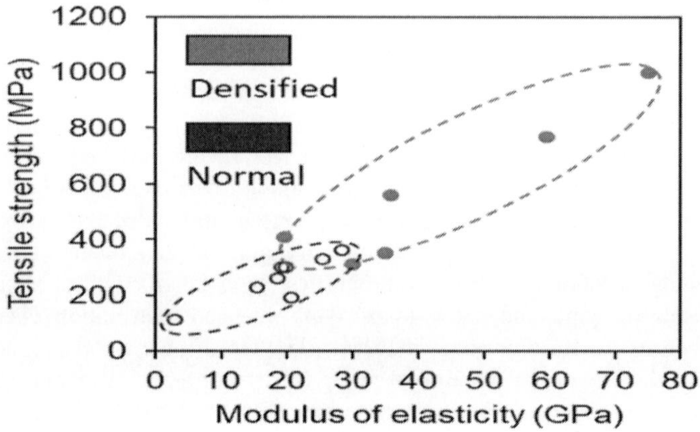

Figure 10. Tensile and tensile strength and modulus of elasticity of normal and dense bamboo (Matthias Jakob et al., 2022)

The bamboo fiber composite under examination consisted of elongated bamboo fibers that had undergone saturation in PF resin. The infusion of resin serves to enhance mechanical attributes and, concurrently, bolster dimensional stability, particularly in the presence of moisture and water [29].

In parallel to wood, subjecting the sample to densification via heat-induced shrinkage utilizing microwave heating yields remarkable results. This process yields a tensile strength of 560 megapascals (MPa) and a modulus of elasticity (MOE) reaching 36 gigapascals (GPa), all achieved at a density of 1 gram per cubic centimeter (g/cm^{-3}) [30]. Notably, bamboo exhibits a substantial density of 1.37 g/cm^{-3}, accompanied by a tensile strength and modulus of approximately 1 and 75 GPa, respectively. These robust mechanical properties vividly underscore the potential of bamboo-based materials for applications demanding high performance. Similar to wood, bamboo scaffolding also presents itself as a candidate for reinforcement within polymer matrices. Impregnating bamboo with epoxy resin, followed by compression, results in the creation of a composite characterized by an estimated fiber content of approximately 22% [31].

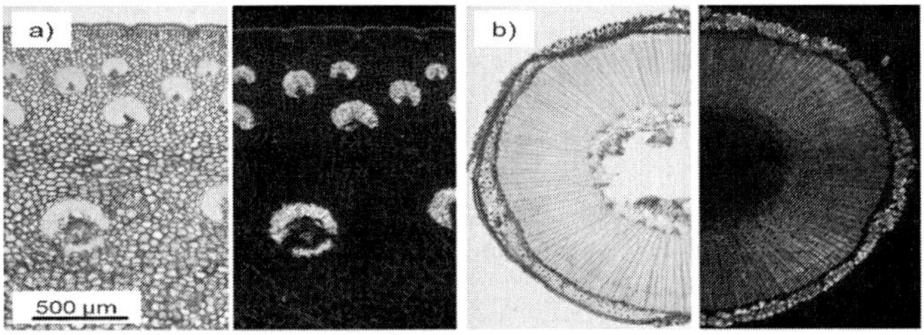

Figure 11. Transmitted light microscopy of a portion of a sisal leaf (a) and a hemp stalk (b) in visible light (left) and polarized visible light highlighting fibers with high cellulose content (Matthias Jakob et al., 2022)

2.3 An Overview of The Basics and Processes of Bamboo Binding

2.3.1 Characteristics Affecting Bamboo and Wood Bonds

a. Chemical Adhesive

This section delves into the foundational chemistry underlying phenol-formaldehyde (PF) adhesives and diphenylmethane diisocyanate (pMDI) polymers, elucidating their impact on the adhesive's interaction with the bamboo structure and the resultant bond strength. Wood adhesives can be categorized into two distinct groups predicated on their chemical composition and mechanical attributes: polymerized adhesives and pre-polymerized adhesives [32].

Within the realm of polymerized resins, one encounters formaldehyde-free PF adhesives, exemplified by pMDI. These resins are typically constituted of diminutive molecules, or monomers, upon application to wood or bamboo substrates. This molecular composition facilitates the resin's wetting action and penetration into the material. As time elapses, often facilitated by elevated temperatures, these monomers undergo a transformation into larger molecules, encompassing oligomers and polymers, engendering a transition from a liquid state to rigid polymers with extensive cross-linkages.

In contrast, pre-polymerized adhesives such as polyurethane (PUR) and polyvinyl acetate (PVAc) exhibit distinct characteristics. These adhesives exhibit limited mobility, restricting their ability to deeply permeate wood or bamboo structures. However, they engender more pliable bonds in comparison to their polymerized counterparts [33].

b. Bamboo Physical Properties

In comparison to softwood counterparts, bamboo presents distinctive characteristics, primarily characterized by its elevated specific gravity (SG) and the variances in SG observed across the culm wall. These properties can significantly influence aspects such as adhesive penetration as well as heat and moisture transfer during the composite pressing process [34]. Notably, specific gravity and shrinkage metrics for bamboo, particularly the tissue wood within bamboo, surpass those of commonly employed hardwood and softwood species by a factor of 1.8 in specific gravity and 5.6 in shrinkage data. These disparities are attributable to the smaller pore structure and the densely structured polylamella configuration present within the bamboo parenchyma and fibers. Recognizing the pronounced specific gravity and heightened shrinkage rate of bamboo becomes imperative when formulating tailored adhesives for bamboo-based applications [35].

Table 2. Specific Gravity and Shrinkage of Bamboo Compared to Wood (William Nguegang Nkeuwa et al., 2022)

Physical Property	Species		Average
Specific Gravity	Bamboo	Moso bamboo (phyllostachys pubescens)	0.29-0.84
		Giant bamboo (Dendrocalamus asper)	0.55-0.90
	Softwood	7 Fir Species	0.30-0.43
		5 Spruce Species	0.35-0.42
	Hardwood	2 Aspen Species	0.38-0.39
		3 Birch Spesies	0.55-065

Shrinkage	Bamboo	Giant bamboo (Dendrocalamus asper)	Radial	6-11
			Tangential	10-20
			Radial	4-9
			Tangential	6-11
	Softwood	7 Fir species	Radial	2.6-4.5
		4 Spruce species	Tangential	6.8-9.2
	Hardwood	2 Aspen Species	Radial	3.3-7.3
		3 Birch species	Tangential	6.7-9.5

2.3.2 Hybrid Bamboo Wood Composite

Numerous investigations have highlighted the advantages of amalgamating bamboo with wood (BW) to engender a diverse range of hybrid structural composites boasting enhanced mechanical strength. Within this context, it's notable that substrate failure incidence stands at 80% for wood, followed by 12.5% for the adhesive phase, and 5% for bamboo. This disparity underscores the fact that the wood component, characterized by its lower density, is relatively more susceptible to shear failure compared to the denser and more robust bamboo network constituting the adhesive phase [35].

Bamboo's higher density, while making it inherently more challenging to bond, also equips it with greater potential for achieving superior bond strength. In the context of hybrid wood and bamboo composites, the presence of the tough and less compressible bamboo web effectively channels compaction stress towards the tie lines and into the adjacent wood web. This, in turn, results in heightened deformation and contributes to the formation of a more robust local connective interface. Regardless of whether bamboo or wood is considered, it's important to note that, for a given density product, increased strand density exhibits an inverse relationship with surface area and yields a greater resin content [36-37].

The synergy between bamboo and wood harnesses the specific strengths of larger wood components, particularly softwoods like Douglas fir. This amalgamation has been empirically established as conferring greater strength and enhanced functionality when compared to exclusive reliance on either bamboo or wood materials. Ultimately, the fusion of bamboo with wood offers the prospect of optimizing wood resource utilization, a critical consideration in addressing the ongoing depletion of global wood supplies [38].

2.4 Characteristics of the Bamboo Fiber Composite

2.4.1 Thermal Characterization of Bamboo Fiber-reinforced Composites

The thermogravimetric analysis of novolac resin treated with alkali-processed bamboo exhibits superior thermal stability in comparison to the untreated composite [39]. This enhanced thermal stability is attributed to the intricate interplay of forces operating among the individual constituents bonded to the novolac resin. Post-alkali treatment, a noteworthy reduction in the fiber's affinity for water absorption, is observed, primarily due to the occupation of hydroxyl functional groups by corresponding soft metal ions. Subsequent neutralization instigates the formation of fresh cellulose bonds within the parent cellulose molecules. These novel bonds render cellulose a hermetically sealed system, thereby elevating the crystallinity index of alkali-treated bamboo, which in turn augments its stability. This newly established cage-like system effectively retains crystallized water, contributing to an elevation in the ultimate temperature during thermal events. Regarding thermal behavior, such as the heat deflection temperature analysis conducted on polypropylene-based bamboo fiber composites, it becomes evident that alkali treatment leads to a discernible enhancement in the thermal properties of bamboo fiber-based composites [40-41].

The composite's fiber content stands out as a pivotal factor directly impacting its thermal properties. In a biodegradation comparative study, researchers elucidated that partial degradation emerges as a salient factor exerting a direct influence on the thermal stability of bamboo composites. Comprehensive insights into the structural attributes of bamboo composites and their interrelation with diverse characteristics, including crystallization and morphology, have been expounded upon [42].

In a separate line of inquiry, scholars have underscored the pivotal role of the filler type utilized, with inorganic fillers serving as effective b-nucleators. This, in turn, precipitates the generation of an abundant array of forms that directly impact various composite properties, notably thermal characteristics [43]. In a distinct investigation, bamboo fiber was posited as a potential source of b-nucleators. Variations in the degree of thermal behavior are attributable to the heightened degree of adhesion forged between the matrix and bamboo fiber, with augmented adhesion correlating with enhanced thermal stability [44].

Exploring the thermal behavior of modified bamboo-reinforced composites with differing fiber loading levels unveiled that the observed weight loss is intricately tied to fiber loading. Furthermore, the maximum peak temperature experiences direct influence contingent upon an increase in inorganic silane content adorning the bamboo surface. An exploration into the kinetics of recycled polypropylene-reinforced bamboo fiber composites revealed that escalating cooling rates correlate with a downward shift in the peak crystallization temperature and initial temperature [45-46].

It is discerned that recycled polypropylene bamboo composites manifest a reduced degree of crystallinity relative to their pristine recycled polypropylene counterparts, illustrating that the composite's relative crystallinity necessitates a slower cooling rate. This phenomenon stems from the "nucleation effect" inherent to bamboo, promoting the crystallization of the polypropylene matrix. It is posited that, under slower cooling rates, the polymer core benefits from an extended duration to reach an active state conducive to crystallization at elevated temperatures. The introduction of bamboo fiber into bamboo composites founded on recycled polypropylene was found to reduce the melting temperature of the matrix while concurrently elevating crystallization, thus confirming the direct influence of bamboo fiber content on the composite's thermal properties [48]. Previous research has ascertained that the presence of fibers within the composite accelerates crystallization rates and augments the percentage of recycled polypropylene crystals. Therefore, bamboo fiber, at varying percentages, plays a constructive role in enhancing the crystallization of recycled resin within composites, typically occurring at higher temperature thresholds characteristic of polypropylene resins. The nucleation effect stands as the principal driving factor behind this intricate behavior [49].

Figure 12. Various applications of bamboo fiber composite

2.4.2 Effect of Heat Treatment on Chemical Properties of Bamboo

a. Change of Crystallinity

In a general context, subjecting bamboo to heat treatment can result in an elevation of its heat-induced crystallinity when compared to bamboo that has not undergone such treatment. Notably, the extent of crystallinity augmentation is contingent upon both the temperature and duration of the heating process [50]. Initially, crystallinity experiences an increment with increasing temperature and duration, but this trend subsequently reverses. The heat treatment process triggers the degradation of hemicellulose into propionic acid and acetic acid, which, in turn, can lead to a partial degradation of microfibrils found in both amorphous and crystalline regions. Consequently, this causes a marginal reduction in the relative crystallinity of cellulose. However, it is worth noting that despite this reduction, the crystallinity of the treated bamboo remains higher than that of untreated bamboo [51].

b. Change of Chemical Components

Through chemical fractionation, the constituents of bamboo, including holocellulose, alpha-cellulose, hemicellulose, and lignin, exhibit resistance to dissolution in acidic environments. Numerous research endeavors have elucidated that, when subjected to heat treatment at relatively low temperatures, specifically below 160°C, the chemical composition remains relatively stable. Conversely, within higher temperature ranges, such as 160°C-180°C and 180°C-220°C, temporal and thermal degradation precipitates a decline in hemicellulose and cellulose content, concomitant with an elevation in lignin content attributable to condensation reactions [52].

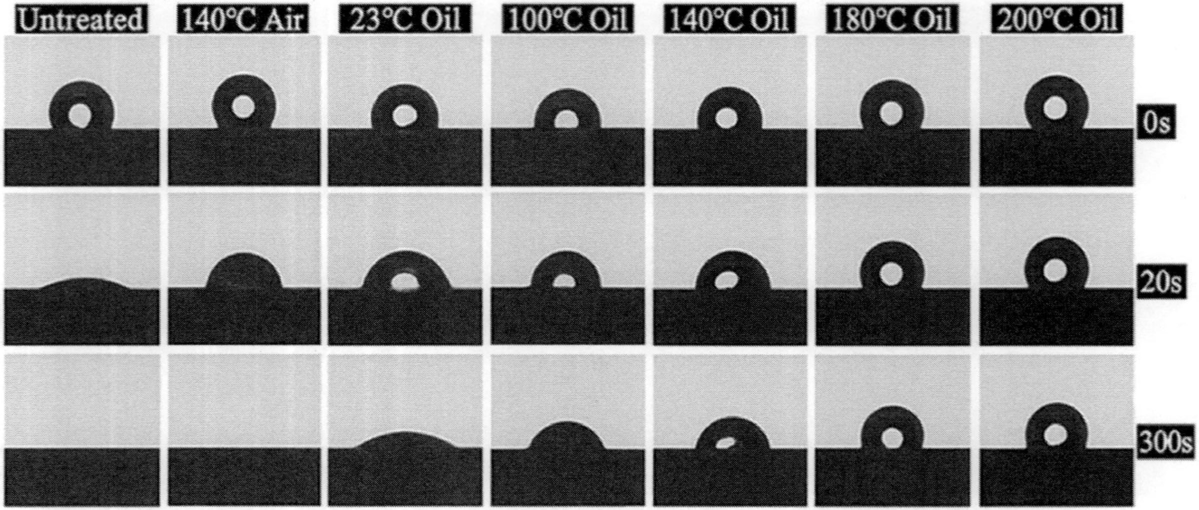

Figure 13. Impact of hot oil treatment on contact angle

c. Change Density

Fiber density assessment can be conducted utilizing a helium pycnometer. As depicted in Figure 15, an augmented concentration leads to a marginal reduction in fiber density. Initially, raw bamboo fiber exhibits a density of 1.47 g/cm^3, which subsequently diminishes to 1.39 g/cm^3 following treatment with an 8% NaOH solution. This reduction in density can be ascribed to the extraction process of fiber solutions, which includes the removal of components like hemicellulose, lignin, and other impurities, as well as corrosion of the cell wall [53].

The helium pycnometer, in its measurement of absolute fiber density, solely considers the denser constituents of the fibers, excluding lumens and pores from the assessment. In the context of alkalization, various natural fibers exhibit diverse responses regarding their density, as exemplified by Mwaikambo's findings. Some fibers undergo a reduction in density, while others experience an increase, and there are instances where no discernible change is observed. Notably, alkalization-induced density reduction has also been observed in sisal and bamboo fibers. Conversely, certain fibers, such as borassus, hemp, and kenaf, exhibit heightened density, potentially attributable to the densification of their cell walls [54-55].

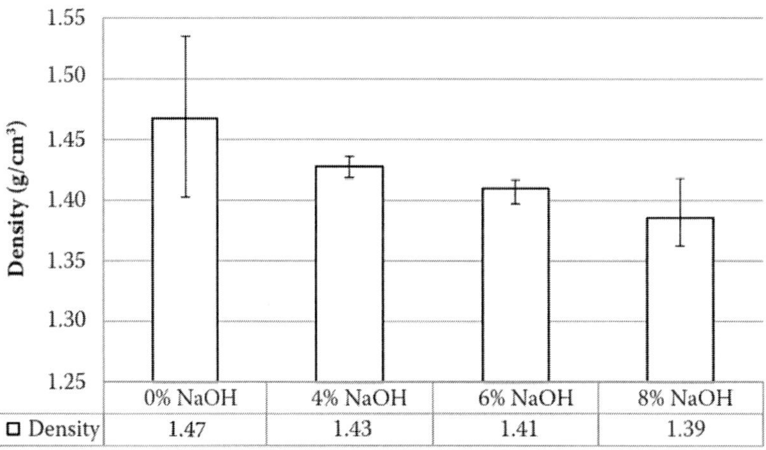

Figure 14. Density of bamboo fiber at various concentrations of alkaline treatment.

2.5 Bamboo Fiber-reinforced Composites and Design Applications

The rapid and sustainable growth of bamboo has spurred significant advancements in both theoretical and applied research related to bamboo-based products. This is particularly evident in sectors like construction, furniture, packaging, and transportation, as illustrated in Figure 15. Products derived from bamboo composites effectively address shortcomings such as dimensional stability, durability, weather resistance, high impact resilience, low maintenance requirements, non-toxicity, and low flame spread [56].

These bamboo-based products have found their place in the market, typically comprising 70% recycled bamboo fiber and 30% recycled high-density polyethylene (HDPE). Multiple companies are actively engaged in the utilization of bamboo composites across a range of everyday applications, encompassing bamboo composite materials, bamboo composite fences, and more [57].

The bamboo composite boasts unique characteristics, including shape retention, resistance to termites, high thermal stability, and ease of handling and installation. End users reap substantial advantages from these bamboo-based composites, particularly in terms of cost-effectiveness [58]. The evolution of methodologies aimed at achieving cost-efficient, environmentally-friendly practices, reducing reliance on finite building materials, curbing pollution, and conserving energy has been significantly influenced by the burgeoning bamboo-based composite industry. Bamboo-based composites tailored for water sports are commonly referred to as surfboards, characterized by their lightweight design and water-resistant surface. Mitsubishi Japan has recently employed a bamboo-based composite in prototyping, using polyurethane resin to develop the prototype. Mitsubishi introduced this product to minimize production costs and reduce CO2 emissions by embracing plant-based materials, as depicted in Figure 17. Despite the potential benefits of advanced composites, they have not garnered significant attention for use in automotive applications. Overhauls from an economic standpoint are imperative for advanced composites, with the primary alternative being the integration of natural plant fibers. Bamboo presents itself as an excellent resource to address the ongoing challenges related to cost and the availability of raw materials. Substantial economic transformations are warranted in the realm of advanced composites, and the utilization of natural plant fibers, particularly bamboo, emerges as a promising solution to alleviate the existing issues concerning cost and raw material accessibility. Significant economic shifts are imperative within the

Figure 15. Application of Composite Furniture Based on Bamboo Fiber

domain of advanced composites, and the singular recourse lies in the adoption of natural plant fibers. Bamboo stands out as a remarkable asset to address the prevailing challenges concerning cost and the accessibility of raw materials. Substantial economic transformations are warranted in the realm of advanced composites, and the utilization of natural plant fibers, particularly bamboo, emerges as a promising solution to alleviate the existing issues regarding cost and raw material availability [59-60].

Figure 16. Composite Based on Bamboo Fiber

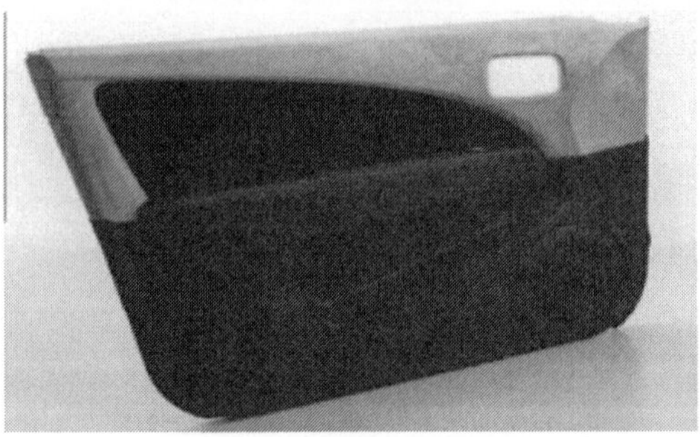

Figure 17. Prototype of door trim that is made of bamboo composite material

The pressing challenge we encounter on a daily basis necessitates a shift toward the creation of sustainable and eco-conscious materials. Among these materials, those derived from rapidly growing plant sources take the spotlight, with bamboo fiber serving as a prime illustration due to its association with swift and sustainable growth. Researchers and industry experts are diligently striving to supplant traditional synthetic materials with environmentally responsible alternatives like bamboo fiber, aiming to establish it as a sustainable and cost-effective resource for the burgeoning field of biocomposites [61]. Biodegradable and sustainable biocomposites crafted from bamboo fiber present versatile solutions suitable for both indoor and outdoor applications, poised to supplant materials traditionally reliant on glass fiber-based and conventional composites. Bamboo fiber-based polymer composites demonstrate a remarkable strength-to-weight ratio, dimensional stability, durability, and the adaptability to be intricately engineered into diverse shapes and sizes at a cost-effective production rate, outperforming their conventional counterparts. Through controlled temperature and pressure processing, bamboo fiber-based hybrid composites can yield product categories boasting superior attributes compared to their individual components. Notably, research indicates that bamboo fiber can seamlessly replace up to 25% of the weight of glass fiber in composites without compromising their mechanical properties [62-63]. This study also highlights that increasing the bamboo fiber content progressively enhances the modulus and strength of the resulting biocomposite. Furthermore, investigations into hybridization involving bamboo and glass fibers reveal a marginal decrease in stiffness with a slight uptick in bamboo fiber content, primarily attributable to suboptimal interfacial bonding [64]. Researchers are diligently exploring resin modifications to facilitate a more comprehensive comparison between bamboo and glass fiber composites. The outcomes are systematically tabulated, delineating the synergistic impact of bamboo and glass fibers. Notably, the thermal properties of hybrid composites receive a significant boost courtesy of bamboo and glass fibers, owing to the hybridization effect and the strengthening of the fiber/matrix interface. Additional scrutiny of the alkaline treatment's influence on bamboo fiber elucidates its role in shaping the tensile properties and chemical resistance of both standalone and mixed bamboo-glass fiber-reinforced composites [65]. The results underscore the profound impact of fiber blending on composite tensile properties, with hybrid composites demonstrating superior qualities compared to bamboo fiber composites. This enhancement is attributed to the low amorphous cellulose component in bamboo fiber, following rigorous acid and base treatments. An earlier study delved into the fabrication of glass-bamboo composites, assessing their flexural and compressive properties [66]. The researchers hypothesized that alkali treatment of bamboo fiber would bolster its binding affinity to the matrix. Indeed, the amalgamation of bamboo fiber with glass fiber directly enhances the mechanical characteristics of bamboo composites. Consequently, these comprehensive investigations underscore the substantial potential for bamboo fiber to supplant or reduce the reliance on glass fiber content in conventional and commercial composites, opening doors for large-scale industrial applications.

2.6 Conclusion

The rapid progress in industrial product manufacturing technology has empowered consumers to make well-informed choices that align with their preferences. A notable and positive development in recent years pertains to natural fiber composites, primarily due to their exceptional performance and long-lasting attributes. The increasing utilization of composite fibers in various applications has spurred advancements in this field. Natural fibers, characterized by their biodegradability, lightweight nature, and renewable origins, can be further categorized based on their source, encompassing animal, plant, and mineral fibers. Bamboo fiber, specifically, comprises predominantly cellulose, hemicellulose, and lignin components. When examining the morphology of bamboo fiber, it exhibits an unblemished surface with surface roughness that enhances the adhesion within the fiber matrix, serving as the principal utility of bamboo fiber. The deployment of bamboo composite products in various applications effectively addresses the shortcomings associated with dimensional stability, extended lifespan, weather resistance, heightened impact resistance, low maintenance requirements, non-toxicity, and reduced flammability. This progress signifies a significant stride in the realm of natural fiber composites.

References

1. Zou, L., Jin, H., Lu, WY, and Li, X. 2009. Nanoscale structural and mechanical characterization of the cell wall of bamboo fibers. Materials Science and Engineering: C. 29(4): p. 1375–79.
2. Tong, J., Ma, Y., Chen, D., Sun, J., and Ren, L. 2005. Effects of vascular ber content on abrasive wear of bamboo. Wear. 259(1–6): p. 78–83.
3. Yao, F., Wu, Q., Lei, Y., Guo, W., and Xu, Y. 2008. Thermal decomposition kinetics of natural fibers: Activation energy with dynamic thermogravimetric analysis. Polymer Degradation and Stability. 93(1): p. 90–8.
4. Shao, S., Wen, G., and Jin, Z. 2008. Changes in chemical characteristics of bamboo (Phyllostachys pubescens) components during steam explosion. Wood Science and Technology. 42(6): p. 439–51.
5. Methacanon, P., Weerawatsophon, U., Sumransin, N., Prahsarn, C., and Bergado, DT. 2010. Properties and potential application of the selected natural fibers as limited life geotextiles. Carbohydrate Polymers. 82(4): p. 1090–6.

6. John, M. and Thomas, S. 2008. Biobres and biocomposites. Carbohydrate Polymers. 71(3): p. 343–64.
7. Wong, KJ, Yousif, BF, and Low, KO. 2010. Effects of alkali treatment on the interfacial adhesion of bamboo bres. Journal of Materials: Design and Applications. 224: p. 139–48.
8. Spinacé, MAS, Lambert, CS, Fermoselli, KKG, and De Paoli, MA. 2009. Characterization of lignocellulosic curaua bres. Carbohydrate Polymers. 77(1): p. 47–53.
9. Aziz, SH and Ansell, MP. 2004. The effect of alkalization and bre alignment on the mechanical and thermal properties of kenaf and hemp bast bre composites: Part 1—polyester resin matrix. Composites Science and Technology. 64(9): p. 1219–30.
10. Sydenstricker, THD, Mochnaz, S., and Amico, SC. 2003. Pull-out and other evaluations on sisal-reinforced polyester biocomposites. Polymer Testing. 22(4): p. 375–80.
11. Boopathi, L., Sampath, PS, and Mylsamy, K. 2012. Investigation of physical, chemical and mechanical properties of raw and alkali treated Borassus fruit ber. Composites Part B: Engineering. 43(8): p. 3044–52.
12. Elenga, RG, Djemia, P., Tingaud, D., Chauveau, T., Maniongui, JG, and Dirras, G. 2013. Effects of alkali treatment on the microstructure, composition, and properties of the Rafa textilis ber. BioResources. 8(2): p. 2934–49.
13. Nam, TH, Ogihara, S., Tung, NH, and Kobayashi, S. 2011. Effect of alkali treatment on interfacial and mechanical properties of coir ber reinforced poly(butylene succinate) biodegradable composites. Composites Part B: Engineering. 42(6): p. 1648–56.
14. Mohanty, A., Misra, M., and Hinrichsen, G., Biobres. 2000. Biodegradable polymers and biocomposites: An overview. Macromolecular Materials and Engineering. 276(1): p. 1–24.
15. Khan, F. and Ahmad, S. 1996. Chemical medication and spectroscopic analysis of jute fiber. Polymer Degradation and Stability. 52(3): p. 335–40.
16. Rout, J., Misra, M., Tripathy, SS, Nayak, SK, and Mohanty, AK. 2001. The influence of fiber treatment on the performance of coir-polyester composites. Composites Science and Technology. 61: p. 1303–10.
17. Ratna Prasad, AV and Mohana Rao, K. 2011. Mechanical properties of natural fiber-reinforced polyester composites: Jowar, sisal and bamboo. Materials and Design. 32(8–9): p. 4658–63.
18. Mwaikambo, L. 2009. Sustainable Composite Materials: Exploitation of Plant Resourced Materials for Industrial Application. VDM Publishing: Saarbrücken, Germany.

19. Rong, MZ, Zhang, MQ, Liu, YL, Yang, GC, and Zeng, HM. 2001. The effect of –ber treatment on the mechanical properties of unidirectional sisal-reinforced epoxy composites. Composites Science and Technology. 61: p. 1437–47.
20. ASTM, ASTM D3379 – 75. 1989.e1 Standard Test Method for Tensile Strength and Young's Modulus for High Modulus Single Filament Materials. ASTM International: West Conshohocken, PA.
21. Bessadok, A., Marais, S., Roudesli, S., Lixon, C., and Métayer, M. 2008. Inuence of chemical medications on water-sorption and mechanical properties of Agave fibers. Composites Part A: Applied Science and Manufacturing. 39(1): p. 29–45.
22. Yousif, BF, Orupabo, C., and Azwa, ZN. 2012. Characteristics of kenaf bre immersed in different solutions. Journal of Natural Fibers. 9(4): p. 207–218.
23. Kalia, S., Kaith, BS, and Kaur, I. 2009. Pretreatments of natural bers and their application as Reinforcing materials in polymer composites—A review. Polymer Engineering Science. 49(7): p. 1253–72.
24. Sawpan, MA, Pickering, KL, and Fernyhough, A. 2011. Effect of various chemical treatments on the fiber structure and tensile properties of industrial hemp fibers. Composites Part A: Applied Science and Manufacturing. 42(8): p. 888–95.
25. Obi Reddy, K., Uma Maheswari, C., Shukla, M., Song, JI, and Varada Rajulu, A. 2013. Tensile and structural characterization of alkali treated Borassus fruit ne bers. Composites Part B: Engineering. 44(1): p. 433–8.
26. Defoirdt, N., Biswas, S., Vriese, LD, Tran, LQN, Acker, JV, Ahsan, Q., et al. 2010. Assessment of the tensile properties of coir, bamboo and jute bre. Composites Part A: Applied Science and Manufacturing. 41(5): p. 588–95.
27. Dhakal, H., Zhang, Z., and Richardson, M. 2007. Effect of water absorption on the mechanical Properties of hemp fiber-reinforced unsaturated polyester composites. Composites Science and Technology 67(7–8): p. 1674–83.
28. Khedari, J., Watsanasathaporn, P., and Hirunlabh, J. 2005. Development of bre-based soil– cement blocks with low thermal conductivity. Cement and Concrete Composites. 27(1): p. 111–6.
29. Kelly, A. and Tyson, W. 1965 Tensile properties of bre-reinforced metals: Copper/tungsten and copper/molybdenum. Journal of the Mechanics and Physics of Solids. 13(6): p. 329–50.

30. Saravanakumar, SS, Kumaravel, A., Nagarajan, T., Sudhakar, P., and Baskaran, R. 2013. Characterization of a novel natural cellulosic ber from Prosopis juliora bark. Carbohydrate Polymers. 92(2): p. 1928–33.
31. Wambua, P., Ivens, J., and Verpoest, I. 2003. Natural bres: can they replace glass in fiber-reinforced plastics? Composites Science and Technology. 63(9): p. 1259–64.
32. Akil, H., Omar, M., Mazuki, A., Saee, S., Ishak, Z., and Abu Bakar, A. 2011. Kenaf fiber-reinforced composites: A review. Materials and Design. 32(8): p. 4107–21.
33. ASTM, ASTM D638. 2010. Standard Test Method for Tensile Properties of Plastics. ASTM International: West Conshohocken, PA.
34. Manalo, AC, Karunasena, W., and Lau, KT. 2012. Mechanical properties of bamboo fiber polyester composites, in 22nd Australasian Conference on the Mechanics of Structures and Materials. Sydney.
35. Lee, SH and Wang, S. 2006. Biodegradable polymers/bamboo have biocomposite with biobased coupling agent. Composites Part A: Applied Science and Manufacturing. 37(1): p. 80–91.
36. Wong, KJ, Zahi, S., Low, KO, and Lim, CC. 2010 Fracture characterization of short bamboo bre reinforced polyester composites. Materials and Design. 31(9): p.4147–54.
37. Krenchel, H. 1964. Fiber Reinforcement Academic Forlag Copenhagen. Denmark.
38. Ku, H., Wang, H., Pattarachaiyakoop, N., and Trada, M. 2011. A review on the tensile properties of Natural Fiber-Reinforced Polymer composites. Composites Part B: Engineering. 42(4): p. 856–73.
39. Paul, AS, Boudenne, A., Ibos, L., Candau, Y., Joseph, K., and Thomas, S. 2008. Effect of fiber loading and chemical treatments on thermophysical properties of banana ber/ polypropylene commingled composite materials. Composites Part A: Applied Science and Manufacturing. 39(9): p. 1582–8.
40. TJ Snyder, M. Andrews, M. Weislogel, P. Moeck, J. Stone-Sundberg, D. Birkes, et al. 2014. 3D systems' technology overview and new applications in manufacturing, engineering, science, and education, 3D Print. addit. Manuf. 1 (2014) 169– https://doi.org/10.1089/3dp.2014.1502.
41. GD Goh, YL Yap, HKJ Tan, SL Sing, GL Goh, WY Yeong. 2019. Process–structure–properties in polymer additive manufacturing via material extrusion: a review, Crit. Rev. Solid State Matter. sci. 1–21. https://doi. org/10.1080/10408436.2018.1549977.
42. V. Shanmugam, O. Das, RE Neisiany, K. Babu, S. Singh, MS Hedenqvist, F. Berto, S. Ramakrishna. 2020. Polymer Recycling in

Additive Manufacturing: an Opportunity for the Circular Economy, Mater. Circ. Econ 2 (11). https://doi.org/10.1007/s42824-020-00012-0.
43. J. Gardan. 2016. Additive manufacturing technologies: state of the art and trends, Int. J. Prod. Res. 54. 3118–3132, https://doi.org/10.1080/00207543.2015.1115909.
44. B. Wendel, D. Rietzel, F. Kühnlein, R. Feulner, G. Hülder, E. Schmachtenberg. 2008. Additive processing of polymers, Macromol. Mater. Eng. 293. 799–809, https://doi.org/10.1002/mame.200800121.
45. M. Vaezi, S. Chianrabutra, B. Mellor, S. Yang. 2013. Multiple material additive manufacturing - Part 1: a review, Virtual Phys. Prototype. 8. 19–50, https:// doi.org/10.1080/17452759.2013.778175.
46. ZX Khoo, JEM Teoh, Y. Liu, CK Chua, S. Yang, J. An, et al. 2015. 3D printing of smart materials: a review on recent progresses in 4D printing, Virtual Phys. Prototype. 10. 103–122, https://doi.org/10.1080/17452759.2015.1097054.
47. V. Shanmugam, O. Das, K. Babu, U. Marimuthu, A. Veerasimman, DJ Johnson, RE Neisiany, MS Hedenqvist, S. Ramakrishna, F. Berto. 2021. Fatigue behavior of FDM-3D printed polymers, polymeric composites and architected cellular materials, Int. J. Fatigue 143. 106007, https://doi.org/10.1016/j.ijfatigue.2020.106007.
48. RE Neisiany, SN Khorasani, M. Naeimirad, JKY Lee, S. Ramakrishna. 2017. Improving mechanical properties of carbon/epoxy composite by incorporating functionalized electrospun polyacrylonitrile nanofibers, Macromol. Mater. Eng. 302. 1600551, https://doi.org/10.1002/mame.201600551.
49. SMF Kabir, K. Mathur, AFM Seyam. 2020. A critical review on 3D printed continuous fiber-reinforced composites: history, mechanism, materials and properties, Compos. structure. 232. 111476, https://doi.org/10.1016/j. compstruct. 2019.111476.
50. TNAT Rahim, AM Abdullah, H. Md Akil. 2019. Recent developments in fused deposition modeling-based 3D printing of polymers and their composites, Polym. Rev. 59. 589–624, https://doi.org/10.1080/15583724.2019.1597883.
51. AS Mangat, S. Singh, M. Gupta, R. Sharma. 2018. Experimental investigations on natural fiber embedded additive manufacturing-based biodegradable structures for biomedical applications, Rapid Prototyp. J. 24. 1221–1234, https://doi.org/10.1108/RPJ-08-2017-0162.
52. TD Ngo, A. Kashani, G. Imbalzano, KTQ Nguyen, D. Hui. 2018. Additive manufacturing (3D printing): a review of materials, methods, applications and challenges, Compos. B Eng. 143. 172–196, https://doi.org/10.1016/j.compositesb.2018.02.012.

53. A. Nasirov, I. Fidan. 2020. Prediction of mechanical properties of fused filament fabricated structures via asymptotic homogenization, Mech. Mater. 145. https://doi.org/10.1016/j.mechmat.2020.103372.
54. C. Wendt, AP Valerga, O. Droste, M. Batista, M. Marcos. 2017. FEM based evaluation of Fused Layer Modeling monolayers in tensile testing, Procedia Manufact. 13. 916–923, https://doi.org/10.1016/j.promfg.2017.09.160.
55. E. Cuan-Urquizo, JI Espinoza-Camacho, A. Alvarez-Trejo, E. Uribe, CD Trevi˜no Quintanilla, SE Crespo-S´anchez, et al. 2019. Elastic response of lattice arc structures fabricated using curved- layered fused deposition modeling, Mech. Adv. Mater. structure. https://doi.org/10.1080/15376494.2019.1682728.
56. SH Ahn, M. Montero, D. Odell, S. Roundy, PK Wright. 2002. Anisotropic material properties of fused deposition modeling ABS, Rapid Prototyp. J. 8. 248–257, https://doi.org/10.1108/13552540210441166.
57. N. Mohan, P. Senthil, S. Vinodh, N. Jayanth. 2017. A review on composite materials and process parameters optimization for the fused deposition modeling process, Virtual Phys. Prototype. 12. 47–59, https://doi.org/10.1080/17452759.2016.1274490.
58. W. Zhong, F. Li, Z. Zhang, L. Song, Z. Li. 2001. Short fiber-reinforced composites for fused deposition modeling, Mater. sci. Eng. 301. 125–130, https://doi. org/10.1016/S0921-5093(00)01810-4.
59. LJ Love, V. Kunc, O. Rios, CE Duty, AM Elliott, BK Post, et al. 2014. The importance of carbon fiber to polymer additive manufacturing, J. Mater. Res. 29. 1893–1898, https://doi.org/10.1557/jmr.2014.212.
60. T. Hofst¨atter, DB Pedersen, G. Tosello, HN Hansen. 2017. State-of-the-art of fiber-reinforced polymers in additive manufacturing technologies, J. Reinforc. plast. Compos. 36. 1061–1073, https://doi.org/10.1177/0731684417695648.
61. Y. Nakagawa, Kichiro Mori, T. Maeno. 2017. 3D printing of carbon fiber-reinforced plastic parts, Int. J. Adv. Manuf. Technol. 91. 2811–2817, https://doi.org/10.1007/s00170-016-9891-7.
62. S. Garzon-Hernandez, D. Garcia-Gonzalez, A. J´erusalem, A. Arias. 2020. Design of FDM 3D printed polymers: an experimental-modeling methodology for the prediction of mechanical properties, Mater. dec. 188. https://doi.org/10.1016/j. matdes. 2019.108414.
63. H. Prüß, T. Vietor. 2015. Design for fiber-reinforced additive manufacturing, J. Mech. Design, Trans. ASME 137. https://doi.org/10.1115/1.4030993, 111409-1-111409–7.

64. J. Wang, H. Xie, Z. Weng, T. Senthil, L. Wu. 2016. A novel approach to improve mechanical properties of parts fabricated by fused deposition modeling, Mater. dec. 105. 152–159, https://doi.org/10.1016/j.matdes.2016.05.078.
65. SMJ Razavi, RE Neisiany, SN Khorasani, S. Ramakrishna, F. Berto. 2018. Effect of neat and reinforced polyacrylonitrile nanofibers incorporation on interlaminar fracture toughness of carbon/epoxy composite, Theoret. appl. Mechanics Lett. 8. 126–131, https://doi.org/10.1016/j.taml.2018.02.008.
66. RE Neisiany, SN Khorasani, JKY Lee, M. Naeimirad, S. Ramakrishna. 2018. Interfacial toughening of carbon/epoxy composite by incorporating styrene acrylonitrile nanofibers, Theor. appl. Fract. Mech. 95. 242–247, https:// doi.org/10.1016/j.tafmec.2018.03.006.

CHAPTER III

MECHANICAL PROPERTIES OF NATURAL AND SYNTHETIC FIBER-REINFORCED HYBRID COMPOSITES

AWANIS ILMI, AGUNG EFRIYO HADI, SYAFRUDDIN

Abstract

In this chapter, we provide good composite materials for mechanical characterization processes. The utilization of characteristic filaments in mechanical applications may be a challenge for analysts to create suitable procedures to get great quality strands for utilization as support in polymer composites. Hybrid composites are an affordable, recyclable, and biodegradable material that can either partially or completely replace synthetic fibers in many applications. Understanding the fundamental structural elements of cellulose fibers and how they affect the mechanical, electrical, thermal, and physical characteristics of hybrid composites is crucial. However, there are several problems that can hinder the use of this fiber as a hybrid composite reinforcement. This review study aims to give a summary of the research that has been done on the cellulose/cellulose and cellulose/synthetic hybrid composites' physical, mechanical, electrical, thermal, and dynamic mechanical properties. A number of scientists have created hybrid composites that have chemical fiber alterations. The characteristics of the resultant cellulose nanocrystals, which will be tested with various test specifications like the TGA test, SEM test, tensile strength test, and various other common types of composites, must be thoroughly examined in order to comprehend all of the many characteristics of hybrid composites made with contemporary machinery. Hybrid composites are reinforced with natural and synthetic fibers: natural fibers such as Wool Fiber, Silk Fiber, Cotton Fiber, Jute Fiber, Hemp Fiber, Jute Fiber, Bamboo Fiber, Kenaf Fiber, Sisal Fiber, Coir Fiber, and Asbestos Fiber, then synthetic fibers such as Rayon Acetate Fiber, Polyurethane Fiber, Cuprammonium Rayon Fiber , Polyester Fiber, Nylon Fiber, Mineral Fiber, and Metal Fiber in the right way.

3.1 Introduction

3.1.1 Composites of natural and synthetic fibers

A. Natural Fibers

Common strands are isolated into three bunches depending on the source: creature, mineral, and plant. This comes about in Table 4.1: creatures, minerals, and plants. Plant strands are the filaments most commonly acknowledged by the fabricating industry for different purposes. There is right now a vital inquiry about the field for scholastics around the world to supplant the world's manufactured chemicals and non-renewable assets. Usually, since they have a brief developing period, they are reasonable, boundless, contain ecologically inviting components, and are more broadly accessible [1].

Table 3.1 Categorization of Natural Fibers (Jawad Ahmad et al., 2022)

Categorization of Natural Fibers			
Natural Fibers	Lignocellulose	Wood	Softwood and Hardwood (eg eucalyptus) Corn and bamboo fiber
		Grass/reed	
			Sabut
		Flower stalk	Rice, wheat, corn and oat fiber
		Bast	Hemp, kenaf etc
		Leaf	Sisal, abaca, pineapple and banana fiber
		Seed	Kapok, cotton, coir, dregs, sugar cane, account folss
	Animals	Silk	Bombax, goat, sheep and horse hair
		Hair/wool	Sheep, angora, alpaca, mohair and camel wool
	Mineral		Asbestos, glass and graphite

Natural fiber can be obtained from a variety of renewable resources, such as minerals, plants, and animal sources. Characteristic fiber is made up of four basic components: lignin, cellulose, pectin, and hemicellulose. Figure 1 illustrates the important chemical makeup of these constituents. Cellulose; Many -D-glucose (C6H11O5) units are found in cellulose, which is the most common organic polymer on the planet [2-5]. Cellulose contains several -D-glucose subunits (C6H11HAI5) linked by glycosidic interactions at the C1 and C4 carbon sites (Figure 1a). It aims to find cellulose in microcrystalline and amorphous forms, with high-order and low-order sites, accordingly [6]. Cellulose is a strong and durable material that can be found in both microcrystalline and amorphous configurations. It is resistant to strong alkalis as well as oxidizing and reducing agents. Nevertheless, it is susceptible to hydrolysis with acids to produce air-soluble and biodegradable sugars when subjected to chemical and environmental treatments [7-8].

a. Hemicellulose is made up of polysaccharide subunits, which are carbon ring sugar branched chain polymers. Numerous branched carbon ring carbohydrates polymerize to form hemicellulose (Figure 1b). Its chains are significantly shorter than those of cellulose and its degree of polymerization is lower. Its hydrophilicity and solubility in alkaline environments are also noteworthy [9-11]. It has the ability to create linkages with the cellulose and lignin components of the cell wall, which will increase the fiber's stiffness and flexibility [12]. Lignin is a multifaceted, nanocrystalline molecule that has functional groups like carbonyl, methoxy, and hydroxyl; nevertheless, its precise molecular structure is yet unknown (Figure 1c). Plant cell walls produce the aromatic units that are the building blocks of lignin. It is a part of cell walls and is indefinite and insoluble in most solvents. This phenolic molecule cannot be broken down into its individual monomers and is often resistant to microbial destruction due to its large molecular weight [13-14]. It has been demonstrated that hydrophobic lignin increases plant stiffness when it acts as a compatibilizer between hydrophilic cellulose and hemicellulose [15].
b. The components of pectin are heteropolysaccharides with different methyl esters called glycosaminoglycans. The fundamental pectin is homogalacturonan (HG), an unbranched polymer of D-galacturonic destructive linkages (-(1-4)), as seen in Figure 1d. In the natural world, it also contains the first significant amount of pectin. Pectin gives the fiber its shine and flavor and acts as a coating for the fiber strands inside the bundle. Wax acts as a shielding agent for the external layer of specialized fibers, consisting of two alcohol

varieties—one that dissolves in water and another that can be harmful. It enables hair to glide smoothly by reducing friction with its gentle coating. This influences both how the material interacts with water (hydrophilicity) and how it adheres to surrounding structures.

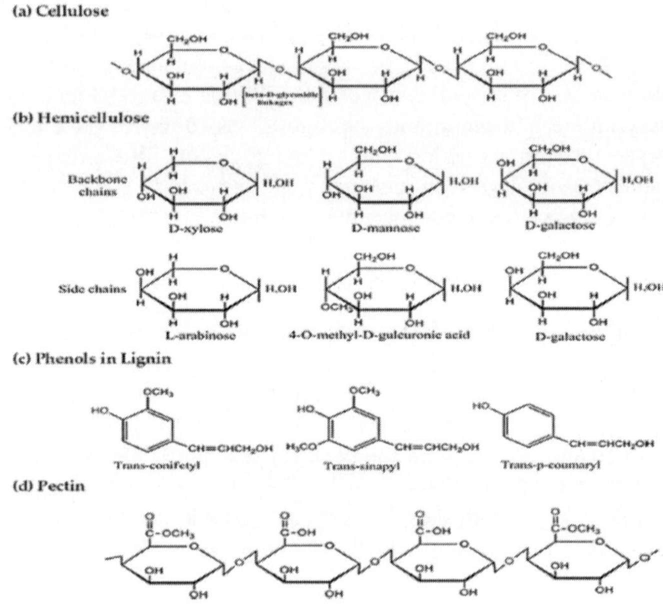

Figure 4.1 Chemical structure of natural fiber barrier blocks (Jawad Ahmad et al., 2022)

B. Fiber Plants

Natural resources provide us with a variety of materials that can be classified as stringy. A variety of plant parts are used to make plant fibers, including seeds (cotton, kapok, milkweed), stems (jute, flax, hemp, kenaf, thorn, bamboo), leaves (sisal, manila, abaca), regular items (coir), and various types of grass fiber. This plant produces fiber that is completely biodegradable and renewable. Plant filaments, which have a long history in human culture, are currently created on a large scale globally and have gained financial respect [16-18].

Table 3.2 Percentage of Pectin, Wax, cellulose, hemicellulose, microfibrillar, and lignin in various natural fiber sources (Zhiguang Zhou et al., 2022)

Fiber	Density (g/cm^3)	Tensile strength (MPa)	Young's modulus (GPa)	Elongation break (%)
Jute	1.23	325-770	37.5-55	2.5
Flax	1.38	700-1000	60-70	2.3
Hemp	1.35	530-1110	45	3
Ramie	1.44	915	23	3.7
Banana	1.35	721.5-910	29	2
Pineapple	1.5	1020-1600	71	0.8
Kenaf	1.2	745-930	41	1.6
Coir	1.2	140.5-175	6	27.5
Sisal	1.2	460-855	15.5	8
Abaca	1.5	410-810	41	3.4
Cotton	1.21	250-500	6-10	7

Numerous useful strands have been gotten from different parts of plants counting takes off, stems (bark strands), natural products, and seeds. The geometric measurements of these filaments, particularly the length of the fiber, depend basically on the area of the fiber within the plant. Fibers obtained from fruits and seeds typically have short lengths, only reaching several centimeters, whereas those extracted from plant stems and leaves tend to be significantly longer, often extending beyond a meter [19, 20].

In addition to a few small non-structural components (such as proteins, extractives, and minerals), vegetable fibers are typically made up of three structural polymers: the polysaccharide cellulose, the aromatic polymers hemicellulose, and lignin [21-25]. Cellulose forms a crystalline structure with high order regions, namely crystalline regions, and low order regions, namely amorphous regions. Long lamellae consisting of pectic polysaccharides connect individual cells in bundles [26].

C. Sources and Properties of Some Natural Fibers

Material filaments are broadly classified as common strands and man-made strands, as appears in Table 4.3. Common strands refer to filaments that occur naturally and are present in minerals (asbestos), organisms (protein strands), and vegetables as individual plants (cellulose filaments) [27-30]. Though they may have distinctive features, manufactured filaments are strands that are not found in nature. They are divided into three main groups: recovered filaments, manufactured strands, and filaments made of inorganic materials such as carbon, glass, and metals. Recovered filaments are formed by changing characteristic polymers. Manufactured strands are made of manufactured polymers. There are many materials found in nature that might be classified as sinewy. Plant filaments are derived from many plant parts, including seeds (cotton, kapok, milkweed), stems (flax, flax, hemp, kenaf, vex, bamboo), and leaves (sisal, manila, abaca). Natural products (coir) and other grass fiber sources are also sources of plant filaments [31]. This plant produces fiber that is entirely renewable and biodegradable. Plant strands, which have a long history in human civilization, have picked up financial esteem and are presently developed on an expansive, all inclusive scale. In Table 4.3 it can be seen that there is a wide range of raw material sources for natural fibers, including cellulose fibers, protein fibers, natural polymers, and synthetic and inorganic polymers. However, in this article the focus is on natural vegetable fiber raw materials or fiber from plants [32-34].

Table 3.3 Classification of Textile Fibers

NATURAL FIBERS	SYNTHETIC FIBERS
Vegetable Origin – Cellulose fiber	**Natural Polymer Based**
Seed Fiber: Cotton, Kapok	Regenerated Cellulose: Viscose, model, lyocell, cupro
Stem Fiber: Jute, Flax, Hemp, Rami, Kenaf	Regeneration Protein: Casein, Arachin Zein
Leaf Fiber: Sisal, Abaca, Henequen	Cellulose Esters: Acetate
	Rubber: Elastomer Alginate
Animal Origin – Protein Fiber	**Synthetic Polymer Based**
Wool	Acrylic, Aramid, Chlorofiber
Animal hair Angora, Mohair, Alpaca	Fluorofiber, Modacrylic, Polyamide
Sutera	Polyester, Polyethylene, Polyimide Polypropylene, Vinylal, Polyimide
Anorganik	**Anorganik**
Asbestos	Asbestos Carbon, Ceramic, Glass, Metal

a. Chemical composition of natural fibers

The chemical structure of distinctive filaments in natural fibers is derived from cellulose microfibrils floating in a lattice of shapeless lignin/hemicellulose composition. It was discovered that lignocellulosic cell dividers are often intricate structures with a spiral form and a variety of biochemical configurations [35-37]. The chemical structure of cellulose filaments is impacted by territorial topographic perspectives and climate varieties [38]. Previous findings show a variety of distinctive chemical elements that are present in a broad range of plant species, as evidenced by their widespread availability. Linear macromolecules are made up of anhydrous D-glucose cycle components connected by b1, a polymer made up of four glycosidic bonds. This provides the plant with its essential hardness, uniformity, and quality [39]. Hemicellulose may be a branched polymer that holds five- and six-carbon chain sugars in a grouping of chemical setups. Lignin is an unstructured, cross-linked polymer complex made up of loosely joined, phenylpropane moieties modified with hydroxy and methoxy that can split in a variety of ways. Since lignin may be a chemical epoxy adherent found in and between the strands of cellulose fibers, it isn't really cellulose.

A polysaccharide with a variety of uses, pectin is mostly composed of polymers refined from rapid rhamnose and glucuronic corrosive, which results in a more complicated polymer. Galactose, rhamnose, arabinose, and arabinose sugars are boundless in their supply of side chains. Calcium particles regularly interface the chains, giving the chain in the pectin-rich zone its basic judgment back. The network is composed of pectin, lignin, and hemicellulose, while the conceptual system of cellulose is joined together inside the most prevalent chemical structure of typical fiber compounds using epoxy tar. The main basis for common strands' suitability and appropriateness in a variety of applications is their chemical makeup, which varies based on plant species, age, climate, and soil conditions. Different forms of sensitivities to characteristic strands can be produced by each structure, including water, chemical, mechanical, thermal, photochemical, and natural helplessness. For instance, cellulose might be a regulating element in the expansion of fiber quality and modulus. Hemicellulose is 2.6 times more notable than lignin for its ability to assimilate tall water and naturally provoke heated corruption; lignin, on the other hand, is mostly concerned with fire and UV damage. The orientation of the microfibrils with respect to the cell axis determines the solidity of the fiber, which is another important feature of normal strands that significantly affects their mechanical properties.

Highlights and barriers of distinctive strands.

Compared to synthetic fibers, natural fibers are more readily available, less expensive, have a low density, a good modulus-weight ratio, great acoustic attenuation, require less industrial fuel, have a lower carbon footprint, and are biodegradable [40-43]. More convincing evidence is provided by other research groups supporting their claims, such as the fact that they are substantially less expensive than alternative materials for reinforced composites and need a lot less energy than traditional reinforcing fibers like carbon and glass [44-46]. However, because produced strands predominate and natural filaments have a lower steadfast quality, typical filaments have distinct confinements. As seen in Table 4, these strands exhibit imprinted inconsistencies in their mechanical and physical attributes, maintaining moisture much better while having weaker strength, less bendable quality, and lower operating temperatures.

Table 3.4 Advantages and disadvantages of concrete reinforcement using natural fiber

Strengths	Weaknesses
Low density	High water absorption
Higher and non-abrasive filling level for printing machines	Dimensional variability
Non-abrasive for processing equipment good thermal and acoustic conductivity	Poor fire resistance
Eco-friendly insulation characteristics, no harmful gas emissions, reduced thermal irritation and zero respiratory CO2 fingerprints	Lower durability and lower thermal resistance
Less energy production	Anisotropic behavior
Renewable	Sensitive to UV attacks, microbes, and fungi
The track is not brittle upon impact	Volatility of characteristics and basic value of origin

b. Improve the properties of natural fibers

The disadvantages of natural fibers, such as poor interfacial relationship of matrix fibers, poor usability, water permeability, and water absorption, are currently one of the difficulties faced by construction engineers and fields to fully apply natural fibers to various combination material products. A factor influencing the weak interfacial contacts between the polymer mixture and the fibers is the hydrophilic character of natural fibers. Therefore, the limits of natural fibers must be addressed by applying chemical management technologies to improve the morphology of the fiber surface and prevent undesired components from adhering to the surface in order to optimize the mechanical characteristics of natural fibers for prime movers [47].

c. Synthetic Fibers

Mineral or polymer bases can be used to create fabricated strands. Mineral fibers are frequently found in carbon and fiberglass. Two examples of artificial polymer surfaces used in orthoses are polyester and polyamide (Nylon and Aramid) [48]. Prosthetic socks, grips for upper member prostheses, tubular stockinettes, straps, liners, and less precise orthotic fittings like fabric spinal bodices are all frequently made of materials that resemble fabric. Stockinettes and imitation socks are the most popular uses of manufactured yarns. The most widely used materials in their enhancement are downy, cotton, or mixes of these distinctive fibers enhanced with nylon, acrylic, or other synthetic surfaces. Elastic textures can be used to create flexible textures that, for example, can be used for the straps in upper appendage prosthetic saddle frameworks. Snare and circular latches (Velcro), which are used for traditional clasps to hold hardware in place more comfortably than buckles and straps, are among the most widely used applications of nylon. Web tape is a solid, flat piece of woven texture that is used in many different applications. It is often made of nylon or polyester [49]. Non-woven textures like felt are created from material strands like fleece or polyester and can provide a smooth surface. They can also be added to coverings as fillers or cushioning. Currently, nylon texture is commonly used as an outer layer of various prosthetic liners and as a covering cloth for neoprene products [50]. To ensure a snug fit, nylon wires and cables are only used in body-operated prosthetic controls and rope suspension systems. The closed-measure prosthesis will progress with the use of a useful nylon stockinet. Since not every intended strand is circular, an assessment of the width can be obtained by determining the fiber's cross-sectional form by watching it longitudinally (a process known as optical isolation) and making estimates in between specific times [51], as was displayed within Figure 2's layout. Fiber synthetics outlast other typical filaments in terms of hardness.

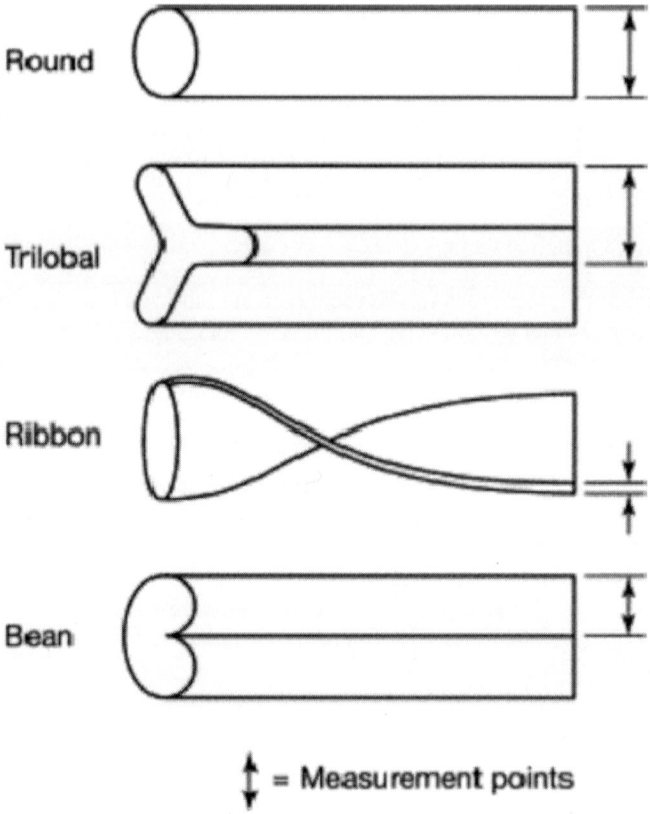

Figure 4.2 Measurement of the cross-sectional area of the fiber along with its thickness through optical cross-section (J. Ahmad dan Z. Zhou, et al., 2022)

The quality of manufactured strands and textures can also be effectively adjusted to match specific requirements by modifying their chemical qualities and the conditions under which they are produced. Depending on their properties, they can absorb more or less water and be solid, hardened, flexible, and adaptive. Compared to natural strands, manufactured filaments are usually more resistant to heat, stains, water, and chemical degradation. Regular strands are frequently more prone to chemical deterioration than synthetic strands since they are biodegradable and vulnerable to attack by a variety of bacteria and organisms. Differentially, while typical filaments decompose easily, most designed filaments are not biodegradable and rip and wear out with time. However, there are issues with produced filaments because of their lower dissolving point and chemical composition. For example, a few engineered filaments are more vulnerable to warm harm when washed in hot water. Other materials also have a more notable propensity to generate electrostatic charges when scratched. Furthermore, some of these components may cause allergic reactions and are not always good for the skin. Some synthetic materials, such as polyester and nylon, don't absorb sweat as quickly or as well as natural fibers. Sometimes, natural fibers like cotton and wool are combined with synthetic fibers to get the desired effect. Sometimes, natural and synthetic fibers are blended on spinning looms to achieve particular blended qualities. For instance, well-maintained cotton-polyester blends have the ability to shrink, withstand tearing and wrinkling, and have an extended lifespan. They also offer greater resilience to water, ripping, and the environment than the majority of fibers made from plants. On the other hand, blends of cotton and polyester are less absorbent than cotton and retain more water when applied to the skin. They might not be resistant to fire and might melt entirely if set ablaze. Table 5 illustrates the many synthetic fiber kinds that are utilized in daily life.

Table 3.5 Classification of Synthetic Fibers

Synthetic fiber	Organic	Synthetic polymer base	Polyethylene
			Aromatic Polyester
			Aramid/Kevlar
		Modified natural polymer	Alginate
			Rubber
			Zein
			Viscose
	Inorganic		Glass
			Carbon
			Boron
			Ceramics

3.2 Analysis of natural fiber mixture process and synthesis

Making bionanocomposites can be done by referring to the process of forming polymer-clay nanocomposites, the interaction mechanism (pressure reduction into the nano gallery (space), miscibility, etc.) between polymer and clay depending on polarity, molecular weight, hydrophilicity, reactive groups, etc.; others are from polymers and solvent types such as water, polar or nonpolar solutions, and mineral types as fillers. Below are several methods for making polymer/layer nanocomposites which can generally be classified into 3 types, namely, intercalated nanocomposites, flocculated nanocomposites, and exfoliated nanocomposites.

3.2.1 In situ polymerization

In situ polymerization is a conventional process for the synthesis of nanocomposites for thermosets and thermoplastics. By using this technique polymer formation can occur in intercalated sheets. The silicate layer expanding in liquid monomer serves as the foundation for the in situ polymerization process, which can be started by radiation or heating and filled with an appropriate initiator. This process has been successfully used to produce nylon/monmorillonite nanocomposites and continues to be developed for other thermoplastics. This method can also be used on thermosets.

3.2.2 Exfoliasi-adsorption

This process uses a solvent where the polymer or prepolymer is dissolved causing the silicate layer to swell. Because of its weak surface bond, the silicate layer is quickly dissolved in solvents like toluene, acetone, chloroform, and water. After the solvent has gone and the sheets are layered, sandwich-like, the polymer is absorbed into the layer and covers it. Epoxy-clay nanocomposites are processed using these methods, however, solvent removal is now crucial. The disadvantage of this process is the use of large amounts of solvent. Toyota Research Group applied this process for the first time to produce polyamide nanocomposites.

3.2.3 Melt Intercalation

The method of melt intercalation was initially documented by Vaia et al. The silicate layer is combined in a molten state with a polymer matrix during this manufacturing process, which does not call for the inclusion of solvents. Conventional techniques like extrusion and injection molding are used to mechanically combine thermoplastic and organophilic polymers at a specific temperature. Nanocomposites are then created by exfoliating or intercalating the polymer chains.

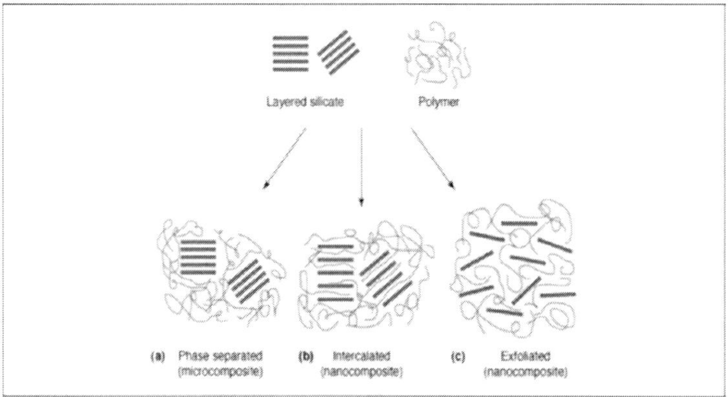

Figure 4.3 Illustration of the different types of composites that can occur as a result of the interaction between the polymer layer silicate (Sozer, 2009)

3.3 Hybrid Composite

The term "hybrid," which has multiple scientific applications, is derived from Greco-Latin. When it comes to polymer composites, hybrid composites are systems where two or more fillers and reinforcing materials are present in one matrix or both approaches combined. One type of reinforcing material is combined in a mixture of different matrices (mixture). Hybrid composites are products of the integration of two or more lignocellulosic fibers into a single matrix. A hybrid composite's behavior is the weighted total of its constituent parts, where the inherent benefits and drawbacks are more favorably balanced. When using hybrid composites containing two or more types of fibers, the advantages of one type of fiber can compensate for the disadvantages of another. As a result, balance in cost and performance can be achieved through appropriate material design. The fibers' characteristics, the aspect ratio of the fiber content, the length of each individual fiber, its orientation, the degree of intermingling of the fibers, the interfacial bonding between the fibers and the matrix, the configuration of the two fibers, and the failure strain of each individual fiber all affect the strength of hybrid composites. When the fibers are highly strain compatible, the hybrid yield is maximized. The mixture rule can be used to anticipate the characteristics of a hybrid system that consists of two components.

PH=P1V1+P2V2 (Thwe & Liao, 2003a)

PH is the property to be investigated, P1 the corresponding property of the first system, and P2 the corresponding property of the second system. V1 and V2 are the relative hybrid volume fractions of the first and second systems and V1+V2= 1. Positive or negative deviations of specific mechanical properties from the hybrid mixing rule are known as positive or negative hybrid effects. The occurrence of a notable synergistic enhancement in the qualities of composites having two or more types of fibers has been dubbed the "hybrid effect." (Jones, 1994). The criteria placed on the material or building being built, as well as the goal of the hybridization, dictate which components go into creating a hybrid composite. In the process of creating and manufacturing hybrid composites, the choice of appropriate fiber kinds and property levels is crucial.

3.3.1 Potential and challenges in development cellulose fiber hybrid composites

a. Hybridization of cellulose fibers

The fiber content, length, orientation, degree of intermingling, fiber to matrix interface, stacking pattern of the two fibers, and failure strain of the individual fibers all affect the properties of hybrid composites. When the fibers are extremely compatible with one another, the hybrid yield is maximized. A positive or negative hybrid effect is characterized by a divergence from the hybrid mixture rule of a certain mechanical property. When two or more types of fibers combine to significantly improve the properties of composite materials, the process is known as the "hybrid effect". The choice of components that form a hybrid composite is determined by the purpose of the hybridization, the requirements imposed on the material or construction being designed. The issue of selecting compatible fiber types and levels of properties is very important when designing and producing hybrid composites. To maximize the favorable aspects of one fiber and minimize its bad features, a number of researchers have experimented with mixing two fibers. Other justifications for combining two natural fibers are to create an opulent effect, lower the finished product's cost, and identify an appropriate blend of natural origin to close the gap between supply and demand. Because of their respective great tensile strengths and hardness, palm and sisal fibers seem like viable materials. Consequently, any composite made of these two fibers will display each constituent's intended qualities. Combining thermoplastics with natural fibers, like hemp and kenaf, helped FlexFormTechnologies become well-known and appear in the door panels of Chrysler's Sebring convertible. In addition, FlexForm wants to manufacture knee bolsters, instrument panel top covers, seat backs, load floors, headliners, and trunk liners for automobiles.

b. Cellulose fiber moisture content

In hybrid composites made of cellulose fiber, the fiber arrangement fills in the spaces created during the composite's production while the hybrid arrangement prevents water from penetrating the composite. This is because cellulose elements are hydrophilic, which enables the composite to absorb a lot of moisture from its surroundings. When the sample is submerged in water, the hydrophilic properties of the cellulose material and capillary action will cause an influx of water, increasing the composite's dimensions. As a result, the fibers will swell, increasing the composite's thickness swelling. The fiber has a moisture content that ranges from 5 to 15%. Water content modifies the composite's dimensions, which in turn impacts the material's mechanical qualities. Humidity during the production process of thermoset-based composites results in empty contents and damages the fiber-matrix bonds, which lowers the mechanical characteristics.

c. Dispersion of cellulose fibers in the matrix

Due to their hydrophilic nature and incompatibility with the hydrophobic polymer matrix, lignocellulosic fibers have poor adherence to the matrix, which makes it challenging for the fibers to disperse throughout the polymer matrix. Researchers studied cellulose fiber composites with polypropylene, polyethylene, and polystyrene, finding that poor fiber dispersion resulted in agglomeration into a complex mass, leading to composites with poor final properties. Several methods have been reported to improve filler dispersion and interfacial interactions between filler and matrix.

d. Fiber-matrix interface

Natural fibers react poorly to moisture, are hydrophilic, and do not work well with hydrophobic polymer matrices. The incompatibility of natural fibers leads to an inadequate fiber/matrix contact, ultimately resulting in a reduction of the composite's mechanical properties. The fiber can be chemically altered to reduce its hydrophilicity. They conducted a thorough analysis of surface modification of cellulose fibers and came to the conclusion that the chemical modification techniques that result in continuous covalent connections between the fiber surface and the matrix seem to be the most promising. Surface modification of natural fibers includes physical treatments, such as solvent extraction, and physicochemical treatments, such as the use of corona and plasma discharge or laser, ray, and UV bombardment; and chemical modification, both by direct condensation of the coupling agent onto the cellulose surface and by its grafting by free radicals.

e. Thermal stability

Exposure to natural environments can readily destroy cellulose fibers. Biological, chemical, mechanical, thermal, photochemical, and aqueous degradation techniques are a few. It is common to present cellulose fibers' biodegradability as a benefit that justifies using them. However, composites need to be functional for a number of years in outdoor applications. Controlling this natural degradation is crucial to extend their service life. Changing the chemistry of the cell wall is one strategy to stop or slow down the process of natural deterioration. This allows for the elimination or slowing down of cellulose fibers' undesirable characteristics, such as dimensional instability, flammability, biodegradability, and chemical degradation.

f. Biodegradability

Exposure to natural environments can readily destroy cellulose fibers. Biological, chemical, mechanical, thermal, photochemical, and aqueous degradation techniques are a few. It is common to present cellulose fibers' biodegradability as a benefit that justifies using them. However, composites need to be functional for a number of years in outdoor applications. Controlling this natural degradation is crucial to extend their service life. Making changes to the chemistry of the cell wall is one method of stopping or reducing the natural breakdown process. Undesirable properties of cellulose fibers such as dimensional instability, flammability, biodegradability, and chemical degradation can be eliminated or slowed down in this way. Chemical treatment can reduce water absorption in the fiber, and therefore can reduce the amount of fiber swelling and biological degradation by blocking the availability of –OH groups on the fiber surface. It has also been reported that encasing natural fibers in thermoplastics reduces water absorption.

3.4 Physical and mechanical properties of hybrid composite

3.4.1 Hybrid composites' mechanical and physical characteristics

Fiber content, fiber length, fiber orientation, individual fiber arrangement, degree of fiber mixing, and interfacial adhesion between fibers and matrix all affect the physical and mechanical properties of hybrid composites. The majority of research on natural fiber hybrid composites focuses on the mechanical properties as a function of the two fibers' arrangement, bonding to the matrix, degree of intermingling, length, and loading, as well as the impact of different chemical treatments on the fibers.

3.4.2 Thermoset hybrid composite

Many researchers have created hybrid composites by fusing fibers with polyester, polyvinyl ester, epoxy resin, phenolic acid, and polyurethane. Table 5 displays the work presented on hybrid composites of cellulose/cellulose fiber and cellulose/synthetic thermoset. The study discussed hybrid thermoset composites made of natural fibers.

3.4.3 Thermoplastic hybrid composites

Thermoplastic hybrid composites, developed by various researchers, combine fibers with polypropylene, polystyrene, polyethylene (low and high density), poly vinyl chloride, polycarbonate, and natural rubber. The reported work on cellulose/cellulose and cellulose/synthetic fiber-reinforced thermoplastic hybrid composites is shown in Table 6.

Table 3.6 Report on research on thermoplastic polymer composites with natural fiber hybridization

Hybrid fiber	Matrix polymer	Reference
TKKS/glass palm oil	Polypropylene (PP)	Rozman et al. (2001a)
Jute/glass	Polypropylene (PP)	Esfandiari (2007)
Sisal/glass	Polypropylene (PP)	Jarukumjorn and Suppakarn (2009), Nayak and Mohanty (2010), Nayak et al. (2009), and Schmidt et al. (2009) Kalaprasad et al. (2004)

	Polyethylene (PE)	Nayak et al. (2010) and Samai et al.
	Polypropylene (PP)	(2009a)
Banana/ glass		
	Polystyrene	Haneefa et al. (2008)
	Polypropylene (PP)	Nayak et al. (2009, 2010a)
Bamboo/glass		Samal et al. (2009b) and Thwe and Liao (2002, 2003a, 2003b)
Hemp/ glass	Polypropylene (PP)	Panthapilakkal and Sain (2007) and Reis et al. (2007) Arbelaiz et al. (2005)
Hemp/ glass	Polypropylene (PP)	Jiang et al. (2003)
Cotton/hemp	Polyethylene (PE)	Tasdemir et al. (2008) Haseena et al. (2004, 2005, 2007) Anuar et al. (2006)
Wood flour/glass	Polyvinyl chloride (PVC)	
Cotton/silk waste	Polycarbonate (PC)	
Coir/sisal	Natural rubber	Wan busu dkk (2010)
TKKS/glass palm oil	Natural rubber	

A. Electrical properties of hybrid composites

It is clear from the literature review that very limited research has been reported on the electrical properties of hybrid composites. The electrical properties of banana/glass hybrid fiber-reinforced composites with various hybrid ratios and coating patterns were analyzed (Joseph and Thomas, 2008). Researchers looked into the hybrid biocomposites made of sisal and oil palm's dielectric properties and volume resistivity. It is seen that the dielectric constant increases with fiber loading at all frequencies and the volume resistivity decreases with frequency and fiber loading, implying that the conductivity increases upon addition of lignocellulosic fibers (Jakub, Francis, et al., 2006; Yakub, Jose, et al., 2006; Jacob, Varughese, et al., 2006). Chemical alteration of the fiber causes the dielectric constant to drop and the volume resistivity to rise. With increased fiber loading, the dissipation factor increases, suggesting that the electrical charge can be sustained for an extended amount of time. Jute/glass fiber-reinforced epoxy composites' electrical characteristics were investigated (Koradiya et al.,

2010). The results showed that the hybrid composite had intermediate electrical properties compared to jute and glass composites. Similar studies on the electrical properties of sisal/glass hybrid composites (treated and untreated) reinforced in a mixture of epoxy and bisphenol-C formaldehyde have been carried out (Patel & Parsania, 2010). Hemp-biomass hybrid composites' electrical strength and volume resistivity have been assessed and contrasted with hemp-bisphenol-C-formaldehyde (BCF) composites. There is little difference in the dielectric breakdown strength (1.21-2.11 kV/mm) between the BCF-jute composite and the hybrid composite. (1.41 kV/mm) was observed but the volume resistivity of hybrid composite was mainly for BCF-jute-husk wheat and BCF-jute-jamun flower peel increased 197–437%. In an interesting work, the researchers studied the electrical properties such as dielectric constant, volume resistivity, and dielectric loss factor of a sisal/coir hybrid fiber-reinforced natural rubber composite in relation to fiber loading, fiber ratio, frequency, fiber chemical modification, and presence of a binding agent. The dielectric constant and volume resistivity values of sisal/coconut fiber hybrids show the same behavior as reported in relation to fiber loading and frequency. Chemically treated fiber composites exhibit a decrease in dielectric constant value as a result of an increase in fiber hydrophobicity. The dielectric constant of the composite is decreased by adding a two-component dry binding agent made of resorcinol and hexamethylene tetramine, which is intended to improve interfacial adhesion between the matrix and fibers.

B. Thermal properties of natural fiber hybrid composites

Researchers studied the thermal properties of banana/glass hybrid composites, they observed that MAPP treated Banana/glass hybrid composites (BSGRP) showed higher thermal stability compared to banana fiber-reinforced polymer composites. This could be as a result of the glass fibers' SiO groups interacting with the MAPP anhydride groups to create a synergistic effect between the glass and banana fibers. The hybrid composite's increased melting point, crystallization temperature, and first thermal degradation temperature were also validated by thermal analysis. Comparable investigations using thermogravimetric analysis and differential scanning calorimetry showed that bananas treated with glass fiber and MAPP raised the temperature at which polypropylene crystallized and maintained its thermal stability. When glass fibers were added, the thermal properties of the sisal/PP composites were found to be enhanced. This was the result of an investigation into the thermal properties of the sisal/glass hybrid PP. The highest Td5 (temperature at which weight loss was 5%) and Td50 (temperature at which weight loss was 50%) were obtained by the

PP/10sisal/20GF/PP-g-MA composite, whereas the lowest Td5 and Td50 were obtained by the PP/sisal composite. When glass fibers were included in the sisal-PP composite, there was an additional rise in the heat distortion temperature (HDT). The thermal properties of sisal/glass hybrid composites were also studied by Nayak and Mohanty (2010) and also confirmed higher thermal stability in the case of hybrid composites. The crystallization temperature (Mongkollapkit, Kositchaiyong, Rosarpitak, & Sombatsompop, 2010) of the 2 wt% MAPP treated sisal/glass hybrid was found to be more than that of the composite without MAPP, indicating further enhancement in the nucleation process in the presence of compatibilizer. Differential scanning calorimetry (DSC) and thermo gravimetric analysis (TGA) demonstrated the enhancement of the thermal stability of the matrix polymer with the incorporation of bamboo and glass fibers, confirming the hybridization effect and efficient interfacial adhesion of the fiber matrix (Nayak et al., 2009). Additionally, it was shown that the PP matrix's melting temperature was unaffected significantly by the inclusion of bamboo, glass, or MPP fiber. Nevertheless, the PP matrix's linear crystallization order was upset and the degree of crystallization was reduced by the addition of fiber and MAPP. In a related investigation, they found that adding bamboo and glass fibers improved the polymer matrix's heat stability, maybe as a result of improved fiber/matrix adhesion (Nayak, Mohanty, & Samal, 2010a). The incorporation of glass fibers resulted in a considerable increase in the thermal stability of both composite systems which may be due to the higher thermal stability of glass fibers than bamboo fibers (Lee & Wang, 2006). Furthermore, the bamboo/glass hybrid composite system also showed maximum charred residue indicating higher fire resistance of the system. Samal, Mohanty, and Nayak (2009b) also studied the thermal properties such as crystallization, melting behavior, and thermal stability of bamboo/glass hybrid composites. TGA demonstrated an increase in the thermal stability of the polymer matrix with the incorporation of bamboo and glass fibers, confirming the hybridization effect and efficient interfacial adhesion of the fiber matrix. The thermal properties of jute fiber composites are improved by hybridization with glass fibers (Panthapulakkal & Sain, 2007). The degradation of hemp fiber composites and hybrids was shown to occur in two steps, and the hybridization of hemp fiber composites with glass fibers caused the degradation temperature to shift to higher values, suggesting that the hybrid composites were more thermally stable. The thermal characteristics of natural rubber composites reinforced with hybrids of sisal and palm palm fibers have been the subject of intriguing investigations (Jakub, Francis et al., 2006; Yakub, Jose et al., 2006; Jacob, Varughese et al., 2006) and the stability Composite thermal properties were seen to increase upon fiber loading and chemical modification. The thermal properties of kapok/sisal hybrid composites with and without alkali treatment were analyzed by differential scanning calorimetry (Venkata Reddy, Shobha Rani et al., 2009; Venkata Reddy, Venkata Naidu et al., 2009). The findings demonstrated that, in comparison to the untreated hybrid composite, the treated kapok/sisal hybrid composite exhibited a somewhat higher melting temperature.

C. Dynamic mechanical properties of natural fiber hybrid composites

Dynamic mechanical analysis (DMA) is one of the most powerful tools for studying the behavior of polymer composite materials and allows quick and easy measurement of material properties (Swaminathan & Shivakumar, 2009). The dynamic properties of a randomly oriented hybrid banana/sisal fiber-reinforced polyester composite were determined (Idicula, Malhotra, Joseph, & Thomas, 2005a). Dynamic characteristics including damping behavior (tansaya) and storage modulus (Ekan) were studied in relation to the relative volume fraction of two fibers and the total fiber volume fraction. The 3:1 banana to sisal volume ratio in the hybrid composite results in tan; at Tg they are the lowest I value and the highest E value. Additionally, they looked at how layering patterns affected the loss modulus (Ekan), storage modulus (Ekan), and damping behavior (tan saya) of composites, trilayers (banana/sisal/banana and sisal/banana/sisal), and bilayers (banana/sisal); intimate combination is achieved by keeping the overall fiber loading at a volume fraction of 0.40 and the relative volume fractions of banana and sisal at 1:1 as a function of temperature and frequency (Idicula, Malhotra, Joseph, & Thomas, 2005b). In contrast to the other patterns, the tightly blended banana/sisal/banana composite showed higher stiffness, whilst the bilayer composite had strong damping qualities. It was discovered that the densely packed composite had the maximum activation energy. Dynamic mechanical analysis was performed on natural rubber reinforced with sisal and palm fiber to ascertain its dynamic properties as a function of temperature (Jakub, Francis et al., 2006; Yakub, Jose et al., 2006; Jacob, Varughese et al., 2006). Because natural fibers provide greater stiffness, it was discovered that storage modulus (Ekan) increased with fiber weight percentage and decreased with loading. Tan, on the other hand, was observed to decrease. Maximum E kan was shown by natural fiber hybrid composites treated with 2% and 4% NaOH, respectively. Tan values of chemically treated composites are obviously lower than those of untreated composites. An unsaturated polyester matrix containing glass fibers and pineapple leaf fiber (PALF) was used to create randomly oriented hybrid composites (PALF/glass/PALF) and glass/PALF hybrid composites (glass/PALF/glass). These hybrid composites' dynamic mechanical properties were examined (Uma Devi, Bhagawan, & Thomas, 2010). In comparison to the PALF/glass/PALF composite, the intimate blend and trilayer hybrid polyester composite (glass/PALF/glass) offer higher storage modulus values. Glass fiber and PALF together result in a significant reduction in the cost of polyester resin. Dynamic mechanical investigation of the glass/banana hybrid polyester composite was performed at three different frequencies and a variety of temperatures (Pothan, George, John, & Thomas, 2010). The E value of the hybrid composite decreases above the Tg temperature where glass is the core material. It was observed that the coating pattern of the composites showed a significant effect on the dynamic properties of the composites and the tightly blended composites showed the highest Ekan values in all the compositions. DMA results showed that the storage modulus and loss modulus of MAPP-treated banana/glass hybrid composites increased over the entire temperature range, indicating better adhesion between fibers/matrix (Samal et al., 2009a).

3.5 Application of natural and synthetic fiber based composites

3.5.1 Mechanical properties of natural and synthetic fibers

a. natural fiber

Since ancient times, mankind has utilized natural fibers. The use of natural fibers in composites has drawn a lot more attention in the last few years [1–15]. Fibers are utilized as reinforcement in fiber-reinforced composites, while polymers serve as the matrix. Animal, vegetable, and mineral fibers are the three primary categories into which natural fibers can be separated, as shown in Figure 1. Research has been done on a variety of plant fiber types for use as reinforcement in composites. Natural fibers are becoming more and more abundant in production, not just because they are easy to work with and have many environmental benefits [17]. Table 1 displays the mechanical characteristics of natural fibers. The chemical makeup of natural fibers is another crucial factor in their use in a variety of applications. Table 2 displays the composition of various natural fibers under investigation [18, 19]. Hemicellulose, cellulose, pectin, lignin, wax, and water-soluble materials are the components of natural fibers. Even among the same plant species, the chemical makeup of fiber might differ due to environmental influences, geographic location, and techniques used to remove the fiber.

The majority of chemical fiber treatments include surface alteration, as [2,7,20-23] explains. The strength of natural fibers and traditional reinforcing fibers, like glass fiber, currently differs by a significant amount. The primary reason natural fibers cannot fully replace glass fibers is the difference in strength. Thus, in an effort to create new materials with superior properties than the original materials, researchers started looking into the hybridization of composites, which is the process of merging two or more types of fiber. A laminate's composite structure is made up of layers of reinforced fibers that are woven together and a matrix that holds the layers together and gives the assembly stiffness as a whole. Because each component of the composite material has maintained its original chemical, physical, and mechanical properties, its properties are distinct from those of metal alloys.

The primary determinants of composites' mechanical characteristics. Tensile tests [24], thermal buckling analysis [25], post-buckling analysis [26,27], bending analysis [28], and bending analysis [29,30] are a few methods of testing mechanical properties on a composite structure. According to Mustapha et al.'s research [31], more observation can be done after the structure has been tested to ascertain whether it has sustained total damage or can still be used for a given amount of time.

b. synthetic fiber

In polymer matrix composites (PMCs), synthetic fibers are specialized reinforcements that are frequently utilized to create lightweight materials with higher stiffness, modulus, and strength. PMC's reinforcement is one of its key characteristics. Retention serves the primary purpose of supporting the load placed on the composite. However, in order for the reinforcement to serve its intended purpose, it must meet a number of fundamental requirements, including compatibility with the matrix, the ability to form a chemical bond or adhesion with the matrix, the possession of superior qualities, the best possible composite orientation, and an appropriate shape. The current research provides an in-depth analysis of the developments made in synthetic fibers for different types of reinforced composites.

The primary characteristics, mechanisms of failure, and applications of synthetic fiber-based composites are described in depth based on the specified parameters and kind of fiber (organic or inorganic). Furthermore, the physical and mechanical qualities of synthetic fibers, together with their synthesis process, play a major role in the classification, applications, and properties that are chosen for them. Lastly, several obstacles and directions for future study will be emphasized. The four steps in the manufacturing process of synthetic fibers are chemistry, spinning, winding, and packaging. Usually, polymerization is the chemical process used to arrange and combine the fiber sections. The process of polymerization involves the recurrence of essential units to create macromolecules. The various components are initially solid and need to be transformed into a liquid in order to be released into fibers. These substances are processed artificially, melted, or crushed to produce a viscous liquid. Next, a viscous liquid is forced into the spinneret during the spinning cycle to deliver the fiber. An instrument with several apertures of a fixed width is called a spinneret. The liquid exits the spinneret as liquid rope strands after being contained by its apertures. The application sets the fiber measurement, which is determined by the spinneret opening. The ejects is emptied until it turns into unwavering fiber. The fiber becomes yarn through the winding cycle. From the spinneret, the fibers fall upward and become entangled in a sizable vacuum jet. As the pressure is packed and circulated around the coil for future use, the vacuum force maintains its availability [21]. However, electrospinning is the method of obtaining fiber that is most frequently utilized. Electrospun fibers are produced by a manufacturing process that uses interactions powered by electrostatic forces. These fibers are spaced out at regular intervals of a few nanometers to a few micrometers. The ability to manipulate the electrospinning process to create filaments with a variety of action possibilities and morphological constructs is one of its key features. Scientists are becoming more interested in concentrating on the remarkable properties of materials at the nanoscale due to the advancement of nanotechnology. Due to its versatility and prospective uses in a variety of industries, electrospinning—a process for creating or melting electrostatic fiber in polymer assemblies—has recently garnered increased interest and attention. Compared to traditional methods, the fibers produced by this process more closely resemble portions of extracellular tissue. These interactions produce twisted filaments in the submicron region, which have several benefits including high surface area to volume ratio, tunable porosity, and controllable nanofiber structure to achieve desired functionalities and attributes [22]. The most sought-after material in the market for fiber-reinforced composites is now SF. Every kind of composite reinforced with SF has comparable significance for different industrial application levels [23]. The reinforced portions or fillers in the composite matrix are discrete components within the overall volume of the composite material, and they play a major role in determining the composite's characteristics [24]. Based on their constituent parts—the kind of matrix material, the orientation of the reinforcement, and the composite production process—all polymer composite materials may only be loosely categorized [25]. Crucial details about SF show that it greatly boosts the load-bearing capacity by enhancing toughness, fatigue resistance, and impact resistance; as a result, it can be used with concrete for tasks involving building [26,27].

3.5.2 Application of Natural and Synthetic Fiber-Based Composites

Researchers are being drawn by environmental consciousness to develop new composites that combine many natural resource enhancements through hybridization. By combining fillers with natural fibers, a process known as hybridization, the composite's mechanical qualities are enhanced [28]. There is a wealth of literature outlining the mechanical characteristics of composites made of natural fibers. Numerous factors, such as the volume or weight percentage of reinforcement, fiber orientation, fiber aspect ratio, fiber-matrix adhesion, fiber alignment, distribution, usage of additives, and chemical treatment of the fibers, can affect the mechanical performance of fiber-reinforced composites. It is noteworthy to mention that the mechanical behavior of composites is impacted by their moisture absorption, which might result in inadequate interfacial bonding between the hydrophobic matrix polymer and the fibers [29]. Various kinds of natural fiber elements for interior components have been actively produced by the automotive and aircraft sectors [30-32]. Natural fibers are also used to make insulation materials for a variety of applications, including impact sound insulation, blown insulation, poured insulation, ceiling panels for thermal insulation, and acoustic soundproofing [33]. Natural fibers offer a vast range of construction materials, shapes, and even better building materials than those often utilized now, indicating a sustainable future for architecture. Natural fiber reinforced polymer composites have showed a lot of promise for use in food packaging, home furnishings, automotive body components, biomedical buildings, and residential applications [34]. To create hybrid composites, natural fibers can be combined with other natural fibers and added to a polymer matrix. This can improve the performance and characteristics of the final material while balancing the cost of natural fibers [35]. High stress can be reached to achieve the ideal tensile strength of natural fiber-reinforced fibers, according to Edhirej et al. [36]. Natural fiber reinforced hybrid composites are the subject of current research as a viable substitute for synthetic fibers like carbon and glass fibers. The mechanical characteristics of natural-natural fibers reinforcing hybrid composites have been studied by numerous researchers [37-39]. One of these researchers, Das [40], created a polyester hybrid composite reinforced using waste paper and woven jute cloth, and he examined the mechanical characteristics of both the composite and the hybrid. Cotton, which is a natural fiber formed from seeds, is commonly used as a raw material for fine textiles. Kapok, on the other hand, is typically used as a filler in the furniture industry and is not suitable for fine textiles due to its fiber length, which is less than 2 inches for spinning. Textiles, including technical and ready-to-wear textiles, are frequently created from stem fibers like hemp and flax. But because stem fiber has a high strength and elongation, it can also be used effectively as reinforcement when creating composites.

a. Composite based on natural fiber textiles

The numerous reinforcing materials that make up composite textiles with reinforcing materials are arranged as textile preforms, which can be made of knitted, woven, or non-woven materials. The pros and cons of each technology must be taken into account in order to select the best one for creating textile preforms. Composite structures are created using a variety of techniques: the base material is consolidated by heating (curing) the layers, creating lamination; the reinforcing materials (textile weaves) are embedded into the matrix. These materials can be macromolecular substances, colloidal solutions, or suspensions with coagulation properties. Textiles can be used on an industrial scale for performance, economy, and ease of design, according to studies on a number of technologies [41]. Although it is the fundamental raw material used in the textile industry, textile fiber is also a vital component of many other industrial sectors. Technical textiles can be made essentially from any kind of fiber, including fibers derived from reusable materials [42]. Reusable textile materials, sometimes referred to as textile waste, are primarily produced by fabrication, machining, and other textile processing industries, as well as by the natural or artificial deterioration of textile goods [43]. During the weaving process, reinforcing bars, strands, or threads that interlock with over-or under-placing are commonly used to build woven structural fabrics. The most popular types of fabrics are satin, keeper, and plain weave. The most popular type of weave is called plain weave, which is produced by alternating each fiber between the longitudinal (warp) and transverse (weft) directions of the weave.

The mass/strength ratio, shorter travel times, elimination of waste from the manufacturing and processing processes, improved control over the product's final shape, and superior quality are all benefits of textiles [44]. Textile materials also provide you with direct control over the properties of the material right from the design phase. Textile materials can also have a three-dimensional structure, and some of the most intricate fittings can be produced without the need for preforms, which can save money and time during the production process and eliminate the potential of regulating the final shape from the material design stage [45,46].

Technological advancements in textiles, including knitting, weaving, and braiding, have produced textile composites with improved mechanical qualities. Because they are highly suitable for structural purposes and offer great integrity, woven fabrics are an appealing option for reinforcing. Improved mechanical qualities, particularly interlaminar or interfacial strength, lower production costs, and superior drapeability are the main reasons why woven fabrics are used more often than non-woven ones. In textile reinforcement, interconnectivity between neighboring fibers adds extra interfacial strength to the comparatively weak fiber-resin interface. Three-dimensional braided composites have a great deal of potential for ballistic protection applications due to their non-delamination properties under ballistic impacts [47]. Creating various textile preforms is a crucial phase in the field of composite technology.

b. Application of synthetic fiber

All synthetic fibers are made in laboratories using non-natural polymers, usually leftovers from the refining of petroleum. Synthetic fibers can be processed via spinning, polymerization, and filament processing. To create filaments from synthetic polymers, the spinning process is further divided into three subcategories, including melt spinning, wet spinning, and dry spinning [48]. To make a filament, the viscous polymer is run through a spinneret, which is a tiny plate that resembles a nozzle and has numerous tiny holes in it. A substance can be polymerized to create a high molecular weight complex with a variety of physical characteristics. The by-product spun from the process is called a polymer [49]. Synthetic filament yarns form an endless number of fine filaments. Static fibers are created by cutting continuous filaments short; these fibers are then spliced and supplied to the textile industry to generate yarn. Yarn, another product of synthetic fibers, comes in three forms and is similarly derived from tow.

Although SF has many drawbacks, including being flammable, not ideal for hot washing, producing microplastic pollution, having a low insulating capacity, melting easily, being vulnerable to heat damage, and not being environmentally friendly, its main benefits are that it is stretchable, waterproof, moisture-proof, durable, and wear-resistant. In order for SF to be employed as effectively as possible, it must be compatible with the matrix, have better mechanical and physical qualities than the matrix, have the best orientation inside the composite, and form a chemical link or adhesion with the matrix.

The non-biodegradable characteristics and lower melting point of SF are the largest obstacles for manufacturers. SF is behind the rest of the market because of safety and environmental issues. There is a fantastic chance to concentrate on science fiction research in order to create additional technologies that offer improved technical and safety aspects. Furthermore, it is imperative that significant consideration be given to incorporating advantageous modifications into SF that have few or no negative environmental effects. Research on SF-reinforced bioplastics, plastic nanocomposites, self-healing polymers, plastic electronics, and smart and reactive polymers should be directed toward addressing these issues. Research across multiple disciplines will enhance the possibilities for biodegradation and biorecycling of SF.

Conclusion

In the journey the type of fiber used and the manufacturing procedure affect the qualities of SF. Because SF has better mechanical, tribological, thermal, and chemical resistance than traditional metals, its use for a variety of applications is growing globally. SF is divided into three categories: inorganic, organic, and other fibers, depending on its chemical makeup. Organic fibers have been a part of our daily lives since prehistoric times. However, inorganic fibers have lately been produced and shown to have exceptional qualities, making them a very promising subject for further investigation compared to organic fibers.

References

1. KS Kene, VS Vairagade, S. Sathawane, Experimental study of the behavior of steel and glass fiber-reinforced concrete composites, Bonfring international journal of industrial engineering and management science 2 (4) (2012) 125-130.
2. KR Kashyzadeh, NA Kesheh, Study of types of Cracks in construction and their control, International Journal of Emerging Technologies and Advanced Engineering 2 (8) (2012) 1-4.
3. H. Carvalho, A. Raposo, I. Ribeiro, J. Kaufmann, U. Götze, P. Peças, E. Henriques, Application of a Life Cycle Engineering approach to

assess the relevance of the use of natural fibers in rocker composites-case study, Procedia CIRP 48 (2016) 364–369.
4. T. Rohan, B. Tushar, G. Mahesha, Review of natural fiber composites. IOP Conference Series: Materials Science and Engineering, 2018.
5. T. Sathishkumar, P. Navaneethakrishnan, S. Shankar, R. Rajasekar, N. Rajini, Characterization of natural fibers and composites–A review, Journal of Reinforced Plastics and Composites 32 (19) (2013) 1457–1476.
6. Muhammad, L., Ansari, MN, Pua, G., Jawaid, M., & Islam, MS (2015). An overview of Natural Fiber-Reinforced Polymer composites and their applications. International Journal of Polymer Science, 2015.
7. G. Musalaiah, BV Suresh, M. Tanniru, P.R. Reddy, Development and Mechanical Characterization of Reinforced Glass Fiber, Polymer Composites 8 (7) (2017) 231–239.
8. TD Tavares, JC Antunes, F. Ferreira, HP Felgueiras, Biofunctionalization of natural fiber-reinforced biocomposites for biomedical applications, Biomolecules 10 (1) (2020) 148.
9. TYG Girijappa, et al., 'Natural Fibers as a Sustainable and Renewable Resource for the Development of Eco-Friendly Composites: A Comprehensive Review', Frontiers in Materials, 6 (September), DOI 10 (2019) 1–14.
10. S. Shahidi, J. Wiener, M. Ghoranneviss, Surface modification methods to increase the dyeability of textile fabrics, Eco-friendly textile dyeing and finishing 10 (2013) 53911.
11. M. Sanjay, G. Arpitha, P. Senthamaraikannan, M. Kathiresan, M. Saibalaji, B. Yogesha, Effect of Jute/Kenaf/E-glass woven fabric epoxy composite hybrids for medium load applications: Impact, inter-laminar strength, and failure surface characterization, Journal of natural fibers. (2018).
12. JA Andrade, AP Duarte, MC Simoes, Education and health: composition and growth of welfare states across country groups, Eastern Journal of European Studies 9 (2) (2018).
13. P. Kaur, M. Talwar, Different types of Fibers used in FRC, International Journal of Advanced Research in Computer Science 8 (4) (2017).
14. Geremew, A., De Winne, P., Demissie, T.A., & De Backer, H. (2021). Treatment of Natural Fibers for Applications to Concrete Pavements. Advances in Civil Engineering, 2021.
15. M. Bahrami, J. Abenojar, M.Á. Martínez, Recent Advances in Hybrid Biocomposites: Mechanical Properties, Water Absorption and Fire Resistance, Materials 13 (22) (2020) 5145.

16. Abral, H., Dalimunthe, M.H., Hartono, J., Efendi, R.P., Asrofi, M., Sugiarti, E., SM Sapuan, S.M., Park, J.W., Kim, H. (2018) Characterization of tapioca starch biopolymer composites reinforced with micro scale water hyacinth fibers. Starch - Stärke, 70(7-8).
17. Aji, T.A., Purwanto, H., & Respati, S. M. B. (2018) Pengaruh ketebalan komposit matrik resin dengan penguat kulit eceng gondok (eichhornia crassipes) yang dianyam terhadap kemampuan balistik, Momentum, Vol. 14(1), 75-79.
18. Asrofi, M., Abral, H., Kasim. A., Pratoto, A., Mahardika, M., & Hafizulhaq, F. (2018) Mechanical Properties of a Water Hyacinth Nanofiber Cellulose Reinforced Thermoplastic Starch Bionanocomposite: Effect of Ultrasonic Vibration during Processing. Fibers, 6(40), 1 – 9.
19. Asrofi, M., Abral, H., Putra, Y.K., Sapuan, S.M. & HyunJoong Kim. (2018) Effect of duration of sonication during gelatinization on properties of tapioca starch water hyacinth fiber biocomposite. International Journal of Biological Macromolecules, 108, 167– 176.
20. Athijayamani, A., Thiruchitrambalam, M., Manikandan, V., & Pazhanivel, D. (2009) Mechanical properties of natural fibers reinforced polyester hybrid composite, Int J Plast Technol, 13(1), 1-12.
21. Bahra, M.S., Gupta, V. K. & Aggarwal, L. (2017). Effect of Fiber Content on Mechanical Properties and Water Absorption Behaviour of Pineapple/HDPE Composite. Materials Today: Proceedings, Vol 4 (2), Part A, 3207-3214.
22. Balachandar, M., Vijaya Ramnath, B., Barath, R. & Bharath Sankar, S., (2019) Mechanical Characterization of Natural Fiber Polymer Composites. Materials Today: Proceedings, Vol. 6 (2), 1006-1012.
23. Baley, C., Gomina, M., Breard, J., Bourmauda, A., & Davies, P. (2019). Variability of mechanical properties of flax fibers for composite reinforcement. A review., Industrial Crops & Products, Article In Press Available online 22 November 2019.
24. Bhuvaneshwari, M. & Sangeetha (2017). Development of Water Hyacinth nonwoven fabrics for thermal insulation. Journal on Future Engineering & Technology, 13(1), 22 – 29.
25. Boca Raton, FL: CRC Press. Bordoloi, S., Kashyap, V., Garg, A., Sreedeep, S., Wei, L. & Andriyas, S. (2018) Measurement of mechanical characteristics of fiber from a novel invasive weed: A comprehensive comparison with fibers from agricultural crops, Measurement, 113, 62–70.
26. Cai, M., Takagi, H., Nakagaito, A.N., Yan Li & Waterhouse, G.I.N., (2016). Effect of alkali treatment on interfacial bonding in abaca fiber-

reinforced composites. Composites Part A: Applied Science and Manufacturing, Vol. 90, 589-597.

27. Chonsakorn, S., Srivorradatpaisan, S., & Mongkholrattanasit, R. (2018) Effects of different extraction methods on some properties of water hyacinth fiber. Journal of Natural Fibers, 1 – 11.
28. Choudhary, A. K., Chelladurai, H., Kannan, C. (2015) Optimization of combustion performance of bioethanol (water hyacinth) diesel blends on diesel engine using response surface methodology. Arabian Journal for Science and Engineering. 40, 3675– 3695.
29. Deb, A., Das., S., Mache, A. & Laishram, R. (2017) A study on the mechanical behaviors of jute-polyester composites, Procedia Engineering 173, 631-638.
30. Elanchezhiana, C., Ramnath, B.V., Ramakrishnan, G., Rajendrakumar, M., V. Naveenkumar, V., & Saravanakumar, M.K. (2018) Review on mechanical properties of natural fiber composites. Materials Today (Proceedings 5), 1785–1790.
31. M. J. Mochane, T. C. Mokhena, T. H. Mokhothu, A. Mtibe, E. R. Sadiku, S. S. Ray, I. D. Ibrahim, O. O. Daramola. Recent progress on natural fiber hybrid composites for advanced applications: A review. eXPRESS Polymer Letter Vol.13, No.2, 159-198 (2019). https://doi.org/10.3144/expresspolymlett.2019.1 5
32. Kiruthika A.: A review on physico-mechanical properties of bast fiber-reinforced polymer composites. Journal of Building Engineering, 9, 91–99 (2017). https://doi.org/10.1016/j.jobe.2016.12.003
33. Ahmad F., Choi H. S., Park M. K.: A review: Natural fiber composites selection in view of mechanical, light weight, and economic properties. Macromolecular Materials and Engineering, 300, 10–24 (2015). https://doi.org/10.1002/mame.201400089
34. Saba N., Paridah M. T., Abdan K., Ibrahim N. A.: Dynamic mechanical properties of oil palm nano filler/ kenaf/epoxy hybrid nanocomposites. Construction and Building Materials, 124, 133– 138 (2016). https://doi.org/10.1016/j.conbuildmat.2016.07.0 59
35. Gurunathan T., Mohanty S., Nayak S. K.: A review of the recent developments in biocomposites based on natural fibres and their application perspectives. Composites Part A: Applied Science and Manufacturing, 77, 1–25 (2015). https://doi.org/10.1016/j.compositesa.2015.06.0 07
36. Pickering K. L., Efendy M. G. A., Le T. M.: A review of recent developments in natural fibre composites and their mechanical performance. Composites Part A: Applied Science and Manufacturing, 83, 98–112 (2016). https://doi.org/10.1016/j.compositesa.

37. Khan, M. A., Hassan, M. M., and Drzal, L. T. Effect of 2-hydroxyethyl methacrylate (HEMA) on the mechanical and thermal properties of jute- polycarbonate composite. Compos. Part A Appl. Sci. Manufact. 36, 71–81 (2005). https://doi:10.1016/S1359-835X(04)00178-2
38. Borba, P. M., Tedesco, A., and Lenz, D. M. Effect of reinforcement nanoparticles addition on mechanical properties of SBS/curauá fiber composites. Mater. Res. 17, 412–419 (2013). https://doi:10.1590/S1516-14392013005000203
39. Biswas, S., Shahinur, S., Hasan, M., and Ahsan, Q. Physical, mechanical and thermal properties of jute and bamboo fiber-reinforced unidirectional epoxy composites. Proc. Eng. 105, 933–939 (2015). doi: 10.1016/j.proeng.2015.05.118
40. Dixit, S., Goel, R., Dubey, A., Shivhare, P. R., and Bhalavi, T. Natural fibre reinforced polymer composite materials - A review. Polym. Renew. Resour. 8, 71–78 (2017). https://doi:10.1177/204124791700800203
41. Pickering, K. L., Efendy, M. G. A., and Le, T. M. A review of recent developments in natural fibre composites and their mechanical performance. Compos. Part A. Appl. Sci. Manuf. 83, 98–112 (2016). https://doi:10.1016/j.compositesa.2015.08.038
42. Arpitha, G. R., and Yogesha, B. Science direct an overview on mechanical property evaluation of natural fiber-reinforced polymers. Mater. Today Proc. 4, 2755–2760 (2017). https://doi:10.1016/j.matpr.2017.02.153
43. M. Jawaid, M. Thariq, and N. Saba (Woodhead Publishing), 103-122 (2019). https://doi:10.1016/B978-0-08102292-4.00006-0
44. Asim, M., Saba, N., Jawaid, M., and Nasir, M. "Potential of natural fiber/biomass fillerreinforced polymer composites in aerospace applications," in Sustainable Composites for Aerospace Applications, eds M. Jawaid and M. Thariq (Elsevier Ltd.), 253–268 (2018). https://doi:10.1016/B978-0-08-102131-6.00012-8
45. Sanjay, M. R., Arpitha, G. R., and Yogesha, B. Study on mechanical properties of natural - glass fiber-reinforced polymer hybrid composites: a review. Mater. Today Proc. 2, 2959–2967 (2015). https://doi:10.1016/j.matpr.2015.07.264
46. Sanjay, M. R., Arpitha, G. R., Senthamaraikannan, P., Kathiresan, M., Saibalaji, M. A., and Yogesha, B. The hybrid effect of jute/kenaf/e-glass woven fabric epoxy composites for medium load applications: impact, inter-laminar strength, and failure surface characterization. J. Nat. Fibers 16, 600–612 (2019). https://doi:10.1080/15440478.2018.1431828

47. Sanjay, M. R., Arpitha, G. R., Naik, L. L., Gopalakrishna, K., and Yogesha, B. Applications of natural fibers and its composites: an overview. Nat. Resour. 7, 108–114 (2016). https://doi:10.4236/nr.2016.73011
48. Boopalan, M., Niranjanaa, M., and Umapathy, M. J. Study on the mechanical properties and thermal properties of jute and banana fiber-reinforced epoxy hybrid composites. Compos. Part B Eng. 51, 54– 57 (2013). https://doi:10.1016/j.compositesb.2013.02.033
49. Madhu, P., Sanjay, M. R., Senthamaraikannan, P., Pradeep, S., Siengchin, S., Jawaid, M., et al. Effect of various chemical treatments of prosopis juliflora fibers as composite reinforcement: physicochemical, thermal, mechanical, and morphological properties. J. Nat. Fibers 00, 1– 12 (2018). https://doi:10.1080/15440478.2018.1534191
50. Pandey, J. K., Ahn, S. H., Lee, C. S., Mohanty, A. K., and Misra, M. Recent advances in the application of natural fiber based composites. Macro Mater. Eng. 295, 975–989 (2010). https://doi:10.1002/mame.201000095
51. Komuraiah, A., Kumar, N. S., and Prasad, B. D. Chemical composition of natural fibers and its influence on their mechanical properties. Mech. Compos. Mat. 50, 359–376 (2014). https://doi:10.1007/s11029-014-9422-2
52. Gurunathan, T., Mohanty, S., and Nayak, S. K. A review of the recent developments in biocomposites based on natural fibers and their application perspectives. Compos Part Appl. Sci. Manuf. 77, 1–25 (2015). https://doi:10.1016/j.compositesa.2015.06.007

CHAPTER IV

CHARACTERIZATION OF CELLULOSE NANOCRYSTALS FROM (ELAEIS GUINEENSIS) EMPTY FRUIT BUNCHES/NANO-HYDROXYAPATITE POLYMER HYDROGEL COMPOSITE FOR POTENTIAL AGRICULTURAL APPLICATION

UMI QHANITA PUTRI SWYA, ATIQAH AIDA, M. YUNUS, TEZARA CIONITA

Abstract

This book discusses the preparation of nanometer cellulose crystals from oil palm empty fruit bunches. The properties of oil palm empty fruit bunches and the process of extracting cellulose from them are explained. In addition, the benefits and applications of nanometer cellulose crystals in various industries are discussed. The synthesis of nano-hydroxyapatite polymer hydrogel composites and their potential use were also presented. This research investigates the use of nano-hydroxyapatite polymer hydrogel composites in agriculture. The composites showed good chemical interaction and improved hydrophilicity. XRD analysis showed a more amorphous pattern compared to single cellulose. These composites have the potential to improve soil conditions, reduce manufacturing costs and environmental contamination, and ensure safe use of nanoparticles in fertilizers. The production process of nanocellulose from agricultural waste involves chemical pretreatment, bleaching, and acid hydrolysis. Hydrogels are substances that can swell without changing their volume or structure, and can be made from natural or synthetic polymers. Hydrogel polymers have a wide range of applications, including in agriculture as soil conditioners and slow-release fertilizers. Nano-hydroxyapatite is a potential source of fertilizer for agriculture, and can be synthesized using different techniques. Testing of natural fiber composites can be done using Fourier Transform Infrared Spectroscopy to analyze their chemical composition.

4.1 Introduction

4.1.1 Empty Palm Oil Bunches

Palm oil is one of the industries that produce large amounts of solid waste. One of the few industries that produce large amounts of solid waste materials. Empty fruit bunches (EFB) are the largest solid waste compared to other palm oil. As per [1], oil palm sawmills produce sawdust consisting of bunch sawdust (16-23%), juice sawdust (11-26%), core meal sawdust (4%), bark sawdust (4-6%), and other solid sawdust, the palm oil sawmill industry produces sawdust consisting of bunch sawdust (16-23%), juice sawdust (11-26%), core meal sawdust (4%), skin sawdust (4-6%), and other solid sawdust (16.50%).

Table 1. Characteristics of Empty Palm Bunches (Windiastuti et al., 2022)

Parameter	Range
C (%)	40.93-68.3
H (%)	2.88-7.33
HAI (%)	26.4-51.78
N (%)	< 0.1-2.18
S (%)	0.04-0.92
Lignin (%)	10-34.37
Cellulose (%)	22.2-65
Hemicellulose (%)	19.5-38.8
pH	7.20-7.80

Improper management of empty bunches from palm oil mills can lead to environmental issues. These bunches are residues left after the sterilization and tipping processes. Table 1 provides details on the properties of empty bunches sourced from various origins. According to the table, carbon is the predominant component of empty bunches. Additionally, they contain oxygen, hydrogen, nitrogen, and sulfur. Similar to other agricultural materials, empty bunches comprise lignin, hemicellulose, and cellulose. The specific composition of each element can vary based on factors such as sample origin, soil fertility, local climate conditions, soil nutrients, the analytical methods utilized, and the preparation of samples before analysis. [1]

4.1.2 Cellulose Nanocrystals

Cellulose, a natural carbohydrate polymer present in numerous plants, animals, and bacteria, is thereby considered a renewable and biodegradable substance. Annually, the global production of cellulose ranges from 75 to 100 billion tons. Alongside hemicellulose and lignin, cellulose constitutes a significant component of plant fiber cell walls, arranged in a fibrous structure through β-1,4-glycosidic bonds linking glucose monomers. Cellulose is extracted or produced by eliminating its non-crystalline components, resulting in the formation of nanocellulose, which is characterized by outstanding physical, chemical, and mechanical properties such as low density, a significant specific surface area, high crystallinity, and excellent tensile strength. These characteristics make cellulose and nanocellulose highly desirable for a wide array of applications, including air purification, energy storage, biomedical devices, strengthening composite matrices, and serving as components in hybrid materials like aerogels and hydrogels. Furthermore, cellulose offers the advantage of being sourced from various abundant natural reservoirs. [2]

Most cell walls in plants consist of cellulose, hemicellulose, and lignin (see Figure 1), where lignin is present at around 10-25% of the dry weight and acts as a binder between the cellulose and hemicellulose components. Lignin provides rigidity and strength through its binding function and protects cell walls. The two main components of plant cell walls, namely cellulose and hemicellulose, respectively contribute about 35-50% and 20-35% of the dry weight of lignocellulosic biomass. [3]

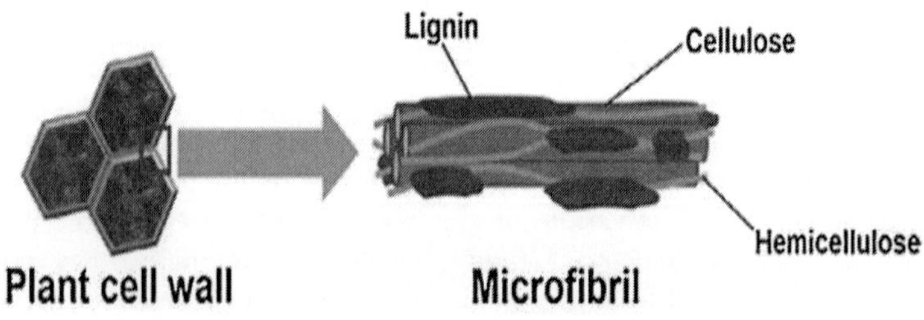

Figure 1. Plant cell wall structure in lignocellulosic biomass consisting of lignin, hemicellulose, and cellulose (Sharmaa amita et al., 2019).

Naturally occurring cellulose fibrils within lignocellulosic biomass typically exhibit both crystalline areas, contributing strength and rigidity, and amorphous regions, providing flexibility to plant cells. Nanocellulose refers to cellulose structured as fibers or crystals, ranging in length and diameter from several micrometers down to less than 100 nm, and can be extracted from natural cellulose fibers (refer to Figure 2). Nanocellulose is biodegradable, lightweight, with a density of approximately 1.6 g/cm^3, and possesses a high tensile strength of up to 10 GPa, comparable to that of cast iron. Moreover, nanocellulose features reactive hydroxyl groups, making it suitable for surface modifications tailored to diverse applications. [3]

4.1.3 Nano-hydroxyapatite polymer hydrogel composite

Hydroxyapatite can be synthesized not only through the reaction of chemical compounds, but can also be reacted using chemical compounds with natural compounds. Natural compounds that can be used in the HAp synthesis process must contain calcium, such as biogenic materials. Biogenic materials are materials derived from living things, such as various types of shells and bones. Bones and shells are known to contain elements dominated by calcium (Ca) and several other elements, such as sodium (Na), magnesium (Mg), oxygen (O_2), and hydrogen (H) in small amounts. [4]

Biogenic-based hydroxyapatite can be synthesized using several methods, such as wet method and solid-state method. The wet method is divided into three, namely hydrolysis, hydrothermal, and precipitation. Hydrothermal method is a method that utilizes high temperature and pressure in the synthesis process. This method is more complex and equipment is expensive. Similar to the hydrothermal method, the hydrolysis method is a reaction method of binding hydroxyl groups derived from water. This hydrolysis method has the disadvantage that it is difficult to hydrolyze phosphate. Different synthesis processes will produce different hydroxyapatite powders, such as particle size, particle size homogeneity, and particle shape. A smaller hydroxyapatite particle size will expand the contact area between the implant and the surrounding tissue so that a better bond can be obtained. Solid-state method is a method of HAp synthesis that can be mass produced with relatively simple procedures. The calcination temperature in this method plays an important role in controlling the size and shape of HAp particles. However, HAp products obtained from this method often show heterogeneity in their phase composition. This occurs due to the small diffusion coefficient of ions in the solid phase. [4]

4.2 Preparation of Cellulose Nanocrystals

4.2.1 Processing Palm Oil Empty Bunches into Cellulose

The oil palm plant (Elaeis guineensis Jacq) was named by Jacquin in 1763 after observing oil palm trees growing in Martinique, West Indies, Central America. The word "Elaeis" comes from Greek which means oil, while "guineensis" was chosen because Jacquin believes that oil palm comes from Guinea (Africa). Oil palm plants generally have an average growth period of 20-25 years. [5]

Figure 1. Empty Fruit Bunches (Praevia Muhammad Farizan et al., 2022)

Oil palm produces waste that can provide great benefits for life, including as organic fertilizer and as activated charcoal. One of the solid wastes of the palm oil industry is palm empty fruit bunches (TKS). Palm oil bunches are one of the solid wastes of the palm oil industry. Palm empty bunches also produce strong fibers as fillers in rubberized fiber products, including car seats, mattresses, and composite boards. Solid waste is characterized by its composition. The largest component in solid waste is cellulose, in addition to other smaller components such as ash, hemicellulose, and lignin. Oil palm bunches are one of the solid wastes of the palm oil industry. Empty palm bunches also produce strong fibers as fillers in rubberized fiber products, including car seats, mattresses, and composite boards. [7]

Empty palm bunches are one of the largest wastes in the process of processing palm oil into palm oil. Empty palm bunches contain 43%

cellulose and 24% hemicellulose. With this high cellulose content of polysaccharides, it has the potential to be processed into cellulose nanofibrils. A mill with a capacity of 60 tons/hour can produce approximately 100 tons/day of waste. Waste production may increase or decrease depending on the FFB (Fresh Fruit Bunches) processed. Oil palm empty fruit bunches contain high fiber content. The main contents of PKS are cellulose and lignin. The cellulose in TKS can be 43-60%, while the lignin content reaches 21-27%.

Cellulose is often used as a reinforcing agent in various polymer composites. Recently, composite materials with nano-sized reinforcements, called nanocomposites, have attracted much attention. Incorporation of nanoelements into a polymer matrix usually results in better properties, such as high surface area, high Young's modulus, high tensile strength, and low coefficient of thermal expansion. Cellulose nanocrystals (NKS) are crystalline nanoparticles made from cellulose and are highly relevant for the development of renewable biomaterials in many fields such as chemistry, food, and pharmaceuticals. NKS is obtained from the hydrolysis process of α-cellulose using acid. During the acid hydrolysis process, the amorphous portion of the cellulose dissolves, producing crystalline nanoparticles with a diameter of 8 to 20 nm and a length of 100 nm to several micrometers, depending on the cellulose source. [9]

4.2.2 How to Prepare Cellulose Nanocrystals

Cellulose, which is the most commonly used material worldwide, has great potential as 'nanocellulose' for a wide array of applications. Cellulose nanocrystals (CNC) are a variant of nanocellulose that have attracted attention due to their ability to form rod-shaped structures with high aspect ratios and can be dispersed evenly in water and other solvents. The CNC manufacturing process generally involves acid hydrolysis, where biomass is reacted to with a strong acid to produce rod-shaped nanocellulose that is highly crystalline and has a high aspect ratio. These CNCs can then be dissolved into a polymer matrix, forming nanocomposites that have enhanced mechanical properties, with increased composite modulus first reported in latex matrices. Research conducted by various groups shows that CNCs can increase the strength of a number of polymer matrices. [10]

Cellulose is a linear polymer of glucose molecules bonded together through (1,4)-glycosidic linkages. In its natural structure, cellulose forms fibrils composed of multiple cellulose molecules connected by glycosidic bonds. The chemical formula of cellulose is (C6H10O5)n, where n denotes the degree of polymerization. The production of cellulose nanocrystals can vary depending on the source of cellulose used, offering researchers various avenues for investigation, including the type of cellulose source, reaction conditions, and processing methodologies. [11]

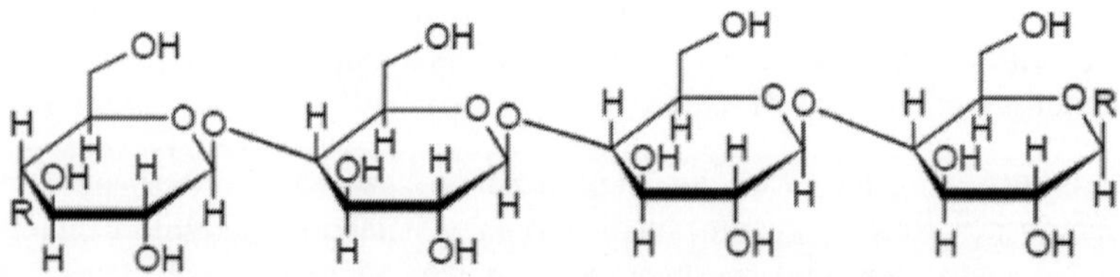

Figure 2. Cellulose structure

Cellulose possesses a distinctive structure characterized by its propensity to form hydrogen bonds, facilitated by its high concentration of hydroxyl groups. These bonds exert strong forces both within and between molecules, significantly impacting cellulose properties. For instance, the hydrophilic nature and swelling behavior of cellulose are profoundly influenced by these hydrogen bonds. Cellulose exhibits high hydrophilicity and swells in environments with elevated humidity levels. In dry conditions, the hydroxyl groups are less accessible as they reinforce each other, whereas in high humidity, absorbed water disrupts the hydrogen bonds between polymer chains, making the hydroxyl groups more accessible. [12]

Cellulose nanocrystals are produced from highly abundant and renewable biomass. Recently, cellulose nanocrystals have attracted much attention as a nanostructured material. Because they are nano-sized, have good mechanical properties (young's modulus 138 GPa and tensile strength 1.7 GPa), a highly reactive surface, and are easily degradable, cellulose nanocrystals are used as reinforcement for biopolymer nanocomposites. Biopolymers are known to have brittleness and poor thermal stability during their use in several applications. The use of cellulose nanocrystals as a filler in biopolymers leads to an improvement in the mechanical and thermal properties of the biopolymers due to cellulose nanocrystals having a very tight network of hydrogen chains in the polymer matrix. [13]

Cellulose fibers exhibit exceptional physical and mechanical properties due to their diverse array of hydroxyl groups and robust network of hydrogen bonds. These fibers consist of both highly organized regions (crystalline structure) and less organized regions (amorphous structure). The crystalline regions feature neatly aligned molecular chains that contribute to the stiffness and strength of cellulose. In contrast, the amorphous regions provide flexibility to the material. This combination of structured and unstructured areas results in cellulose fibers typically having diameters ranging from 3 to 100 micrometers and lengths from 1 to 4 millimeters, characteristics commonly found in lignocellulosic biomass. Cellulose nanocrystals, shaped like rods, have a width ranging from 5 to 70 nm and a length of approximately 100 nm. These nanocrystals consist entirely of cellulose with a cellulose content of 100% and exhibit a high crystallinity ranging from 54% to 88%. Extracting cellulose nanocrystals from the lignocellulosic composition of empty fruit bunches involves a series of processes typically divided into two main stages: pretreatment and subsequent isolation of cellulose nanocrystals. [14]

The initial step in preparing Palm Oil Empty Fruit Bunches (EPOFB) begins with the preparation stage, where the EPOFB is washed and dried for about two days. After that, EPOFB were reduced through a grinding and size adjustment process to produce a fine sample measuring 200 mesh. At this stage, the moisture content of fine EPOFB and EPOFB fibers was 9.18% and 6.9%, respectively. Before use, the processed EPOFB are dried until the water content is less than 10%. [15]

During the pretreatment phase, delignification and bleaching processes are employed to isolate ÿ-cellulose. Delignification involves the removal of lignin between cells and cell walls using a 17.5% (w/v) NaOH solution, which includes stirring and heating at 80°C for 30 minutes. This alkaline treatment induces delignification by causing regular surface swelling and increasing the formation of cracks and voids in the fibers. The color difference observed between the filtrate and residue confirms the effectiveness of the delignification process. Subsequently, the residue is mixed with water and allowed to settle for 24 hours, resulting in three distinct layers: a light brown foam on the top, a dark brown middle layer containing lignin, and a light brown sediment at the bottom. Following this, the residue undergoes continuous rinsing with distilled water until it reaches a neutral pH, followed by filtration and drying for subsequent processing. [15]

However, after completing the chemical pretreatment step, the process is only halfway done. To achieve the desired level of brightness and remove any remaining extractives, a bleaching process using chlorine dioxide, sodium percarbonate, or hydrogen peroxide is necessary. The use of sodium hypochlorite for bleaching is generally not recommended due to the formation of dioxins, which are toxic organic compounds [16]. This bleaching process aims to increase the cellulose content while reducing the levels of lignin and hemicellulose. Anions generated by hydrogen peroxide in an alkaline solution effectively remove most of the chromophoric groups (such as carbonyl and conjugated carbonyl groups and quinones) from the lignin molecule. The presence of lignin contributes to a darker film layer, resulting in undesirable physical and optical properties. Additionally, hemicellulose reduces the resistance of individual fibers, thereby impairing the mechanical properties of nanocellulose. Therefore, lignin and hemicellulose must be removed from cellulose [17]. A 10% (v/v) H2O2 solution was added to the treated solution and heated with stirring at 70°C for 15 minutes to produce a lighter residue. The resulting residue was washed with distilled water through a neutralization process until it achieved a neutral pH. [11]

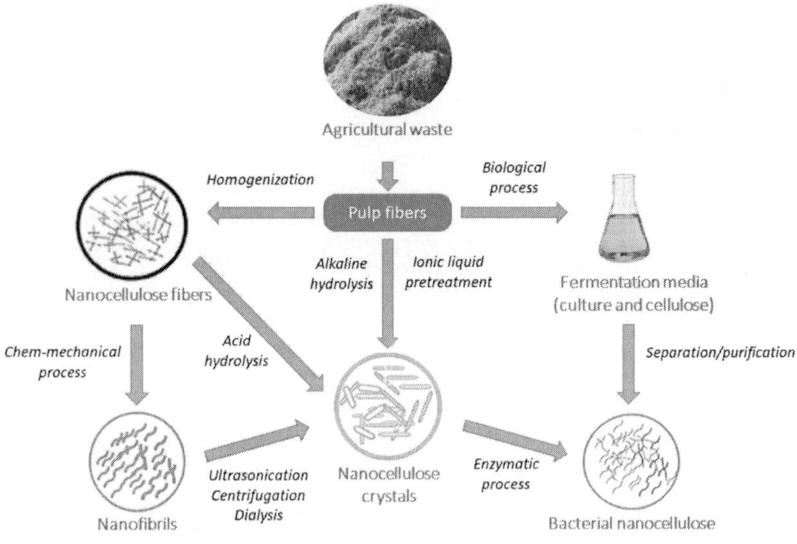

Figure 3. Schematic representation of nanocellulose production from agricultural waste (Mateo Soledad et al., 2021)

Moreover, cellulose nanocrystals can be obtained through hydrolysis conducted under controlled conditions. Mineral acids are commonly employed as hydrolysis agents. The amorphous regions of cellulose, which are relatively weaker, are susceptible to acid attack and degradation, resulting in cellulose with predominantly crystalline segments remaining

[18]. During the acid hydrolysis process, the extracted cellulose nanocrystals take the form of needle-shaped nanoparticles characterized by high surface area, aspect ratio, and crystallinity. Subsequent homogenization is typically necessary to achieve uniform dispersion of the cellulose nanocrystals within the suspension, as these particles tend to agglomerate. Finally, the suspension of cellulose nanocrystals undergoes freeze-drying to yield a bright white solid sample. [19]

4.3. Composite Hydrogel Polymer Nanohydroxyapatite

4.3.1 Definition of Hydrogel

A substance called hydrogel expands when water is added because of linked polymer chains. Hydrogel is defined as a substance that has a high capacity for swelling without changing its volume, form, or structure. Depending on the use, hydrogels can be easily molded and made flexible [20]. These materials can be made from synthetic or natural polymers, and depending on a number of variables including pH, temperature, UV light, and ionic strength, they may gel. A distinct three-dimensional configuration arises as a result of the polymer's elevated hydrophilicity [21]. Depending on the pH, temperature, ionic strength, and electromagnetic radiation in the surrounding environment, hydrogels exhibit swelling behavior [22]. These factors can also be changed to change the form, mechanical flexibility, opacity, and porosity of hydrogels.

There are two primary categories of hydrogels: those made of natural and those made of synthetic polymers. Each kind of hydrogel has benefits and drawbacks of its own. However, it can be presumed that natural materials are biocompatible, biodegradable, and non-toxic to the human body given the medicinal applications of hydrogels. They are a superior option than synthetic chemicals because of these qualities. Conversely, hydrogels derived from synthetic materials possess superior mechanical properties, which augment their potential for industrial utilization. Furthermore, composite materials—a blend of natural and man-made polymers—are widely used.

Numerous varieties of hydrogels exist:

1. Homopolymer hydrogels – these networks are a class of monomers, the fundamental building blocks of a structure. The cross-linking and polymer type can determine whether or not they have a cross-linked framework structure.

2. Hydrogels made of copolymers, which are composed of two or more monomer species, at least one of which needs to be hydrophilic. They are able to be set up in many ways.

3. Multipolymer hydrogels, which consist of two separate synthetic or natural polymers that are cross-linked to form a network [23].

Hydrogels can also be categorized according to their charge (nonionic, ionic, amphoteric electrolyte, zwitterionic), cross-linking type (chemical or physical), physical appearance (matrix, film, or microsphere), and configuration (amorphous, semicrystalline, and crystalline). Due to their many benefits, hydrogels based on natural polymers have recently supplanted those based on synthetic polymers. They are biocompatible in the first place and offer superior cellular and tissue responses when applied. For the goal of tissue engineering, novel hydrogels can be created using proteins and polysaccharides. Secondly, hydrogels derived from naturally occurring polymers exhibit biodegradability, with no harmful by-products emerging from the degradation process. [24]

4.3.2 Hydrogel Polymer

The three-dimensional tissue substance known as hydrogel polymer (PHG) is able to absorb and hold a significant volume of water because it contains hydrophilic groups like -COOH, -CONH2, and -JADI3H, as well as cross-links made of covalent, electrostatic, hydrophobic connections that are resistant to dissolution [25–27]. This property makes it possible to use the product widely in the food industry as a thickening agent, in biomedicine as artificial organs, in agriculture as soil conditioners, in pharmaceuticals for controlled drug delivery, and in personal hygiene products like diapers and sanitary napkins, among other applications [26, 28]. The majority of these polymers are derived from petrochemicals, and there is a growing push to make them using naturally occurring renewable resources including starch, cellulose, chitosan, alginate, pectin, carrageenan, and collagen because monomers are so expensive.

For instance, despite their rate of degradation in soil being less than 10% annually, linear and cross-linked polyacrylamides (PHGs) are utilized as soil conditioners to enhance moisture retention and stabilize soil aggregates to reduce crust formation and soil erosion [31]. By combining vinyl monomers with biodegradable substances like cellulose, these difficulties can be solved. Because of its abundance, affordability, and exceptional qualities—such as hydrophilicity, biodegradability, biocompatibility, and non-toxicity—cellulose is highly sought-after. Using free radical polymerization procedures, cellulose derivatives, such as carboxymethyl cellulose, are frequently used to create cellulose-based PHG [32, 33]. Acrylamide, methacrylamide, and acrylic acid and its salts are examples of frequently used monomers [34, 35].

PGH's water absorption and retention qualities have attracted a lot of attention for their potential application in agriculture as soil conditioners and slow-release fertilizer (SRF) carriers [8]. Conventional fertilizers lose between 40 and 70 percent of their nitrogen, 80 and 90 percent of their phosphorus, and 50 and 70 percent of their potassium to evaporation, surface runoff, and leaching, which causes significant financial losses and degradation of the environment [26, 36]. By using PGH that can biodegrade as a delivery system to encapsulate nutrients and synchronize nutrient release with plant uptake, these losses can be minimized and the detrimental effects of using traditional fertilizers on the environment can be decreased. Additionally, PGH can improve soil moisture retention, which lowers the need for irrigation—particularly in arid and semi-arid areas. Unfortunately, PGH's high production costs prevent it from being used in many industries, including agriculture. Several inorganic fillers, or "clay minerals," including kaolinite [31], montmorillonite [37,38], bentonite [39], attapulgite (Palygorskite) [40], and clinoptilolite [41], have been used in the development of polymer nanocomposites in an attempt to overcome this difficulty. Because of their abundance in the natural world and other exceptional qualities including high specific surface area, high cation exchange capacity, physical-chemical stability, and high intercalation, these clay minerals are appealing.

4.3.3 Nano-hidroksiapatit

Kenyan farmlands' natural soil fertility has decreased over time [42–43] as a result of ongoing farming, insufficient nutrient input, and other restricting factors. Farmers use conventional fertilizers like calcium ammonium nitrate (CAN), urea, mono-ammonium phosphate (MAP), single superphosphate (SSP), diammonium phosphate (DAP), triple superphosphate (TSP), nitrogen, phosphorus, and potassium (NPK), and to some extent, organic fertilizers, to supply the most limited nutrients (NPK). However, leaching causes low nutrient usage efficiency (NUE) when fertilizers are heavily used in sub-humid zones. Although top dressing applied separately is known to increase NUE, small-scale farmers with limited resources may view this as an extravagance or apply it less frequently than suggested [43], which can result in subpar crop yields.

However, its big size, poor nutrient quality, low nutrient mineralization, and additional effort limit its exclusive usage [45]. Alternatively, less expensive sources of fertilizer are being looked for since the high producer prices in Kenya [46] pose a major obstacle to the benefits of fertilizer use in agriculture. A new type of fertilizer known as a "Smart" fertilizer was created in an effort to lessen these problems. Slow Release Fertilizer, or SRF, is one of the fertilizers that has been created. Only the efficient application of contemporary technology can lead to agricultural progress in the face of limited resources and a constantly expanding population. The incorporation of nanotechnology into the formulation or production of fertilizers has been the subject of extensive research. Nano fertilizers are anticipated to outperform traditional polymer-coated SRFs in terms of NUE, toxicity, and potential side effects from overuse, such as groundwater pollution, because of their high surface area-to-volume ratio [47]. One of the primary prospective agricultural nutrient sources is thought to be nano-hydroxyapatite (HA) particles [48]. Though there are still few agricultural uses, the majority of the data on HA that are currently accessible are concentrated on biomedical applications [49, 50, 51].

4.3.4 Synthesis of Nano-Hydroxyapatite

Three techniques exist for the synthesis of nano-hydroxyapatite:

1. The synthesis of nano-HA is carried out in accordance with the protocol used by Rop et al. [52], in which Ca(OH)2 and H3PO4 are agitated mechanically (at 1000 rpm) to react with or without a surfactant (TX-100). Scheme 1 states that the stoichiometric ratio of Ca:P is kept at 1.67. After achieving a consistent weight, the resultant nanoparticles are dried at 105 °C in an oven and then ground into a fine powder. The nano-HA powder is cleaned of the surfactant by running it through methanol.

$$6H_3PO_4 + 10Ca(OH)_2 \rightarrow Ca_{10}(PO_4)_6(OH)_2 + 18\ jam_2\ O$$

Figure 4. Reaction between orthophosphoricaci and calcium hydroxideto yield nano HA

2. Aqueous solution of nanohydroxyapatite (nHap) is generated by a twofold decomposition technique. In a reactor, calcium nitrate Ca(NO3)2,4H2O (23.6 g/mole) is introduced and heated for approximately 60 minutes. After that, over the course of three hours, a dropwise addition of ammonium dihydrogen phosphate (NH4)2HPO4 (7.92 g/mole) is made.

To reduce fluid loss due to evaporation, a reflux column is employed. A gentle addition of 200 ml of concentrated ammonia is made once the precipitation is finished. Throughout the synthesis process, there is continuous mechanical stirring of the reactor. Throughout the maturation process, the suspended precipitate is kept in contact with the stock solution while stirring and heating it continuously for 30 minutes. Hot filtering is carried out using a Büchner funnel, and the resultant solid is allowed to dry for the entire night at 80°C in the oven. [53]

3. With minor adjustments, the synthesis's technique was taken from Kottegoda et al. (2011). A 100 mL beaker containing 7.716 grams of Ca(OH)2 was filled with a 0.22 mM TX-100 (non-ionic surfactant) solution. Using a motorized stirrer, the mixture was swirled for thirty minutes. Dropwise (15 mL/min), 100 mL of 0.6 M H3PO4 was added to the Ca(OH)2 suspension from a buret while it was being rapidly agitated at 1000 rpm. Following the reaction, the dispersion was given a 10-minute agitation before being left to stand for two hours. After reaching a consistent weight, it was dried at 105°C in an oven and then milled into a fine powder. By washing, the surfactant was eliminated [54].

4.4. Testing of Natural Fiber Composites

4.4.1. Fourier Transform Infrared Spectroscopy Testing

Figure 3 displays the FTIR spectra of the polymer hydrocompositenano-HA and cellulose/hydrogel polymer nano-HA. The stretching vibration of alcohol –OH is responsible for the spectral band detected at 3600–3000 and 1647 cm−1. This is cellulose. Other peaks arise because of the stretching vibration of C—O—C for glucosidic bonds, which occurs between 1300 and 1000 cm−1. For both primary and secondary alcohol groups, the stretching vibration of C-O is represented by a very strong peak at 1018 cm−1. The peaks observed in nano-HA at 1419 and 875 cm−1 are ascribed to CO2, an ion thought to arise from the interplay between nano-HA and CO2 [28]. The weak peaks at 1635 cm−1 correspond to the –OH stretching vibration within the nano-HA lattice and the H-O-H water lattice, while broad and weak peaks span from 3600 to 3000 cm−1 [29–31]. PO3−4 stretching vibrations are associated with bands at 1022 and 964 cm−1, whereas bending vibrations are associated with bands at 601 and 563 cm−1.

i) As seen in the image below, the polymer nano-composite's spectra display a strong, broad band between 3600 and 2500 cm−1. This could suggest that multiple functional groups overlap.
ii) The principal amine-induced stretching vibration of N\\H is responsible for 3600–3300 cm−1.
iii) The stretching vibration of alcohol --OH, caused by certain unreacted cellulose and nano-HA, is 3600–3000 cm−1.
iv) The C--H vibration mode is linked to 3000–2800 cm−1. A carboxylate—OH proportion of 3000–2500 cm−1 is attributed to unneutralized acrylic acid. The axial deformation of C-O for the amino groups is responsible for a rather strong band at 1674–1651 cm−1, which indicates the conversion of polyammonium acrylate to polyacrylamide during heating or dehydration.

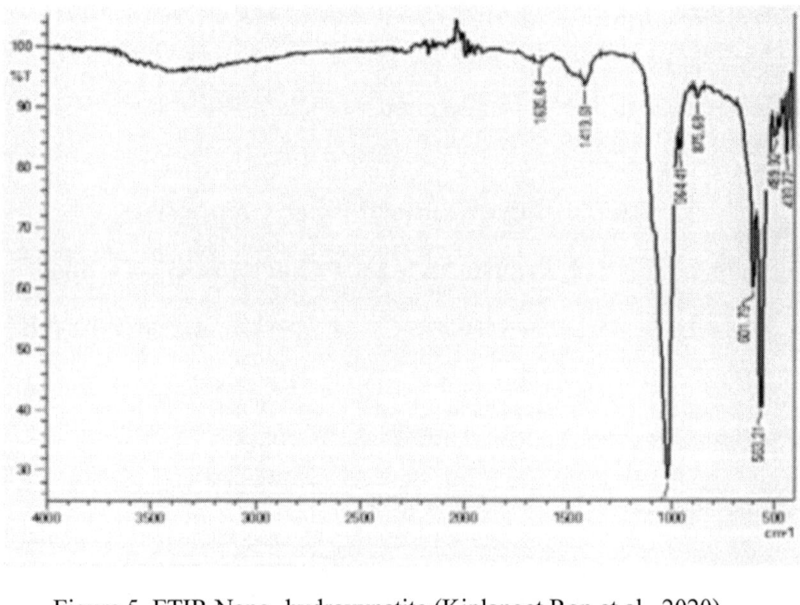

Figure 5. FTIR Nano- hydroxypatite (Kiplangat Rop et al., 2020)

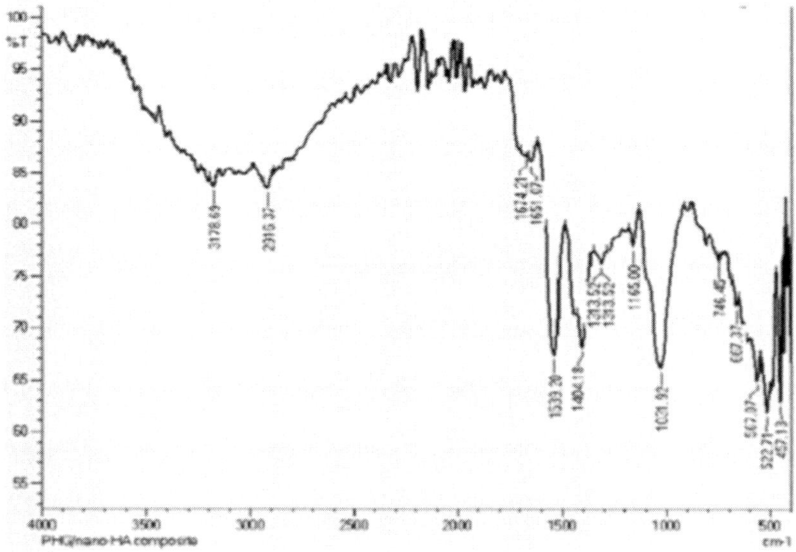

Figure 6. FTIR Polymer Hydrogel Nano-hydroxypatite (Kiplangat Rop et al., 2020)

4.4.2 Electron Microscopy Testing

Figure 3 displays the transmission electron microscopy (TEM) images of polymer hydrogel nanocomposites with nano-hydroxyapatite. In comparison to those synthesized without it, the nano-HA image generated with TX-100 (non-ionic surfactant) displays more agglomerated rod-shaped nanoparticles (Figures 3a & b). Similar findings were reported by [55], who linked the development of rod-shaped nano-HA to the ion-dipole interactions between Ca2+ and the polyethylene oxide group of TX-100 that result in the production of a hydrophobic ring complex (Scheme 4).

Under regulated conditions, the creation of this combination is thought to slow down the pace at which Ca2+ is transferred to the developing crystal, which in turn reduces the size of the nanoparticles. The adsorption of TX-100 onto particular planes of the produced crystal also contributes to the production of nano-rods by driving the growth of the nanoparticles in the appropriate direction. High specific surface energy is linked to nanoparticle agglomeration (Figures 3a & b), which causes aggregation [56, 57].

The dispersion of nano-HA in the copolymer is seen in the TEM image of the nano-composite (Figure 3c). Similar to this, Ragu and Sakthivel [57] found that the agglomeration of nanoHA in polymethyl methacrylate/nano-HA composites decreased to a diameter of around 20 to 50 nm. This reduction in agglomeration is explained by the polymer's ability to reduce the surface energy of nano-HA through its PO independent groups.

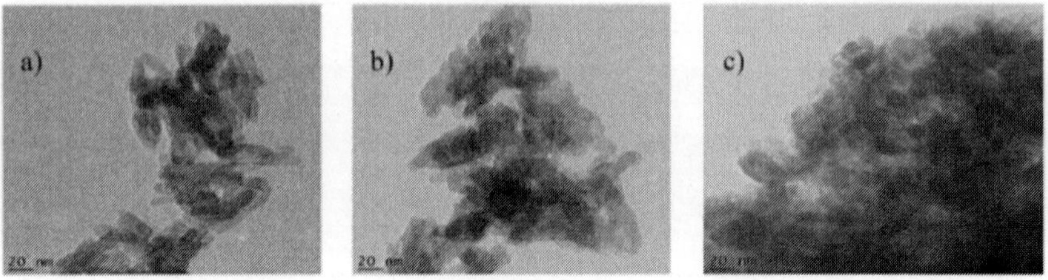

Figure 7. TEM images of (a) nano-HA synthesized with TX-100, (b) nano-HA synthesized without TX-100, (c) cellulose/nano-HA composite, at a scale of 20 nm (Kiplangat Rop et al., 2020)

4.4.3 X-ray Diffraction (XRD) Analysis Testing

An examination of X-ray diffraction (XRD) testing for a polymer nanocomposite, nano-HA, and copolymer-grafted cellulose is presented in the text. In comparison to copolymer without cellulose, the authors found that cellulose-grafted copolymer degraded more quickly in soil microbial culture. During a 16-week incubation period, soil modified with cellulose-grafted co-polymer showed a considerable increase in soil N minerals, most likely as a result of urea-N hydrolysis. The pH of the soil varied from 5.15 to 5.33 in the second week of incubation and rose to 5.71 to 5.97 in the sixteenth week, mostly as a result of the neutralizing effect of Ca2+ ions. The action of microbials may have aided in the release of soluble P from nano-HA by the nanocomposite polymer. In order to solubilize insoluble phosphates like apatite, soil bacteria like Pseudomonas and Anthrobacter as well as fungi like Aspergillus and Penicillium secreted low molecular weight organic acids like citric, gluconic, lactic, oxalic, and acetic acids that lowered pH and chelated mineral ions. According to Scheme 5, it was thought that microbial soil

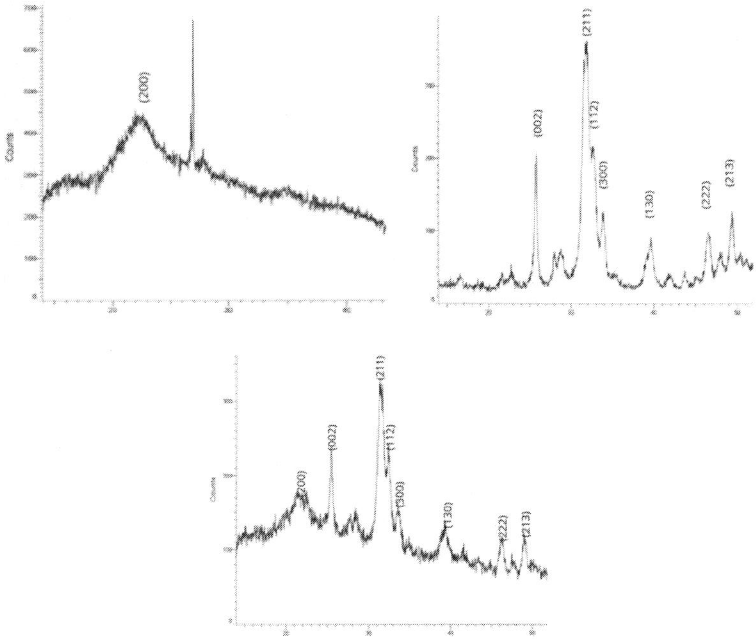

Figure 8. TEM images of (a) nano-HA synthesized with TX-100, (b) nano-HA synthesized without TX-100, (c) cellulose/nano-HA composite, at a scale of 20 nm (Kiplangat Rop et al., 2020)

acidity and the rhizosphere environment created by plant root exudates undcr phosphorus limits would dissolvc nano-IIA. Thc law of mass action states that prolonged mining of Ca2+ and H2PO4 together with an increase in H concentration are what cause the equilibrium to shift to the right, or dissolve.

4.5 Potential Applications of Composites for Agriculture

These "building" materials are employed to manufacture nanomaterials (NMs) utilized across diverse applications. Nanocellulose, carbon-based materials at the nanoscale, and nanosilica are currently employed as constituents in nanocomposites (NCs) used across various sectors including industry, agriculture, medicine, and environmental remediation. Agricultural residues such as banana or orange peels, wheat whiskers, straw, cotton or corn stalks, coconut or almond shells, corn silk, rice husks, palm empty fruit bunches, bagasse, peanut husks, and ginger rhizomes have also proven effective for this purpose. [68]

Figure 9. This type of agricultural waste is applied to make nanocomposites. (Jampilek Josef et al., 2020)

Nanocellulose comprises cellulose nanocrystals (CNC) and cellulose nanofibers (CNF), with cellulose nanowhiskers derived from various cellulose sources, including agricultural waste. The mechanical breakdown of cellulose fibers, typically 20-40 nm wide and several micrometers long, containing both crystalline and amorphous regions, yields CNFs characterized by a high aspect ratio of length to diameter. When employed as reinforcing fillers in composites, CNFs have the potential to significantly enhance composite quality. [68]

Traditional agricultural practices, often referred to as industrial agriculture, are deemed unsustainable depending on their methods and management practices. This system is characterized by the widespread use of synthetic fertilizers, herbicides, and pesticides, as well as the implementation of crop rotation. While fertilizers boost crop yields in the short term, their prolonged use has resulted in non-point source environmental pollution. Moreover, conventional farming methods are inefficient due to excessive chemical inputs, soil degradation, large-scale farming operations, susceptibility to pest and disease outbreaks, reduced profitability, and risks to human and environmental health. These practices also contribute to water quality deterioration and significant water consumption, with irrigation drawing water from multiple sources carrying salts and contaminants from soil treated with fungicides, pesticides, and fertilizers. The return flow of water from irrigated agricultural lands into rivers, compounded by urbanization and expansion of settlements, increases water salinity, waterlogging, reduces river discharge, and diminishes soil moisture. Efforts to ensure agricultural sustainability have spurred interest in sustainable and biodegradable materials as alternatives.

Natural polysaccharides like cellulose, pectin, alginate, starch, chitosan, and xanthan gum have been employed as biopolymers for centuries, offering numerous benefits to humanity. These complex biopolymers, composed of 20 to 60,000 monosaccharide units, are abundant in the cell walls of plants, algae, animal products, and bacterial sources. Their intricate molecular structure allows polysaccharides to interact effectively with other components such as proteins, making them valuable in a wide range of applications. Polysaccharides are highly valued for their sustainability, cost-effectiveness, and edible nature, making them popular biomaterials used in coatings, films, hydrogels, and fibers. In recent times, there has been a growing use of polysaccharide-based products in agriculture as a modern solution to the challenges posed by conventional farming practices. Polysaccharides possess desirable properties such as hydroxyl groups and the ability to form hydrogen bonds between these groups, which contribute to their water solubility, rigidity, high tensile strength, and viscosity. For example, chitosan films find applications in agriculture for controlling soil microbiota, facilitating remediation, creating fertilizer release systems, and improving water absorption capacities. [69]

4.5.1 Excellent Water Saver for Plantation Media

Several advanced farming techniques have been developed, such as aeroponics, hydroponics, nutrient film technique (NFT), floating systems, wicking systems, drip irrigation, flow systems, and aquaponics. These methods of farming without soil offer opportunities to reduce environmental impact and represent viable paths towards achieving sustainable agriculture. Key elements supporting soil-less farming systems include nutrient recycling to reduce fertilizer runoff and the adoption of vertical or stacked planting techniques to maximize space efficiency, both contributing to its promising future. Recently, soil-less farming systems integrating renewable carbon-based (RC-based) products have gained significant attention for their potential applications in agriculture. With ongoing efforts to meet future agricultural demands, soil-less farming systems are continuously evolving and being adopted. [69]

The advancement of nanocellulose hydrogel products, characterized by their amphipathic properties, excellent biocompatibility, and biodegradability over time, has consistently motivated researchers to enhance these products further. Hydrogels have demonstrated numerous agricultural benefits, including (1) enhancing moisture retention, particularly beneficial in arid regions and during cold winters, thereby reducing water usage in traditional irrigation methods, (2) serving as nutrient-rich mediums capable of controlled release, (3) contributing to soil health by minimizing erosion and binding loose soil particles to enhance root adhesion and soil structure.

The increasing demand for food commodities is prompting the expansion of agricultural land and higher water consumption through conventional farming practices reliant on irrigation systems. This trend has led to substantial global water usage, particularly affecting arid regions where water resources are under immense strain, resulting in water scarcity. RC hydrogels have emerged as a promising growing medium due to their high porosity, water retention capacity, and developmental properties that enhance oxygenation of plant roots. Hydrogels, by effectively retaining water, can also absorb and store rainwater or irrigation water, thereby contributing to water conservation efforts. [69]

Cellulose is composed of both crystalline (highly ordered) and amorphous (less ordered) domains embedded in a matrix of lignin and hemicellulose, the composition of which varies depending on the source. Reagents face challenges in accessing cellulose due to its biphasic nature, with greater accessibility to the amorphous regions. Various approaches are being explored to enhance cellulose accessibility. For instance, the process of converting cellulose I to cellulose II or amorphous cellulose has shown improved arrangements that facilitate reagent penetration, particularly water in agricultural contexts. The hydroxyl groups present along the amorphous cellulose chain enable water molecules to penetrate cellulose fibers through hydrogen bonding. Additionally, the anisotropic van der Waals forces acting on cellulose chains enhance strong adsorption energy, aiding in the retention of water molecules through electrostatic interactions. [69]

Interest in the prospective use of hydrogels in agriculture primarily focuses on their integration with soil for water conservation purposes, particularly in uncontaminated plantations. Hydrogels with superabsorbent properties are under development to enhance water conservation efforts. Numerous studies aim to modify hydrogels to improve their robustness, given their high hydrophilic chains that enhance water absorption capacity and porosity while reducing soil erosion. However, these modifications often compromise the structural integrity of the hydrogel, making it less conducive to supporting root systems. Hydrogels derived from rice straw, utilizing cellulose as a source, exhibit significant water absorption capabilities, thereby reducing irrigation frequency and enhancing soil aggregation. Research indicates that soil blends incorporating chemically cross-linked cellulose hydrogel powders demonstrate threefold increased water retention capacity, as demonstrated in cherry tomato cultivation trials. Anionic carboxylic cellulose hydrogels, characterized by higher carboxyl content and a homogeneous macroporous structure, exhibit excellent water retention properties and support seedling growth. Additionally, blending cellulose with other natural polysaccharides forms composites that reinforce hydrogels while maintaining microporosity and water absorption capacity. The strength imparted by cellulose renders hydrogels suitable as growth media for sustainable water conservation practices. Biodegradable synthetic polymers, such as poly(acrylic) acid grafted with cellulose, are also employed to create superabsorbent hydrogels with enhanced development potential, facilitated by a large pore volume resulting from electrostatic repulsion. [69]

Conclusion

Agriculture can benefit from the application of fertilizer containing cellulose nanocrystals and nano-hydroxyapatite polymer hydrogel composite. This is so that the polymer hydrogel's hydrophilicity can be enhanced by the nano-hydroxyapatite content. Good chemical interaction between nano-hydroxyapatite and the polymer structure is indicated by FTIR measurement. When compared to cellulose alone, the composite displays a more amorphous pattern, according to XRD examination. The polymer component increases mechanical strength by trapping nano-hydroxyapatite inside the tissue structure. This fertilizer has the potential to improve soil conditioning through increased moisture retention capacity, lower manufacturing costs and environmental contamination, and safe use of nanoparticles.

References

1. Windiastuti, E., Suprihatin., Y, Bindar., and u, Hasanudin. 2024. Identification of potential application of oil palm empty fruit bunches (EFB): a review.IOP Conference series: Earth and environmental science. 1063(2022)012024.
2. Nasrudin, Raimi Fariz. et al. 2022. Production of cellulose nanocrystals from oil palm empty fruit bunch and pineapple leaf fiber using double oxidation approach, 84:5(2022)73-81.
3. Sharma, Amita., Thakur, Manisha., Bhattacharya, Muna, Mandala, Tamal., Goswami, Saswata. 2019. Commercial application of cellulose nano-composites- A riview, 2029. 2215-017X.
4. Yusuf, Yusril., Khasanah, DU., Syafaat, FY. et al. 2024. Biogenic based hydroxyapatite. Book chapter gadjah mada university press, 978-602-386-421-8.
5. Cherlina, DI., Gea, S., Nainggolan, H. (2007). Preparation of polyvinyl alcohol/cellulose nanocomposites isolated from empty oil palm bunches usingthe steam explosion method. 14(2),120-126.
6. Praevia, Muhammad Farizan., Widayat. 2022. Analysis of the utilization of empty palm oil bunches waste as cofiring at coal power plant, Vol.3, No 1, PP 28-37.
7. Aulia, fenny., Marpongahtun., Gea, Saharman. 2023. Study of providing cellulose nanocrystals from bunches palm oil. Journal saintia chemical. Vol.1, No.2, 2013.
8. Habibi, Y.; Goffin, AL; Schiltz, N.; Duquesne, E.; Dubois, P.; Dufresne, A. 2008. Bionanocomposites based on poly(ε-caprolactone)-grafted cellulose nanocrystals by ring-opening polymerization. J. Materi. Kimia, 2008,18, 5002.
9. Putri, Elisa, Gea, Sharman. 2018. Isolation and characterization of nanoastrals cellulose from palm bunch *(Elaeis guineensis jack)*, Journal of islamic science and technology, Vol.4, No.1.
10. Ichwan, M., Onyianta, AS., Trask, RS., Etale, A., & Eichhorn, SJ. 2023. Production and characterization of cellulose nanocrystals with different allomorphs from empty oil palm bunches to increase toughness, Carbohydrate Polymer Technology and Applications, 5(2023).100272.
11. Ng, Lawyong., Wong, TJ., Ng, Ching Y., Amelia, Chiang KM. 2021. A Review on cellulose nanocrystals production and characterization methods from *Elaeis guineensis* empty fruit bunches, Arabian Journal of Chemistry, 14,103339.

12. F, Anjana. 2016. Study of making cellulose nanocrystals from empty palm oil bunches using the sonication hydro thermal process (Doctoral dissertation), Sepuluh Nopember Institute of Technology Journal.
13. Houyong Yu, Zongyi Qin, dkk. 2013. "Facile extraction of thermally stable cellulose nanocrystals with a high yield of 93% through hydrochloric acid hydrolysis under hydrothermal condition". Jurnal of Materials Chemistry, A Vol.1, hal 3938-3944.
14. Han, Apichart., Gobwute, R., & Randorn, C. 2017. Preparation of hydroxyapatite hydrogel for bone-like materials via novel self-initiated photocatalytic polymerization, Materials letters,193 (2017) 142-145.
15. Hidayatulloh, I., Emmanuela, MW., Kusumawati, E., & Elizabeth, L. 2021. Nanocellulose productionfrom empty palm oil fruit bunches (EPOFB) using Hydrolysis followed by freeze drying, ASEAN Journal of chemical engineering, Vol. X21, No 1, 52-61.
16. Taflick, T., Schwendler, L.A., Rosa, S.M.L., Bica, C.I.D., Nachtigall, S.M.B. 2017. Cellulose nanocrystals from acacia bark–Influence of solvent extraction. Int. J. Biol. Macromol. 101, 553–561. https://doi.org/10.1016/J.IJBIOMAC.2017.03.076.
17. Mussatto, S. I., Rocha, G. J. M., & Roberto, I. C. 2008. "Hydrogen peroxide bleaching of cellulose pulps obtained from brewer's spent grain". Cellulose, 15(4), 641–649.
18. El Achaby, M., Kassab, Z., Barakat, A., Aboulkas, A. 2018. Alfa fibers as viable sustainable source for cellulose nanocrystals extraction: Application for improving the tensile properties of biopolymer nanocomposite films. Ind. Crops Prod. 112, 499–510. https://doi.org/10.1016/J.INDCROP.2017.12.049.
19. Soledad, Mateo., Silvia, Peinado., Francisca, Morillas-Gutiérrez., M. Dolores La Rubi., & Alberto J. Moya. 2021. Nanocellulose from Agricultural Waste: Products and Applications—An Overview, Department of Chemical, Environmental and Materials Engineering, University of Jaén, 9, 1594. https://doi.org/10.3390/pr9091594.
20. YX Liu, L. Chen, ZG Zhao, RC Fang, MJ Liu. 2018. Desain dan sintesis hidrogel multiskala yang diilhami secara biologis: dari antarmuka hingga jaringan tiga dimensi, Acta Polym. Dosa, 91155-1174.
21. K. Yang, Q. Han, B. Chen, Y. Zheng, K. Zhang, Q. Li, et al. 2018. Hidrogel antimikroba: bahan yang menjanjikan untuk aplikasi medis, Int. J. Nanomed. 13 (2018) 2217-2263.
22. DE Ciolacu, DM Suflet. 2018. Hidrogel berbasis selulosa untuk aplikasi medis/farmasi, dalam: V. Popa, I. Volf (Eds.), Biomassa

Sebagai Bahan Baku Terbarukan untuk Memperoleh Bioproduk Bernilai Teknologi Tinggi, Elsevier, hlm. 401-439.
23. E.M. Ahmed. 2015. Hydrogel: preparation, characterization, and applications: a review, J. Adv. Res. 6 (2015) 105-121.
24. M.C.G. Pella, M.K. Lima-Tenorio, E.T. Tenorio-Neto, M.R. Guilherme, E.C. Muniz, A.F. Rubira. 2018. Chitosanbased hydrogels: from preparation to biomedical applications, Carbohyd. Polym. 196 (2018) 233-245.
25. M.S. Hashem, A.M. El-Hady, A. Hebeish. 2013. Synthesis and characterization of novel carboxymethyl cellulose hydrogels and carboxymethyl cellulolse-hydrogel-ZnOnanocomposites, Carbohy Polym 95 (2013) 421–427, https://doi.org/10.1016/j. carbpol.2013.03.013.
26. M. Baki, J. Abedi-Koupai. 2017. Preparation and characterization of a super absorbent slow release fertilizer with sodium alginate and biochar, J. Appl. Polym. Sci. 135 (2017), 45966. https://doi.org/10.1002/app.45966.
27. P. Milani, D. França, A.G. Balieiro, R. Faez. 2017. Polymers and its applications in agriculture, Polímeros 27 (3) (2017) 256–266, https://doi.org/10.1590/0104-1428.09316.
28. M. Tally, Y. Atassi. 2015. Optimized synthesis and swelling properties of a pH-sensitive semi-IPN superabsorbent polymer based on sodium alginate-g-poly(acrylic acidco-acrylamide) and polyvinylpyrrolidone and obtained via microwave irradiation, J. Polym. Res. 22 181, https://doi.org/10.1007/s10965-015-0822-3.
29. G. Mahdavinia, A. Afzali, H. Etemadi, H. Hosseinzadeh. 2017. Magnetic/pH-sensitive nanocomposite hydrogel based carboxymethyl cellulose-g-poly acrylamide/ montmorillonite for colon targeted drug delivery, Nanomed Res J 2 (2) (2017) 111–122, https://doi.org/10.22034/nmrj.2017.58964.1058.
30. A. Martinez-Ruvalcaba, J.C. Sanchez-Diaz, A. Gonzalez-Alvarez. 2009. Polyacrylamidegelatin polymeric networks: effect of pH and gelatin concentration on the swelling kinetics and mechanical properties, PolymBull62539–548, https://doi.org/10.1007/s00289-008-0037-4.
31. R.A. Hussien, A.M. Donia, A.A. Atia, O.F. El-Sedfy, A.R. Abd El-Hamid, R.T. Rashad, et al. 2012. Studying some hydro-physical properties of two soils amended with kaolinite-modified cross-linked polyacrylamides. Catena, No.92: 172–178. https://doi.org/10.1016/j.catena.2011.12.010.

32. M. Sadeghi, S. Safari, H. Shahsavari, H. Sadeghi, F. Soleimani. 2013. Effective parameters onto graft copolymer based on carboxymethyl with acrylic monomer. Asian Journal of Chemistry. 25 (9): 5029–5032.
33. C.Demitri, F. Scalera, M. Madaghiele, A. Sannino, A. Maffezzoli. 2013. Potentialofcellulosebased superabsorbent hydrogels as water reservoir in agriculture. Internasional Journal of Polymer Science. https://doi.org/10.1155/2013/435073.
34. W.A. Laftah, S. Hashim, A.N. Ibrahim. 2011. Polymer hydrogels: a review. Polymer-Plastics Technology and Materials. 50 (14): 1475–1486.
35. E.M. Ahmed. 2015. Hydrogel: preparation, characterization and applications: a review. Journal of Advanced Research. 6 (2): 105–121, https://doi.org/10.1016/j.jare.2013.07.006.
36. 36.A.S. Giroto, G.F. Guimarães, M. Foschini, C. Ribeiro. 2017. Role of slow-release nanocomposite fertilizers on nitrogen and phosphate availability in soil. Science Report. 7, 46032.
37. G.B. Marandi, G.R. Mahdavinia, S. Ghafary. 2011. Collagen-g-poly(sodium acrylate-coacrylamide)/sodium montmorillonite superabsorbent nanocomposites: synthesis and swelling behavior. Journal of Polymer Research. No. 18: 1487–1499, https://doi.org/10.1007/s10965-010-9554-6.
38. 38.A. Bortolin, F.A. Aouada, L.H.C. Mattoso, C. Ribeiro. 2013. Nanocomposite PAAm/methyl cellulose/montmorillonite hydrogel: evidence of synergistic effects for the slow release of fertilizers. Journal of Agricultural and Food Chemistry. 61 (31): 7431–7439, https://doi.org/ 10.1021/jf401273n.
39. S.R. Shirsath, A.P. Hage, M. Zhou, S.H. Sonawane, M. Ashokkumar. 2011. Ultrasound assisted preparation of nanoclay bentonite-FeCo nanocomposite hybrid hydrogel: a potential responsive sorbent for removal of organic pollutant from water. Desal. No. 281: 429–437, https://doi.org/10.1016/j.desal.2011.08.031.
40. Y. Deng, L. Wang, X. Hu, B. Liu, Z. Wei, S. Yang, et al. 2012. Highly efficient removal of tannic acid from aqueous solution by chitosan coated attapulgite. Chemical Engineering Journal. 181–182: 300–306, https://doi.org/10.1016/j.cej.2011.11.082.
41. A. Rashidzadeh, A. Olad, D. Salari, A. Reyhanitabar. 2014. On the preparation and swelling properties of hydrogel nanocomposite based on sodium alginate-g-poly (acrylic acid-co-acrylamide)/clinoptilolite and its application as slow release fertilizer. Journal Polymer Research. No. 21: 344, https://doi.org/10.1007/s10965-013-0344-9.

42. Kimetu, J.M., Mugendi, D.N., Bationo, A., Palm, C.A., Mutuo, P. K., Kihara, J., Nandwa, S., Giller, K. 2007. In: Advances in integrated soil fertility management in Sub-Saharan Africa: challenges and opportunities. 521–530.
43. Mucheru-Muna, M., Mugendi, D., Pypers, P., Kungu, J., Vanlauwe, B., Merckx, R. 2013. Enhancing maize productivity and profitability using organic inputs and mineral fertilizer in Central Kenya small-holdfarms. ExperimentAgriculture. No. 50: 250–269.
44. Mathenge, M. 2009. Towards increase duse of fertilizer and improved seed for food security and economic growth. 6thNatlFertil. Conf. KenyaAgri. Res. Inst. Nairobi, Kenya.
45. Makokha, S., Kimani, S., Mwangi, W., Verkuij, H., Musembi, F. 2001. Determinants of fertilizer and manure use in maize production in Kiambu district Kenya. Int. Maize WheatImprov. Cent. Kenya Agri. Res. Inst, Mexico, D.F.
46. Druilhe, Z., Barreiro-Hurlé, J. 2012. In: Fertilizer subsidies in sub-Saharan Africa. FAO, Rome. 12–104.
47. DeRosa, C., Monreal, C., Schnitzer, M., Walsh, R., Sultan, Y. 2010. Nano technology in fertilizers. Nat. Nanotechnol. No. 5: 91.
48. Kottegoda, N., Manaweera, I., Madusanka, N., Karunaretne, V. 2011. A greenslow-release fertilizer composition based on urea-modified hydroxyapatitenano-particles encapsulated wood. CurrentScience.No. 101: 73–78.
49. Mateus, P.A., Barrias, C.C., Ribeiro, C., Ferraz, M.P., Monteiro, F.J. 2007. Comparative study of nano-hydroxyapatite microspheres for medical applications. Journal of Biomedical Materials Research. 86A: 483–493.
50. Pang, X., Zeng, H., Liu, J., Wei, S., Zheng, Y. 2010.The properties of nanohydroxyapatite materials and its biological effects. Materials Sciences and Applications. No. 1: 81–90.
51. Pataquiva Mateus, A.Y., Ferraz, M.P., Monteiro, F.J. 2013. Nanoparticles of hydroxyapatite: Preparation characterization and cellular approach- an overview. Revista Mutis. No. 3: 43–57.
52. K. Rop, G.N. Karuku, D. Mbui, I. Michira, N. Njomo. 2018. Formulation of slow release NPK fertilizer (cellulose-graft-poly (acrylamide)/nano-hydroxyapatite/soluble fertilizer) composite and evaluating its N mineralization potential. Annals of Agricultural Sciences. No.63: 163–172, https://doi.org/10.1016/j.aoas.2018.11.001.
53. Ezzouine, N., Billah, R. El K., Soufiane, A., Esserti, S., Belfaiza, M., etc. 2022. Protection of Solanum lycopesicum induced by chitosan and

chitosan nano-hydroxyapatite against Pepino mosaic virus and Verticillium dahlia. Journal of Biocatalysis and Agricultural Biotechnology.

54. Kottegoda, N., Manaweera, I., Madusanka, N., Karunaretne, V. 2011. A greenslow-release fertilizer composition based on urea-modified hydroxyapatite nano-particles encapsulated wood. Current Science. No. 101: 73–78.
55. E. Iyyappan, P. Wilson. 2013. Synthesis of nanoscale hydroxyapatite particles using Triton X-100 as an organic modifier. Ceramics International. No. 39: 771–777, https://doi.org/10. 1016/j.ceramint.2012.06.090.
56. N. Pramanik, D. Mishra, I. Banerjee, T.K. Maiti, P. Bhargava, P. Pramanik. 2009. Chemical synthesis, characterization, and biocompatibility study of hydroxyapatite/chitosan phosphate nanocomposite for bone tissue engineering applications. International Journal of Biomaterials. https://doi.org/10.1155/2009/512417Article ID 512417.
57. A. Ragu, P. Sakthivel. 2014.Synthesis and characterization of nanohydroxyapatite with polymethyl methacrylate nanocomposites for bone tissue regeneration. International Journal of Science and Research. No. 3 (11): 2282–2285.
58. K. Rop, D. Mbui, N. Njomo, G.N. Karuku, I. Michira, R.F. Ajayi. 2019. Biodegradable water hyacinth cellulose-graft-poly(ammonium acrylate-co-acrylic acid) polymer hydrogel for potential agricultural application. Heliyon. No. 5.e01416. https://doi. org/10.1016/j.heliyon.2019.e01416.
59. K. Rop, G.N. Karuku, D. Mbui, I. Michira, N. Njomo. 2018. Formulation of slow release NPK fertilizer (cellulose-graft-poly (acrylamide)/nano-hydroxyapatite/soluble fertilizer) composite and evaluating its N mineralization potential. Annals of Agricultural Sciences. No. 63: 163–172, https://doi.org/10.1016/j.aoas.2018.11.001.
60. A. Costescu, I. Pasuk, F. Ungureanu, A. Dinischiotu, M. Costache, F. Huneau, et al. 2010. Physico-chemical properties of nano-sized hexagonal hydroxyapatite powder synthesized by sol-gel. Digest J Nanomat Biostruct. No. 5 (4): 989–1000.
61. B. Gayathri, N. Muthukumarasamy, D. Velauthapillai, S. B. Santhosh, V. Asokan. 2018. Magnesium incorporated hydroxyapatite nanoparticles: preparation, characterization, Antibacterial and Larvicidal Activity. Arabian Journal of Chemistry. No. 11: 645–654, https://doi. org/10.1016/j.arabjc.2016.05.010.

62. A. Mechay, H.L. Feki, F. Schoenstein, N. Jouini. 2012. Nanocrystalline hydroxyapatite ceramics prepared by hydrolysis in polyol medium. Chemical Physics Letters. No. 541: 75–80, https://doi.org/10.1016/j.cplett.2012.05.047.
63. R.K. Brundavanam, G.E.J. Poinern, D. Fawcett. 2013. Modelling the crystal structure of a 30 nmsized particle based hydroxyapatite powder synthesised under the influence of ultrasound irradiation from X-ray powder diffraction data. American Journal of Materials Science. No. 3 (4): 84–90. https://doi.org/10.5923/j.materials.20130304.04.
64. K. Rop, G.N. Karuku, D. Mbui, N. Njomo, I. Michira. 2019. Evaluating the effects of formulated nano-NPK slow release fertilizer composite on the performance and yield of maize, kale and capsicum. Annals of Agricultural Sciences. No. 64: 9–19. https://doi.org/10.1016/j.aoas.2019.05.010.
65. M.S. Khan, A. Zaidi, E. Ahmad. 2014. Mechanism of phosphate solubilization and physiological functions of phosphate-solubilizing microorganisms, in: M.S. Khan, A. Zaidi, J. Musarrat (Eds.), Phosphate Solubilizing Microorganisms, Springer, Cham. https://doi.org/10.1007/978-3-319-08216-5_2.
66. E.T. Alori, B.R. Glick, O.O. Babalola. 2017. Microbial phosphorus solubilization and its potential for use in sustainable agriculture. Frontiers in *Microbiology*. No. 8: 971, https:// doi.org/10.3389/fmicb.2017.00971.

CHAPTER V

SYNTHESIS, CHARACTERIZATION, AND DEVELOPMENT OF ENVIRONMENTALLY FRIENDLY POLYMER BLENDS WITH VARIOUS TYPES OF ACEH FIBERS

TEUKU RIHAYAT, AGUNG EFRIYO HADI, JANUAR PARLAUNGAN SIREGAR

Abstract

Natural fiber-reinforced composites have a good ability to replace synthetic fiber composites; apart from being derived from renewable resources they are also environmentally friendly. Many fields benefit from the use of natural fiber-reinforced composites such as in building construction, locomotives, furniture, and the packaging industry. Natural fibers such as sisal, jute, coconut, wood fiber, bamboo, and pineapple have recently been used to strengthen polymer matrices. In Indonesia, in Aceh to be precise, these plants thrive and are widespread. The weak adhesion of natural fibers to the polymer matrix motivated researchers to improve the properties of these composites through various methods with a series of characterization studies. Several ways of modifying fiber are usually by adding additives and physical treatment. In this chapter, the fabrication and synthesis process for the development of natural fiber-reinforced composites will be discussed. A summary of the processing technology, interface modification, assembly pattern design, material properties, and pilot applications of natural fiber-reinforced composites is presented. Then the study of environmentally friendly polymers combined with natural fibers will also be investigated. A processing overview including the properties of these natural fiber-reinforced composites in the related fields is also described. This research on the development of natural fiber composites provides an effective approach for the purpose of protecting the environment.

1. Introduction

The development of renewable energy and the use of materials based on renewable resources is becoming increasingly important as energy sources such as petroleum are increasingly depleted, which dominate various industrial sectors for daily human needs. Fossil resources as well as non-degradable plastic waste cause environmental pollution and dangerous carbon dioxide is trapped in the atmosphere. Biodegradable materials are considered more environmentally friendly because they will eventually decompose and do not cause long-term damage to the earth [1]. Another benefit of biodegradable materials is that they often come from renewable resources that will not run out and can be continuously renewed.

Polymers are long repeating chains of atoms, formed from binders in the form of identical molecules called monomers, which are then joined together to form polymers, which includes natural polymers such as proteins, nucleic acids, cellulose (polysaccharides), clay, sand, sol-gel, silica, siloxa, and so on. Meanwhile, synthetic polymers are polymers that are formed through chemical reactions such as rubber fiber, nylon, polyester, polysterene, polyethylene, and others [1,2]. Polymers, especially those that form plastic products, have become one of the materials in the world of industry and chemical industry that dominate the manufacture of daily household materials and equipment, the needs of the packaging industry, the automotive industry, the electronics industry, and the medical field. Polymer applications are often found, for example in plastic food and drinks, household cleaning equipment (brooms, mops, buckets, and so on) and eating utensils (plates, spoons, bowls, cups, and so on), and automotive materials such as car bodies, motorbikes, tires, and so on. Assembled components can also be found in electronic materials ranging from fan body components, TV refrigerators, cable insulation, battery components, battery containers, and so on. There are several examples of use in the medical field such as infusion containers, tubes, medicine capsules, surgical equipment and clothing, even bone tissue materials. Some of the advantages of polymers that are widely used are because they are easy to process for various products at low temperatures at low costs [3]. Polymers are light and their weight or volume ratio is small. Corrosion and damage resistance properties against aggressive environments are also advantages of using polymers. In general, most polymers are good insulators against heat and electricity, and are useful for special component materials because of their elastic and plastic properties and their large molecular density so their dimensional stability is high. However, for some applications that require high mechanical properties, single natural polymers when processed have weaknesses, namely their mechanical and physical properties are weak and brittle.

Polymer mixtures in the industrial world are predominantly related to what are called composites. Polymers with a mixture of fillers increase their superior mechanical and physical properties. Various industries are adapting and exploring innovative environmentally friendly and renewable materials to meet the environmental problems of end use because most synthetic polymers which have advantages over natural polymers can increase the risk of environmental pollution because they are made from petroleum materials.

Composite materials are heterogeneous combinations of two or more different phase constituents. Composites can be formed from a combination of a polymer with filling or reinforcing materials such as fiber, minerals, carbon, lime, mica, and others. Currently, the use and utilization of natural fiber-reinforced materials continues to develop and is increasingly in demand by the industrial world. Fibers are small filaments of natural or synthetic materials that generally resemble threads in long or short pieces that come together to form a network or can be separated. Fibers can be grouped into natural fibers and synthetic fibers [4]. Natural fibers are very interesting materials to study. Natural fibers are generally made from various plants that are directly obtained in nature.

Natural fibers offer many benefits. The synergistic combination of natural fibers and plastic polymers can produce materials that have enormous potential and prospects for implementation in various industries as well as the properties, quality, and performance of fiber at cost-effective production levels. In the field of materials technology, natural fiber materials used as reinforcing materials are expected to produce mixed materials that are light, strong, environmentally friendly, and economical. This is because natural fibers have low density, are biodegradable, easy to recycle, require low energy production, have good mechanical properties, and can be renewed because they come from nature. Natural fibers are used in the textile, paper, crafts, accessories, decoration, and biocomposite material industries. Some of the fibers included in natural fibers include: hemp, sisal, palm fiber, jute, coconut fiber, pineapple leaf fiber, wood fiber, fiber from empty palm oil bunches, water hyacinth fiber, sugar palm fiber, and so on, which are starting to be used as reinforcing materials for materials.

Natural fibers have been tried to replace the use of synthetic fibers, such as E-Glass, Kevlar-49, Carbon Graphite, Silicone Carbide, Aluminum Oxide, and Boron [4,5]. In fact, asbestos, which used to be a synthetic fiber that was only used in Indonesia, has now been abandoned because it has a negative impact on the environment. Even though it doesn't completely change, the use of natural fibers to replace synthetic fibers is a wise step in saving the environment from waste produced and limited natural resources that cannot be renewed. Various types of fiber plants grow abundantly in Indonesia, such as kenaf (Hibiscus canabinus), alang-alang (Imperata cylindrical), and purun Tikus (Eleocharis dulcis).

Fiber reinforced polymer composites have been widely used over the past few decades and their market share continues to grow in applications such as automobiles, engine parts, sporting goods, consumer durables, and the electrical industry.

2. Potential of Natural Fibers in Indonesia

The amount of rainfall and its distribution as well as the average annual temperature are important climatic factors in evaluating land suitability for plants that produce natural fiber from plants. The amount and distribution of rainfall is related to the availability of water for plant evapotranspiration. Indonesia has natural resources in the form of forests spread throughout the archipelago. Indonesia's natural conditions are rich in natural fiber materials, such as cotton, jute, sisal, kenaf, banana, coconut, palm oil, coarse hemp (flax), fine hemp (hemp), and many others. With several advantages that natural fibers have, innovation and product development can be carried out in various industrial applications such as building construction, locomotives, automotives, furniture, packaging, geotextiles, and many others.

In general, plant fibers are almost the same or similar in that they are composed of three main components, namely cellulose, hemicellulose, lignin, plus other ingredients. The general properties of cellulose fibers are that they absorb water easily (hygroscopic), tangle easily, and if a combustion test is carried out, they produce a burning odor and charcoal. Plant fibers are used as materials for making paper and textiles. Cellulose fibers or what are called regenerated fibers are fibers made from natural fiber polymers, for example made from wood pulp or cotton seed fluff [6]. The raw material is reshaped to produce fibers or filaments that are suitable for making the desired material, for example spun into thread. Cellulose fiber can come from: stems (such as: flax or linen fiber, henep, jute, kenaf, sunn, hemp, purun Tikus, etc.), fruit (such as: coconut fiber), leaves (such as: Abaca or Manilla, henequen and sisal), and seeds (such as: cotton fiber and kapok). Fibers derived from seeds consist of cotton and kapok fibers.

2.1 Several Types of Natural Fibers Widely Spread in Aceh

In Aceh Province the average annual rainfall varies from almost 5000 mm per year on the western slopes of Bukit Barisan to less than 1500 mm per year on the eastern slopes of Bukit Barisan. On the western slopes of Bukit Barisan (including West Aceh district) rain is abundant throughout the year [6]. Bimodal rainfall with peaks in April, May, and around October and November is sufficient to meet the water needs of fiber producing crops in this region. The average monthly temperature fluctuations within a year are small. The temperature difference between the hottest months (around May and November-December) and the coldest months (January and June) is less than 20°C. Variations in air temperature are generally related to the height above sea level. This causes variations in the growth of fiber-producing plants which are widely distributed in the Aceh region itself [7].

Simeulu Regency, Aceh, including Zone A, is found in many fiber-producing plants such as banana abaca fiber. The further east you go, the lower the rainfall. Around Meulaboh, almost every month has an average monthly rainfall of >200 mm with the peak rainy season occurring in April and October. The eastern zone occupies the area around North Aceh and Lhokseumawe where there are many coconut plants which produce coconut fiber (cocofiber). Meanwhile, in the Central Aceh area, pineapple leaf fiber is often found and is widely distributed in the area. Another fiber that stands out is palm oil empty bunch fiber in the Nagan Raya district, which is one of the production areas that produces the most palm oil in Aceh [8].

2.1.1 Abaca Banana Fiber

Abaca bananas, which in Acehnese are called "Pisang Bue", have long been known and used as a vegetable that is widely known in society. Abaca banana plants have long grown abundantly in Aceh Province, especially in plantation areas that grow wild, especially in the Simeulu district. In the results of research [9] abaca fiber (Musa textillis) is another type of natural fiber that can be used as a composite reinforcing material. This fiber is widely used for materials that require strength, durability, and flexibility requirements, such as: ship ropes, submarine cable wrapping, hoist (lift) and power transmission ropes or cables, drilling cables, and high-quality paper. The advantages of abaca fiber lie in its strength and flexibility, as well as being resistant to the influence of microorganisms found in sea water [10].

Along with the development of material technology in the era of advanced materials (green materials), the potential contained in banana abaca fiber is discovered to be extraordinary. Non-structural uses are widely used, such as raw materials: high quality paper, tea bag packaging, baby diapers, and so on. The potential for banana fiber to be applied in making structural materials is because the fiber can be made up to eight meters long and depends on the height of the growing banana stem.

Another term for abaca fiber is Musa textilis or Manila hemp which comes from the Philippine Islands. Apart from the Philippine Islands, abaca fiber is also produced in other places such as Guatemala, Honduras, Panama, and Costa Rica (Central America), as well as in the Netherlands. Judging from its chemical composition as researched by Muller [9], abaca fiber contains: cellulose = 64.72%, moisture = 11.85%, ash = 1.02%, aqueos extract = 0.97%, fat and wax = 0, 63%, and incrusting and pectic matter = 21.83%. Research conducted showed that abaca banana stem fiber powder could increase the tensile strength and hardness of PVA – CaO_3 composites [10,11].

Figure 1. Abaca Banana Fiber [6]

Like other natural fibers, when used as a composite reinforcing material, abaca fiber needs to be given special treatment to increase its interfacial adhesion ability and resistance to absorbing water vapor so that good composite mechanical properties are obtained. Usually, the treatment medium used is a certain suitable chemical solution so that the pro-water (hydrophilic) nature of natural fibers can be reduced and the compatibility of the fibers with a water-repellent (hydrophobic) matrix can be increased. One chemical solution that is widely used to treat natural fibers is alkali (NaOH), to determine the characteristics of abaca fiber both untreated and treated with alkali.

2.1.2 Coconut Fiber

North Aceh Regency is one of the main coconut producers in Aceh Province. Of the total land area of 60,696 Ha spread across 22 districts/cities in Aceh, the area of coconut plantation land in North Aceh reached 32.63% or 19,808 Ha [12]. Coconut fiber (coir) can be developed into various products, including cocopeat, cocofiber, cocomesh, cocopot, coco fiber board, and cococoir. This material is a raw material for the mattress, pot, dry compost industry, and so on. Apart from these applications, coconut coir fiber can also be used as a polymer filling material. The application of coconut fiber composites can be used in industry, for example helmet shells. This study mentioned that the tensile strength of the composite decreases with fiber loading due to the weakest part of the composite and poor compatibility between the fiber and the polymer matrix. However, the flexural strength increases with increasing fiber content, due to the high stiffness of the fibers. Increasing the fiber volume fraction in the coconut fiber composite will increase the tensile strength as well as the tensile strain. The value of the modulus of elasticity increases up to around 70 GPa (gigapascals) — for example, flax fiber can reach values near 60–70 GPa under optimal conditions. The fracture cross-section of the tensile test is dominated by single fractures with an increase in the volume fraction experiencing fiber pull-out. Increasing the fiber volume fraction of coconut fiber composites can increase bending strength [13].

2.1.3 Pineapple Leaf Fiber

Pegasing District, Takengon, is known as a local pineapple producing area in Central Aceh Regency, which is a tropical fruit rich in water content. Not only is it enjoyed directly, pineapple fruit is also processed into various types of food and drinks which have added value from an economic perspective. But not only from fruit, apparently pineapple leaves can have a high selling value which until now has only been considered waste.

Figure 2. Pineapple Leaf Fiber

The pineapple plant (Ananas cosmosus), belonging to the Bromeliaceae family, is a tropical and subtropical plant that is widely found in the Philippines, Brazil, Hawaii, India, and Indonesia. In Indonesia, this plant is widely available in the Central Aceh region, which is one of the natural resources that has potential.

Pineapple fiber is a type of fiber that comes from plants (vegetable fiber) which is obtained from the pineapple plant. Pineapple fiber is starting to be widely used as an additional component in composites because it is known for its strength, like aluminum. Meanwhile, natural fibers are used as fillers or reinforcements in composite materials. From research conducted by Mujiono and Didik, it was found that pineapple leaf fiber has a tensile strength that is almost twice as high as fiberglass, namely 42.33 kg/mm2 for pineapple fiber, and 21.65 kg/mm2 for fiberglass. Thus, pineapple leaf fiber has the potential to be used as a reinforcement in composite materials [13].

Seeing its high economic value, the government is fully committed to supporting the development of pineapple leaf waste processing because it provides greater benefits. Apart from being used as woven fabric, pineapple leaves can also be used for handicrafts, animal feed, fertilizer, paper, and as a filler for composite materials for body materials for electronics, automotive, furniture, and medical equipment.

2.1.4 Palm Oil Empty Bunch Fiber

In Indonesia, empty palm oil bunches are used as paper pulp, fiberboard, and volume filler for furniture. In the Aceh region, oil palm is often found around Nagan Raya district [12,14]. Empty oil palm fruit bunches that are used have various characteristics that need to be researched, therefore research is needed that examines the characteristics of empty oil palm bunch fiber. Processing of palm oil empty bunch fiber begins with the process of taking samples of empty palm fruit bunches, samples of empty palm fruit bunches are then weighed before being given boiling and steaming treatment, empty palm fruit bunches are washed with clean water so that unwanted impurities are reduced.

One solution for its use is to use it as a filler to replace wood in the process of making plastic composite boards. The high cellulose content means it has good mechanical properties such as tensile strength values of 100-400 MPa, elastic modulus of 1-9 GPa, and total strain of 9-18% [14]. So, the properties of palm fiber make it suitable for composite applications. The only very frequent uses of cellulose-rich materials today are as boiler fuel and in the manufacture of potash fertilizers. To date, no systematic work has been carried out to evaluate the morphological and physical properties of palm oil fiber. EFB fiber contains 19% lignin, 65% cellulose, and 2% ash [15].

2.2 How to Prepare and Modify Natural Fibers

Research and use of natural fibers is developing very rapidly nowadays because natural fibers have many advantages compared to engineered fibers, the advantages of natural fibers such as lighter weight, easy to obtain materials, relatively cheap prices, and most importantly environmentally friendly, especially as Indonesia has natural wealth, which is so abundant. Like artificial fibers, natural fibers can also be used as modifications of artificial fibers. In general, the greater the ratio between fiber length and fiber diameter, the better the properties, and a small fiber diameter can reduce surface defects that cause brittleness. The properties of the fiber cannot be separated from the load applied. Optimum strength and stiffness are achieved when the fibers are in the same direction and the load is in the same direction as the fibers.

Even though natural fibers have been used in various applications, intensive research must still be carried out to further understand the forms of treatment given and optimize the potential of natural fibers and obtain new types of fibers. Various types of natural fibers have been explored to produce composite materials with commercial value. The advantages of using polymer composites include corrosion resistance, water resistance, and increased performance. Various composite industries in Indonesia still use glass fiber or metal as reinforcement for composite material products, such as PT. INKA[16]. However, in line with the many scientific findings and studies related to the development of natural fibers, it is hoped that the use of fiber composites in industry will be able to reduce the use of imported metal materials which are more expensive and easily corroded as well as glass materials which do not decompose in nature.

To use banana stem fiber as a PVA-CaO composite filler, the banana stem fiber is first treated with alkali. Treatment with alkali (NaOH) is expected to have an effect on the resulting composite, because the alkali function can remove existing lignin. Providing alkaline treatment to lignin-cellulose materials can change the chemical and physical structure of the fiber surface.

In research [16], it was proven that the addition of abaca banana stem fiber to the PVA-CaO_3 composite could increase the tensile strength, hardness, and flash point of the composite. The thickness of the banana stem fiber affects the compressive strength and maximum tensile strength of the polyester-natural fiber composite board. The use of banana stem fiber as a natural rubber composite mixture can strengthen the mechanical strength of the composite in the form of compressive strength and tensile strength. Maximum strength is obtained when the fiber length is 15 mm.

Another report stated that tensile strength and modulus gradually increase with increasing volume fraction. Reinforced polypropylene coconut coir fiber produces higher impact strength than hemp fiber and kenaf fiber, but its tensile strength and modulus of elasticity are slightly lower. The tensile strength of coconut fiber composite which has a higher flexural strength has the potential to be used in non-structural buildings.

Through research on analysis of the direction of the sieve fibers and the hardener ratio on the physical and mechanical properties of sieve/epoxy composites by comparing NaOH and $KMnO_4$ treatments, it was found that with 2% $KMnO_4$ treatment for 15 minutes with a fiber direction of 450 had the highest value on the mechanical properties of the composite, variations in the percentage of NaOH and KMnO4 in the fiber treatment process have a significant influence on the tensile strength and bending strength of fiber composites [17].

Other related research mentions the mechanical properties of empty oil palm fruit bunch fiber composites with polyester resin. After testing and SEM photos, it was found that the optimal fiber volume fraction of the EFB fiber composite was able to resist crack propagation.

In making fiber composites, the main thing that must be paid attention to is the preparation of the fiber itself. Several techniques and methods of pre-treatment of fibers are carried out to improve the performance of fibers when mixed with polymer matrices for various applications and are adapted to the basic properties of the fiber.

2.2.1 Steam explosion method

This technique requires low energy consumption to separate plant cell walls and produce pulp. It is believed that the steam explosion method is the optimal way for the separation of lignin from plant surfaces. This happens in the case of lignin separation in the pulp industry, for example, which produces final fibers that are stiff and dark in color. One study carried out fiber extraction using the steam explosion method where the bundles were not separated into single fibers. The fiber bundles are first sieved using a sieve and screening machine at mesh sizes to produce fiber sizes of 125–210 mm. Next, drying is carried out at a temperature of 210°C for approximately 2 hours. It turns out that lignin cannot be completely removed from the fiber because this method is more suitable for removing lignin from wood. Therefore, the remaining lignin from the fiber for producing bamboo fiber cotton is removed by a mixing machine. For example, the steam explosion process for bamboo fiber is first cutting raw bamboo and heating it in an autoclave at a temperature of 175°C and 0.7–0.8 MPa for 60 minutes. Then the evaporation process was carried out for 5 minutes which was repeated several times with the aim of cracking the cell walls. The ash content is removed by washing the fibers using hot water at 95°C with the help of soap. Finally, it was dried for a day at 105°C in the oven. During the steam explosion process, the fiber cell walls crack and the bamboo fiber becomes soft, allowing extraction. In this method, the soft cell walls are crushed and attached to the surface of the bamboo fiber, which has low shear resistance. The researchers used an ultrasonic method to wash the fibers. Then unused cells need to be removed by steam blasting using silane isocyanate. According to the analysis results, it was found that the tensile strength of bamboo using the boiling process was higher than the washing method with silane. Appropriate initial treatment is necessary to obtain strong adhesion between the matrix and fiber [19].

2.2.2 Retting Method

In this method, fiber, for example bamboo fiber, is usually prepared by peeling the bamboo skin first and then cutting the bamboo stems into pieces. The pieces are soaked in water for 3 days. Next, the bamboo can be scraped with a knife and combed to obtain fiber. In this method, the fiber surface abrasion process has a strong influence on the fiber quality, and the fiber breaks less along the fiber length. Several other studies only used the technique of cutting raw bamboo into several elongated sections without removing the segments and epidermis of the bamboo by scraping. The bamboo pieces are simply cleaned with water, then fermented for 2 months by soaking in room temperature water. There are 2 types of retting, namely aerobic and anaerobic, this technique is able to separate lignin from the stem with the help of microorganism enzymes without destroying the cellulose content.

2.2.3 Crushing

In the case of coconut fiber, it is extracted by first separating the fiber from old coconuts, characterized by the fiber being brown in color and having very little water content or being dry. Then the coir is crushed (chopped) on a defiber machine which separates the coconut fiber into fine pieces. Before crude fiber can be processed, it is first soaked in water for several days to remove dyes and impurities for approximately 24 hours. After that, the fiber is dried in the heat of the sun or through an automatic drying machine to remove the water content.

2.2.4 Rolling mills

In the case of bamboo stem fiber, the bamboo is cut into pieces measuring 1 mm thick. Bamboo strips are soaked in water for 1 hour to make it easier to separate the fibers. Then a rolling mill is used at low speed and small pressure to soften the stem. Next, the bamboo strips are soaked again for 30 minutes and the fibers are separated manually with a knife. The resulting fiber size is around 220-270 mm, dried in the sun for 14 days [20].

2.2.5 Degumming

In previous research, the degumming method was used for the fiber extraction process. Degumming functions were used to remove gummy and pectin content from decorticated fiber pieces such as bamboo and hemp. [21]. Degumming is a process to remove as much as possible of the gum compounds that are still present between the fiber strands. In crude hemp fiber the gum content ranges from 25-30%. The chemical degumming process is carried out by cooking the fiber in an alkaline solution for several hours. Chemicals that are often used in the chemical degumming process include NaOH 0.5%; a mixture of 32% NaOH and soap; a mixture of Na_2CO_3, soap and $Ca(OH)_2$; a mixture of 3% NaOH, 3% Na-sulfite, 3% Na-tripolyphosphate, and 3% teepol [21].

The chemical degumming process only takes one day. However, the degumming process has several weaknesses, such as producing chemical waste, requiring efforts to neutralize/control chemical contamination, which is a complicated process. One effort that can be made to overcome this is by applying an enzymatic degumming process or biodegumming. The degumming process using enzymes has already been carried out. Among them is the hemp degumming process using the enzymes pectinase, protease, and pectate lyase. Controlling the degumming rate of bamboo is critical to producing long single fibers.

2.2.6 Alkali Treatment

The alkalization process is one method used to provide surface modification to natural fibers. The alkalization process is carried out by immersing the fiber in an alkaline solution (NaOH) with different treatments at time and temperature. The alkalization process will reveal cellulose and remove impurities on the surface of natural fibers such as lignin and pectin so that it can improve the mechanical properties of the fiber.

In the case of pineapple leaves, the initial process of pineapple leaves is then soaked using NaOH to remove dirt and lignin attached to the leaves, then after that the fiber is dried in the outside air.

After the extraction process, namely separating or taking pineapple fiber from the leaves (fiber extraction), washing and drying in the sun can be done in other ways. Next, the fibers that are dry and ready to be used are arranged neatly and tied so that they don't fall apart when you want to use them. In the alkaline treatment procedure, pieces of material, for example bamboo, are first heated in a stainless-steel container, using 1.5 N NaOH with a heating temperature of 70°C for 5 hours. Then a pressing process is carried out on the bamboo pieces to produce fibers which are separated using steel nails arranged like a comb. It is necessary to wash the fiber with water and also dry it in the oven. This treatment minimizes the potential for fiber damage.

In EFB preparation, making empty oil palm fruit bunch fiber begins with drying the fiber for ± 2 days to remove moisture. After that, the empty coconut bunch fibers are cleaned from impurities with clean water. Then the fiber is dried in an oven at a temperature of 90° C. Once clean and dry, the empty palm fruit bunch fiber is reduced in size using an organic chopping machine. The empty oil palm bunch fibers that have been chopped are then meshed to obtain a homogeneous size with a size range between 280 to 900 microns. Alkalization treatment of empty oil palm bunch fiber was carried out with 2% NaOH for 3 hours at a temperature of 70° C using a magnetic stirrer. Washing the alkalization sample is carried out with distilled water until the impurities on the surface are removed. The water content of empty oil palm fruit bunch fiber is removed by baking it in an oven at a temperature of 105° C for 12 hours [22].

2.2.7 Bleaching

Bleaching is a whitening process carried out by placing the fibers that have been alkalized into a solution of H2O2 plus NaOH in general. The empty palm fruit bunch fibers are chopped and sieved to a size of 112-224 μm. The bleaching stage aims to remove residual alkali lignin. Bleaching was carried out by soaking in NaOH 4% by weight and H2O2 7.2% by weight at 550°C for 2 hours using a magnetic stirrer.

3. Making Polymer Composites with Natural Fibers

There are many major advantages of using natural fibers in composite production. First, natural fibers are renewable, biodegradable, and have minimal CO2 emissions so they are environmentally friendly. In addition, fiber-reinforced biopolymer composites can reduce dependence on petroleum-based materials, where up to 50% of the matrix weight can be replaced with natural fibers in some applications. Second, it is more economical because natural fibers are widely available and can continue to exist, thereby reducing the cost of using the matrix. Third, certain countries have implemented restrictions on products made from glass fiber and asbestos, therefore the presence of fiber composites can be an alternative replacement. Apart from that, another advantage is that fiber composites do not produce dangerous residues and other heavy metals when burned.

Several types of fibers, especially hemp and kenaf, which are included in the bast fiber category, have high specific strength and modulus, as well as low density compared to synthetic E-glass fibers, therefore they are lighter. This advantage means that using NFC saves weight and saves fuel. Apart from that, in the building industry, NFC has good acoustic and sound dampening properties. Apart from that, from a health perspective, the NFC production process is safer; it reduces skin and respiratory irritation.

3.1 Several Types of Bio-based Polymers

Biopolymers or polymers based on bio (renewable) materials, especially plastic polymers, are polymers made from materials originating from plants or animals. Some of them come from plants that contain starch because starch itself is a branched structure, namely amylopectin, and a straight structure called amylose. The development of this bio polymer has the impact of minimizing the use of fossil sources and is more environmentally friendly, making it more sustainable.

Bio-based polymers are sustainable polymers synthesized from renewable resources such as biomass instead of conventional petroleum raw materials such as oil and gas. Bio-based polymers are generally synthesized using biological or biochemical processes that produce minimal carbon, thereby reducing the impact of CO2 emissions on the environment.

If polymer production continues using fossil resources in the long term it could result in climate change. Greenhouse gas, namely CO2, which is produced from burning petroleum materials, will thin the ozone layer which will disrupt the layer's function to reduce the intensity of the sun's heat entering the earth, which is why the earth is getting hotter.

Polyethylene furanoate (PEF) is a bio-based polymer that is 100% recyclable. It is made from renewable vegetable sugar. PEF production polymerizes furan dicarboxylic acid and ethylene glycol. The process resembles PET production, but PEF replaces terephthalic acid with 2,5 furan dicarboxylic acid (FDCA).

PLA is a bioplyer as a basic material for bioplastics made from corn starch and can also be made from cassava starch, which has the advantage of a fast decomposition time and is compostable. PLA is an abbreviation for Poly

Lactic Acid, where this polymer is composed of lactic acid monomers obtained from the hydrolysis process of starch into glucose and then fermented using bacteria. The various advantages and benefits of PLA make it one of the most widely processed polymers, not only for industrial products, but also in the agricultural and medical fields. For example, in the medical field, several journal studies state that PLA can be used as medicine capsules, bone fixation devices, surgical implants, and bioscaffolds to repair damaged body tissue [23].

Polyhydroxyalkanoate (PHA) is a bio-polymer produced by microorganisms and has great potential to replace conventional plastic. Industrial wastewater processing to produce PHA using a sequencing batch reactor (SBR) is one of the reactor modifications in the activated sludge process. Tapioca industrial wastewater has a high organic compound content so it can be used as a carbon source for microorganisms.

PHB is a linear aliphatic polyester with short chain branching, which is highly biodegradable. Aliphatic polyesters have received special attention because they are sensitive to hydrolytic degradation while some of them are not enzymatically degraded. Poly-beta-hydroxybutyrate (PHB) is a bioplastic raw material that can be synthesized by microorganisms. For example, glucose as a raw material for PHB production comes from cassava starch, sago, corn, and others.

3.2 Current Manufacturing Techniques Incorporating Polymer Composites with Natural Fibers

Composite materials are a combination of two or more materials that have a number of properties that cannot be possessed by their individual components which are then arranged in a symmetrical combination to obtain certain properties. It can also be defined as a combination of two or more materials on a macroscopic scale that differ in chemical composition and do not dissolve in each other, where one material functions as a reinforcing material and the other material as a binder to maintain the unity of its elements, namely reinforcing reinforcement, which has less ductile but more rigid and stronger properties. What is used is natural fiber and matrix, which is generally more ductile but has lower rigidity strength.

The field of study of fiber composites along with the latest methods and technology for processing them is very broad. Likewise, renewable sources have been widely researched in several studies in various countries which prove that fiber composites have broad benefits and potential both from a technical, environmental, and cost perspective. In general, fiber composite processing techniques can be classified into 4 categories which are described below.

3.2.1 Cold Press

In research conducted by [24], they carried out the process of mixing bamboo fiber with 15% modified phenolic formaldehyde resin. The process begins by soaking the bamboo fiber at room temperature for approximately 6 minutes, then the resin is removed until the mass fraction becomes 13%. Next, the fiber is dried in an oven at 55°C to reduce the water content and weighed as needed. The glued bamboo was put into a mold and pressed using a cold press technique at 8 MPa for 10 hours at a temperature of 132°C. Once formed, the material is removed from the plate. After analysis, it was found that there was a greater increase in the tensile and compressive strength of the composite compared to raw bamboo. Meanwhile, the research carried out the manufacture of bamboo composites using resorcinol glue. The composite was made using a cold press technique at room temperature for 12 hours and using a pressure of 1.9 MPa.

3.2.2 Hot Press

The concept of the hot press method is similar to the cold press, only it uses a high temperature for a mixture of certain materials according to their properties. An example occurs in a mixture of bamboo pieces and Polyvinyl Chloride (PVC) which is pressed at 6 MPa for 3 minutes at a temperature of 175°C. Previously this process began with 5 minutes of heating with constant compression. This is done with the aim of reducing the water content, because if the water content is above 10% it can cause a decrease in the elastic modulus because water vapor cannot evaporate when pressed [24].

In other research, bamboo fiber was extracted using a biodegradable emulsion method, namely spraying steam and spreading randomly. Next, resin is added and printed on a plate container. The sheets are made starting with drying at a temperature of 105°C for 2 hours, then pressed at a pressure of 20 MPa with a temperature of 130°C for just 5 minutes. The result is poor interfacial adhesion between the fiber and the matrix which results in decreased strength.

3.2.3 Molding Process

This method is a method of molding composite materials using a rigid mold (mold). For example, in the process of making a bamboo fiber composite and a polypropylene matrix, it is mixed using a melt mixing procedure with a hot temperature adjusted to the melting point of the mixed material in a mold. Before molding and pressing, the inside of the mold can be sprayed with a thin layer of Semi Permanent Mold Release Agent (SPMRA) or aluminum foil to prevent the mixture from sticking and a thin steel plate placed on top of the mold to remove trapped air bubbles from the composite. The mixture of matrix and fiber is then pressed in a compression molding machine with an adjustable pressure load. This molding process is intended for standard testing purposes.

3.2.4 Injection Molding

Basically, injection molding is the process of inserting melted plastic into the cavity of a mold. The injection molding process begins by inserting the plastic material into a hot melt injection funnel (extruder). The plastic is pushed under high pressure into the mold. Once the plastic fills the mold, it is allowed to cool and harden. The finished part is then extracted (usually automatically) from the mold. For example, in the process of mixing lactic acid pellets and bamboo fiber, it is supplied from the inlet to the extruder, then the injection is carried out at 50-60 MPa and a temperature of 180°C [25]. Then, the extrudate is cooled and granulated in water at room temperature and forms pellets.

3.3 Classification of Samples Applications in Various Industrial Fields

The mechanical performance of fiber reinforced composites can be influenced by many factors including volume or weight fraction of reinforcement, fiber orientation, fiber aspect ratio, fiber-matrix adhesion, fiber alignment, distribution, use of additives, and chemical treatment of the fibers. The automotive and aircraft industries have been actively producing various types of natural fiber parts for their interior components. Insulation materials are also made from natural fibers for different application areas. In the automotive sector, natural fibers are widely used as car panels, seat backs and brake pads, seat backs, side and door panels, boot linings, hat racks, spare tire linings, dashboards, business tables, piller cover panels, under-body protection trim, instrumental panels, and headliner panels.

Natural fibers represent a sustainable future in architecture, with a wide variety of building materials, shapes, and even improving upon materials commonly used today. As a building material, for example, sisal fiber is often used as a composite substitute for wood, frames, doors, roofs, and even in earthquake-resistant buildings because it is durable and cheap.

In the world of textiles, natural fibers derived from seeds (cotton) are widely used as raw materials for fine textiles. Stem fibers such as flax and hemp are widely used for textiles, both ready-made clothing and technical textiles.

In the world of packaging and household equipment industries, fiber composites have begun to be used in making bowls, cups, spoons and forks, chopsticks, and food lunch boxes.

4. Study of Mechanical, Microstructural, and Physical Characterization

4.1 Results of Mechanical Behavior Test

The following are the tensile test results of PLA composite samples with fiber filler modified with NaOH and 1 composite sample without fiber modification at the best volume fraction and holding time as in the picture.

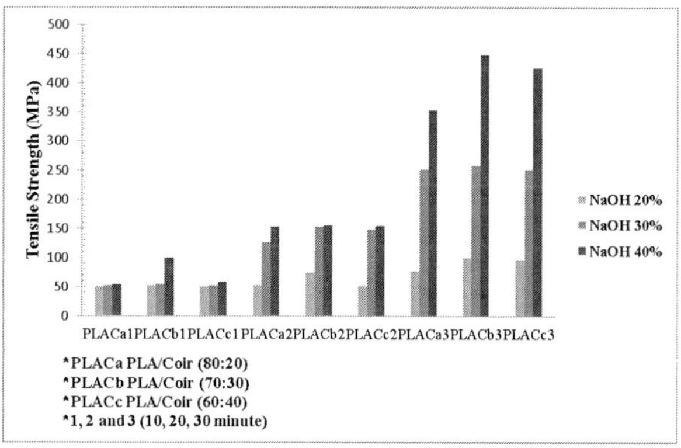

Figure 3. The results of the tensile test of PLA-Coir and % NaOH [24]

Based on the graph, it can be seen that the addition of NaOH to coconut fiber filler before it is used to form a composite with polymer shows a better increase in tensile strength compared to composites without the use of NaOH as a filler preparation material. The tensile strength value of the PLA-

Coconut Fiber composite sample with modified fibers is 55 to 440 MPa, while the tensile strength value of the PLA-Coconut Fiber composite with unmodified fibers is 340 MPa. This shows that apart from the fiber itself, the fiber pre-treatment process can also influence the final product [25].

The greater the concentration of NaOH, the better the effect on the tensile strength of the composite. This proves that alkaline NaOH is successful in modifying the fiber, where NaOH cleans the fiber from dirt and reduces surface tension and increases good interfacial adhesion so that the fiber will be compatible when mixed with a polymer matrix, resulting in better quality composite strength.

It was also reported that the use of urea formaldehyde as an adhesive also contributed to improving the mechanical properties of the composite. The presence of this adhesive will help the bond between the matrix and filler to be maximized. The use of urea formaldehyde as an adhesive also contributes to improving the mechanical properties of the composite. The presence of this adhesive will help the bond between the matrix and filler to be maximized. The use of urea formaldehyde as an adhesive also contributes to improving the mechanical properties of the composite. The average tensile strength test value is directly proportional to the amount of filler used. The greater the volume fraction of PLA polymer coir, the greater the resulting tensile test value. Apart from that, the tensile strength of the material is also not immune to the influence of holding time which is supported by optimal temperature. A holding time that is too short will cause the bond between the matrix and the filler to be weak or not even occur, resulting in a small tensile strength value.

In other research the increase in strength of hemp-reinforced composites is due to the good load-bearing ability of the fibers due to the transfer of higher loads from the matrix to the fibers. There is an effect resulting from the process of soaking the fiber before it is mixed with the matrix where it is known that the tensile strength of PE with hemp fiber which is soaked for approximately 12 days causes the fiber to swell due to the absorption of water which fills the voids during melting, this causes a higher tensile strength compared to the same sample without water immersion (dry).

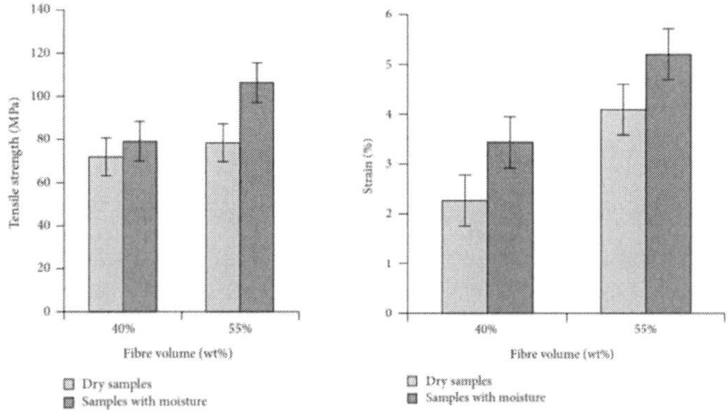

Figure 4: Tensile strain versus fiber volume [24]

During the composite manufacturing process, what might happen is poor impregnation, causing the appearance of voids that are not solidly bound between the fiber and the matrix. This can be prevented by filling with resin to increase its mechanical properties and tensile strength.

In another report, the water hyacinth fiber was treated by soaking the water hyacinth fiber in a toluene solution containing 5% isocyanate for 30 minutes at a temperature of 50°C. Then the fiber is dried and mixed with toluene containing 6% polyethylene glycol. The results of this treatment show that this process results in a decrease in the hydrophilic properties of the fiber, increasing the tensile strength and modulus of elasticity in the LDPE/water hyacinth composite. Composite ductility in non-treated fibers is higher than in treated fibers. Pre-treatment of water hyacinth fiber with silane showed that treatment with silane reduced the contact angle between water hyacinth fiber and epoxy-resin, resulting in a stronger composite. Treatment with silane coupling agent increases the tensile strength and flexural strength of the resulting composite material.

4.2 Results of Microstructural Tests

Figures 5 and 6 show the morphology of the fiber surface before and after treatment with NaOH solution. SEM results generally show that the morphology of palm fiber is cylindrical in shape and has pores on the surface.

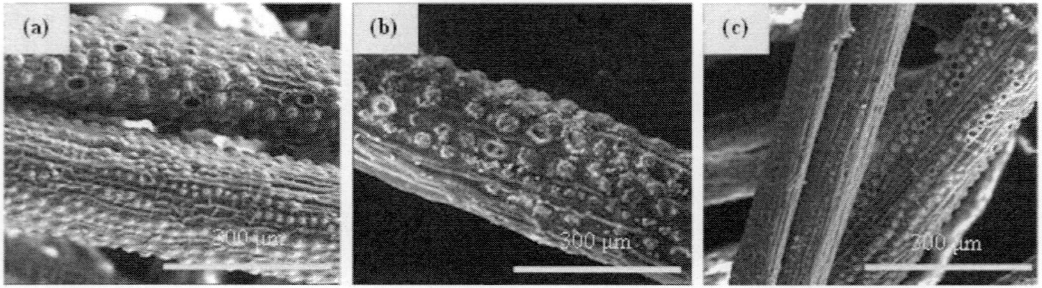

Figure 5. SEM of Palm Oil Fiber (a) Without Treatment (b) 5% NaOH Treatment (c) 10% NaOH Treatment [24]

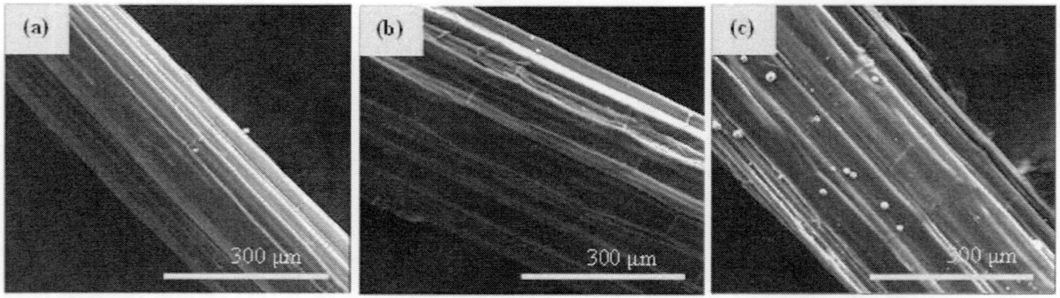

Figure 6. SEM Banana Leaf Fiber (a) Without Treatment (b) 5% NaOH Treatment (c) 10% NaOH Treatment [35]

Fiber without treatment with NaOH solution has a rougher surface compared to fiber treated with NaOH solution. Apart from that, palm fiber without alkali treatment has a surface layer that contains lignin and other impurities such as wax and oil. The layer removal process leads to increased bond stability between the polymer matrix and fiber. Chemical surface modification has an important role in cleaning surfaces from dirt and fibers that grow imperfectly. Several previous studies have reported that chemical fiber treatment can improve the level of surface roughness and reduce surface polarity [26]. In fact, partial removal of the lignin layer causes the breaking of the bonds in the fibrils leading to the protrusion or exposure of some of them. These protrusions will produce mechanical bonding of the fibers so that they can increase the interaction of the fibers with the matrix.

In research conducted by [26], SEM results showed that the NaOH concentration had an effect on the level of surface roughness of the palm oil fiber produced. Increasing the concentration of NaOH causes greater loss of the lignin layer and hemicellulose so that the level of fiber surface roughness becomes lower. This can lead to increased interfacial bonding between fiber and matrix so as to improve the mechanical properties of the composite. It has been reported that the higher the NaOH concentration used in the fiber treatment process, the higher the amount of lignin, hemisulose, and other impurities dissolved. [30] reported that alkalization modified the surface topography and made the fiber surface appear smoother than the surface without NaOH treatment.

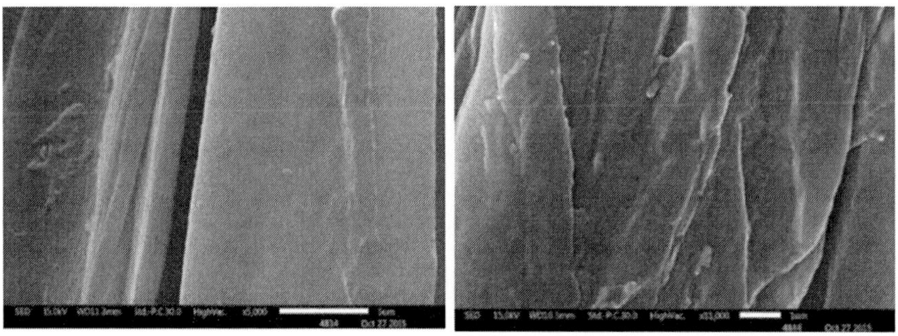

Figure 7. SEM test results for water hyacinth and epoxy resin [25]

Treatment with silane has the effect of increasing the adhesion bond between water hyacinth fiber and epoxy resin. The alkaline treatment process causes the surface of the water hyacinth fiber to become rougher as shown in Figure 7. The surface changes result from the loss of some of the pectin, lignin, wax, hydroxyl groups, and other impurities. Changes in surface roughness result in better interlocking forces between the fiber and the matrix. Modification of the properties of water hyacinth fiber can also be done using polyvinyl alcohol (PVA). Water hyacinth fiber was soaked in a mixture of ethanol with 6% PVA powder. Soaking is done for 24 hours. Test results show that up to a mass fraction of 25%, modification with PVA increases the tensile strength, modulus of elasticity, and ductility of water hyacinth fiber composites with LDPE matrix. Treatment of water hyacinth fiber can also be done by soaking the fiber in a toluene solution containing isophorone diisocyanate for 30 minutes at a temperature of 50° C. When compared with no treatment, this treatment succeeded in increasing the tensile strength of the composite by up to 100%.

4.3 Results of Physical Tests

Figure 8a shows the results of thermogravimetric analysis of palm fiber without alkali treatment and with alkali treatment. At a temperature of 50°C-100°C, it shows a mass reduction of 5.5% in both fiber conditions due to the loss of water content and volatile components.

The TG-DTA results in Figure 8 show that at a temperature of 125-225°C there was a mass reduction of 18.87% in fibers without the alkalization process and 17.44% in fibers with alkalization caused by hemicellulose decomposition. Hemicellulose decomposition can be detected from Differential Thermal Analysis (DTA) which shows the appearance of a low peak at a temperature of around 170°C.

At a temperature of 250-350°C there was a decrease in the mass of palm oil fiber without alkalization and with 10% NaOH alkalization amounting to 47.81% and 44.87%, respectively. This is due to the decomposition process of cellulose contained in palm fiber. The peak of cellulose decomposition was detected on the DTA graph at a temperature of around 332°C [28]. At a temperature of 425-475°C it shows characteristics of lignin decomposition. Fibers without an alkalization process showed a higher intensity decomposition peak at a temperature of 450oC compared to fibers with alkalization. This shows that the alkalization process with NaOH causes the loss of the lignin component from the fiber. The percentage of final mass remaining in the fiber after heating to a temperature of 600°C is 1.6% for fiber without treatment and 11.80% for fiber with treatment. The TG-DTA results show that palm oil fiber with the NaOH alkalization process shows relatively high heat resistance compared to fiber without alkalization.

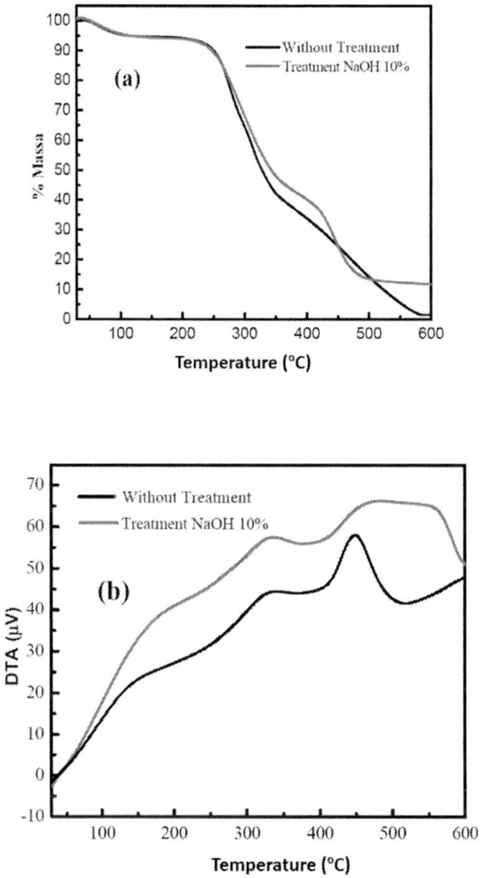

Figure 8. TGA and DTA results of palm oil shabu fiber [27]

5. Conclusion

Natural fiber-reinforced composites have a good ability to replace synthetic fiber composites, apart from being derived from renewable resources they are also environmentally friendly. Many fields benefit from the use of natural fiber-reinforced composites such as in building construction, locomotives, furniture, and the packaging industry. Natural fibers such as sisal, jute, coconut, wood fiber, bamboo, and pineapple have recently been used to strengthen polymer matrices. In Indonesia, in Aceh to be precise, these plants thrive and are widespread. The weak adhesion of natural fibers to the polymer matrix motivated researchers to improve the properties of these composites through various methods with a series of characterization studies. Ways of modifying fiber usually include adding additives and physical treatment. In this chapter, the fabrication and synthesis process for the development of natural fiber-reinforced composites will be discussed. A summary of the processing technology, interface modification, assembly pattern design, material properties, and pilot applications of natural fiber-reinforced composites is presented. Then the study of environmentally friendly polymers combined with natural fibers will also be investigated. A processing overview including the properties of these natural fiber-reinforced composites in the related fields is also described. This research on the development of natural fiber composites provides an effective approach for the purpose of protecting the environment.

References

1 YL Li, SJ Wang, B. Arash, Q. Wang. 2016. A study on tribology of nitrile-butadiene rubber composites by incorporation of carbon nanotubes: molecular dynamics simulations. Carbon,10, 145-150.

2 Navin Chand, Mohammed Fahim. 2020. Tribology of Natural Fiber Polymer Composites, Second Edition: Woodhead Publishing Series in Composites Science and Engineering.

3 J. Liu, J. Shen, Z. Zheng, Y. Wu, L. Zhang. 2015. Revealing the toughening mechanism of graphene polymer nanocomposite through molecular dynamics simulation. Nanotechnology, 26, (29), 291003.

4 Xiao-Su Yi, Shanyi Du, Litong Zhang. 2018. Composite Materials Engineering. Vol.1, China: Springer. 2014. Compos. Part B Eng, 56, 691-697.

5 QX Pei, YW Zhang, VB Shenoy, A molecular dynamics study of the mechanical properties of hydrogen functionalized graphene. 2010. Carbon 48, 898-904.

6 Y. Li, Q. Wang, S. Wang. 2018. A review on enhancement of mechanical and tribological properties of polymer composites reinforced by carbon nanotubes and graphene sheet: molecular dynamics simulations. Compos. Part B: Eng. 160, 348-361.

7 Jackson Singh, Thingujam. 2014. Characterization of Natural Fiber-reinforced Composites-Bamboo And Sisal: A Review. International Journal of Research in Engineering and Technology, 03,(19), 187-195.

8 Y Yolanda and Praswasti P D K Wulan. 2020. Comparison of Natural Fiber Types as Reinforcement Material on Composite Mechanical Properties via Carbon Nanotubes Addition.IOP Conference Series: Materials Science and Engineering, 1041, 012050.

9 Parnia Zakikhani, R. Zahari, M.T.H. Sultan, D.L. Majid. 2014. Extraction and preparation of bamboo fiber-reinforced composites. Materials & Design, 63, 820-828.

10 K. Cahya, H. Faikhul, K. Wawan. 2022Experimental Study of Sandwich Strength Materials Made from Banana Fiber, as Advance Materials of the Unmanned Mine Sweeper (UMS) Hull. 2022. Key Engineering Material. 941, 249-255.

11 Mirza Pramudia, Mahrus Khoirul Umami and Aji Prihantoko. 2022. Effect of fiber volume fraction on tensile strength and fracture analysis of corn husk reinforced epoxy resin composite. MATEC Web Conf, 372, 02005.

12 Yongsheng Zhang, Zhongshun Yuan, Chunbao Xu. 2017. Bio-based resins for fiber-reinforced polymer composites. Natural Fiber-Reinforced Biodegradable and Bioresorbable Polymer Composites, 2017, 137-162.

13 Wiley. 2017. Woodhead Publishing Series in Composites Science and Engineering. Vol.6: Scrivener Publishing.

14 M. Dhinesh Kumar, C. Senthamaraikannan, S. Jayasrinivasan, S. Aushwin. 2017. Study on static and dynamic behavior of jute/sisal fiber-reinforced epoxy composites. Materials Today: Proceedings,6, (19), 2021, 9425-9428.

15 M. Al-Oqla Faris, Salit Mohd. 2017. Materials Selection for Natural Fiber Composites. Duxford: Wood Head Publishing.

16 Muchlisinalahuddin, Dahlan Hendery, Mahardika Melbi, Rusli Meifal. Cellulose-based Material for Sound Absorption and Its Application – A Short Review. 77, 0100.

17 M. Dhinesh Kumar, C. Senthamaraikannan, S. Jayasrinivasan, S. Aushwin. 2021. Study on static and dynamic behavior of jute/sisal fiber-reinforced epoxy composites. Materials Today: Proceedings,46, 19, 9425-9428.

18 Muchlisinalahuddin, Dahlan Hendery, Mahardika Melbi, Rusli Meifal. 2023. Cellulose-based Material for Sound Absorption And Its Application – A Short Review. BIO Web of Conferences, 77, 01003.

19 Champilho. 2016. Natural Fiber Composites. New York: Taylor & Francis Group.

20 Setiawan, Dedi. 2015. Karakterisasi Serat Abaca SebagaiAlternatif Material PenguatKomposit Ramah Lingkungan. INDEPT, 4, 1, 1-7.

21 Macaulay M. Owen, Emmanuel O. Achukwu, Solehuddin Bin Shuib, Zafir Romli Ahmad, Abdul Halim Abdullah & Umaru S. Ishiaku. 2023. Effects of high-temperature optimization and resin coating treatment on the mechanical, thermal, and morphological properties of natural kenaf fiber-filled engineering plastic composites. Polymer Composites, 44(4),2512-2529.
22 Mohammad Farid, Agung Purniawan, Diah Susanti, Mavindra Ramadhani, Ula Charir Anjani. 2021. Characterization of nanocellulose and barium hexaferrite reinforced silicone rubber composites. AIP Conf. Proc. 2384, 050006.
23 Ninis Hadi Haryanti1, Henry Wardhana2 and Suryajaya. 2021. The Effect of NaOH Immersion Time to Physical Properties of Purun Tikus (Eleocharis dulcis) Cement Board Composite. IOP Conf. Ser.: Mater. Sci. Eng. 599 012013.
24 Zhao, W., Zou, Y., Zhang, W., Chen, H., Zhang, J., Wu, A., Zhao, Y. 2024. Optimization of Process Parameters for Bamboo Fiber Extraction by Steam Explosion. Journal of Natural Fibers, 21(1).
25 Debes Bhattacharyya, Stoyko Fakirow. 2012.Synthetic Polymer-Polymer Composites. New Zealenad: Hanser Publisher.
26 Chawla, Krishan. 2012. Composite Materials. Science and Engineering-3rd Edition. New York: Springer.
27 Mohanty Amar, Misra Manjusri, T. Drzal Lawrence. 2005. Natural Fibers, Biopolymers, And Biocomposites. New York: Taylor & Francis.

CHAPTER VI

CHARACTERIZATION OF NATURAL FIBER COMPOSITES FROM HYBRID POLYESTER REINFORCED WITH BAMBOO FIBER, PINEAPPLE LEAVES, AND COCONUT FIBER OF ACEH

RIMA DHINTA DEWI ASTUTI, ISRA ADELYA IZZATI, SARIADI, JANUAR PARLAUNGAN SIREGAR

Abstract

Natural fiber composites have garnered significant attention from researchers due to their environmentally friendly nature, in contrast to synthetic fibers, which can contribute to environmental pollution. Currently, composites are regarded as the most promising materials in society because of their unique mechanical properties, biodegradability, and the abundant availability of raw materials. Researchers have increasingly focused on the development of biodegradable fiber-reinforced polymer (BFRP) composites as alternatives to conventional materials, especially in the automotive, marine, packaging, furniture, and building construction industries.The production of natural fibers as components of composite materials has gained significant interest and is being widely applied due to concerns about global warming and the decreasing availability of petroleum for synthetic fiber production. Natural fibers play a crucial role in developing biodegradable composites that address current ecological and environmental challenges. Studies have shown that natural fibers such as bamboo, coconut, flax, sisal, pineapple, abaca, and coir, when used as reinforcements and fillers in composites, exhibit strong mechanical properties and resistance to bending. These properties make them comparable to, if not superior to, synthetic fibers and other conventional materials.

1. Introduction

Over the past three decades, functionally graded materials have been extensively developed. These materials have been widely used in systems that enhance driving comfort and reduce energy consumption. Polymer-based composite materials have rapidly evolved and have been applied to various purposes. Environmental issues have sparked a significant interest in developing new composite materials sourced from biodegradable resources. There is a strong push to develop innovative materials that are lighter, cheaper, and renewable from local resources.

Natural fibers have garnered more attention as component reinforcements in recent years. India currently produces about 14.5 million tons of natural fibers annually, compared to 45.5 million tons produced worldwide. These fibers are utilized in place of synthetic fibers and reinforced with a variety of matrices, including vinyl ester, polyester, epoxy, and phenol-formaldehyde. There are many different kinds of natural fibers, including wood, fruit, seed, bark, and leaf fibers. These fibers have historically been utilized for a variety of goods, including bandages, roofs, ropes, and household utensils. Because these materials are biodegradable, there are intrinsic environmental benefits to this revolution. Natural fiber-reinforced composites provide excellent mechanical and thermal characteristics. Natural fiber composites have taken on the role of synthetic polymers in the building, aerospace, and automobile sectors. These natural fiber-reinforced composites have several benefits, including high specific strength and modulus, availability, affordability, light weight, biodegradability, recyclability, lack of health risks, and non-abrasive qualities; they include bandages, roofing, and household tools. Because these materials are biodegradable, there are intrinsic environmental benefits to this revolution.

2. Extraction of Natural Fibers

Naturally available plants contain significant amounts of lignocellulosic fibers, which can be considered natural composites. These fibers primarily consist of lignocellulosic fibrils embedded in a lignin matrix (resin). These cellulose fibrils are aligned to form long fibers, regardless of the plant's origin and structure [1].

Sisal fibers are robust fibers found in the leaves of the sisal plant (Agave sisalana family). Originally native to North and South America, sisal is now widely cultivated in Africa, the West Indies, and the Far East [2]. Sisal fibers are extracted from the leaves using mechanical decorticators. In this process, leaf bundles are fed between rotating rollers. The hard surface is scraped off, leaving the sisal fibers intact. These fiber bundles are then thoroughly washed with water to clean their surfaces. This mechanical process yields 2-4% of long fiber bundles [3]. QuillBot's paraphraser takes your sentences and makes changes, helping you to rework and rephrase your content quickly and efficiently.

In a similar manner, Sansevieria cylindrica fibers are extracted from the plant's leaves. The moisture is subsequently removed from the decorticated fibers by sun-drying them for a whole day [4]. Water retting is the method used to remove sisal and roselle fibers. For three to four days, bundles of sisal leaves and roselle stems are submerged in water. After that, the hard sections of the saturated leaves are carefully rinsed under running water, leaving only the sisal and roselle fibers [5].

Agave Americana leaves are cut from the Agave Americana plant and boiled in water until they become soft. The fibers are then extracted by squeezing the softened leaves in running water to dissolve the soft tissues [6]. Additionally, Zostera marina (seaweed) plants are boiled to extract seaweed fibers [7]. Mature coconut husks are soaked in water for five months to decompose the pulp, making it easier to remove the husk. After soaking, the wet husks are beaten with hammers to separate the coir fibers from the shell. These coir fibers are then combed in a carding frame to separate them into individual fibers [8].

Fibers from the green coconut husk are mechanically extracted. Before extraction, the bundles of green husk are soaked in water for three to seven days to soften the outer skin and remove any dye substances [9]. Coconut leaf sheath fibers originate from the palm family (Arecaceae). The mat-like form of the leaf sheath is collected from coconut trees and immersed in water for a week, then thoroughly washed with freshwater and distilled water. These fibers are then sun-dried for a week to obtain pure leaf sheath fibers [10].

Leaves from the Curaua plant are picked from the Curaua tree, and the fibers are extracted using a simple water retting process [11]. Vakka Fibers come from the fallen leaves of the tree, which are dried for two to three days in the shade. These leaves are soaked in a water tank for 15 days, then the loose upper layer is removed. Afterward, the leaves are soaked again in another retting water tank for three days to separate the fibers from the sheath. Bamboo fibers are extracted within three days, while date palm fibers are extracted from pinnate leaves. The surface layer of the stalk is cut off with a knife, and the stalks are dried in the shade for five days [12].

The date pit fibers are pounded into a powder and then sifted to an average size of 0.15 mm using a sieve tray with a shaker. After that, these fibers are dried out on the outside by being placed in a vacuum oven for a full day [13]. One way to harvest okra fiber is from monocotyledonous plants. For fifteen to twenty days, young plants are housed in a tight space until the stems have broken down enough to make it easier to remove fiber. To obtain clean fibers, the fibers are separated from the microbially damaged stems and triple-washed in deionized water [14]. Elephant grass stems are cut at the base, and the nodes are trimmed. These stems are dried in the shade for a week and then soaked in water for the retting process for 10 days. A plastic hammer is then used to beat the stems to loosen and separate the fibers. These stems are also soaked in water containing 1 N NaOH to extract the fibers [15].

The Agavaceae family of succulent plants is used to obtain agave fibers. To eliminate extra waste such as chlorophyll, leaf juice, and solid adhesives, agave leaves are soaked in a retting water tank for two weeks in order to release the upper fiber layers. Afterwards, the fibers are washed in water. Following the extraction procedure, the fibers are extensively cleaned in water to get rid of superfluous waste materials such as leaf juice, solid glue, and chlorophyll. To get rid of the moisture, these fibers are sun-dried for two to three days [16].

3. Study Related to Preparation of Natural Fiber-Reinforced Polymer Matrix Composites

Using a straightforward hand lay-up technique, Jayabal and Natarajan [17] created polyester composites bonded with fibers from coconut coir. Prior to the molding process, they applied a polyvinyl acetate release agent to the mold surface. Short, randomly oriented fibers that were not woven were impregnated with a matrix made of cobalt octoate accelerator, unsaturated polyester resin, and Methyl Ethyl Ketone Peroxide (MEKP) catalyst in a certain ratio. Following impregnation, the fiber layer was put into a 360 x 360 mm mold and compressed for an hour using weight. After carefully removing the composite from the mold and compacting it to a thickness of 3 mm, it was allowed to cure for a full day at room temperature.

Short fibers of Sansevieria cylindrica were evenly distributed between two light steel plates using a steel roller to ensure uniform fiber distribution in the mold. The prepared matrix solution was degassed before being poured into a closed steel mold and placed under hydraulic pressure for 3 hours. The composite sheets were then compacted in a mold measuring 300 mm x 150 mm x 3 mm [18].

In order to create a mat, chopped sisal strands (30 mm in length) were put at random in a mold and compressed using a hydraulic press. Resin Transfer Molding (RTM) was used in a closed mold with isophthalic polyester resin, cobalt naphthenate, and methyl ethyl ketone peroxide mixed in. The pressure applied was 1 kg/cm2. Plate-shaped composites were produced by curing these samples for 12 hours and post-curing them for 4 hours at 80 °C [19]. Using an aluminum mold that measured 250 mm by 250 mm, De Rosa et al. [19] created epoxy composites with random orientation reinforced with short Phormium tenax leaf fibers. Prior to construction, a small layer of polyvinyl alcohol release agent was applied to the mold, and short fibers were inserted into it at random.

Murali Mohan Rao et al. [20] created polyester matrix composites reinforced with unidirectional fiber composites using the hand lay-up method. They minimized fiber deformation and movement to achieve high-quality composite materials. During the curing process, a compression pressure of 0.05 MPa was applied to the mold, and the composite specimens were cured for 24 hours. The specimens also underwent post-curing at 70°C for 2 hours after removal from the mold. Coconut coir fibers and polypropylene filament were cut to a length of 10 mm, mixed, and carded to form a mesh. This mesh was laminated in a mold under a pressure of 15 MPa at a heating temperature of 190°C using a heat press to create the composite material. Mechanical properties of various natural fibers are explained in Table 1.

Table 1. Mechanical Properties of Various Natural Fibers [40]

Fiber name	Gauge length (mm)	Tensile strength (MPa)	Tensile modulus (GPa)	Elongation at break (%)	Reference
Cocount Husk Fiber	10	126–148	3.1–4.3	-	[19]
Bamboo	100	503	35.91	1.40	[12,22]
Palm	100	377	2.75	13.71	[21]
Coconut Or Coir	100	500	2.5	20	[20]
Date Palm	-	97-196	2.5-5.4	2-4.5	[22]
Pineapple Leaf Fiber	10	609-700, 400-627	23.7-30.3, 14.8	0.8-1.6	[19, 24]
Banana	150	600	17.85	3.36	[12, 23]
Oil Palm	-	248	3.2	25	[22]
Cotton	-	400	5.5-12.6	7.0-8.0	[25]
Coconut Tree Leaf Sheath	50	46.4	2.3	2.84	[26]

The coconut fiber's volume percentage was meticulously regulated when the two ingredients were being mixed. The composite is 300 mm by 300 mm by 3 mm in size. Particles of high-density polymer pellets, including polystyrene and polyethylene, were ground. A powdered date palm pit was combined with these particles. Then, several predefined weight percentages were combined with this dry mixture. According to Alsewailem and Binkhder [27], the mixture was heated and blended for ten minutes at 160°C before being poured into a vacuum mold that had already dried. This resulted in a composite made of polymer and date palm pit.

Crushed kenaf and rice husk fibers were sifted to sizes ranging from 70 to 250 mm. The sifted fibers were stored in a desiccator to minimize their moisture content. After drying, these particulate fibers were carefully added to the biodegradable polymer and mixed in a mixer for 15 minutes to achieve homogeneity. This mixture was then fed into an extruder hopper and extruded using a screw extruder to inject it into a mold to prepare the composite [28].

4. Kind of Fibers

4.1 Bamboo Fiber

The bamboo species belong to the subfamily of grasses, Bambusoideae, further divided into Arundinarieae, Bambuseae, and Olyreae. There are approximately 1600 different bamboo species known, with the majority found across Asia, Africa, the Americas, and Oceania [29]. One of the most famous species is Bambusa vulgaris, found in Brazil. This bamboo can reach heights of up to 15 m, with stem diameters ranging from 5-10 cm and internodes measuring 24-40 cm in length [30]. A highly favored bamboo species is Phyllostachys pubescens, also known as Phyllostachys edulis (Moso bamboo) [29], extensively referenced in bamboo studies. Kao Zhu (Bambusa pervariabilis) and Mao Zhu (Phyllostachys heterocycla) are two other species [31] renowned for their structural bamboo qualities, including impressive compressive strength and flexibility [32].

4.1.1 Extraction, Composition, and Properties of Bamboo Fibers

This section will discuss the influence of fiber length in the extraction process on the properties of bamboo fibers. The internodal length of bamboo affects the maximum fiber length that can be extracted, which is often preferred for use in unidirectional composites that have a more homogeneous structure compared to short fibers. The discussion will also cover the extraction process, types of bamboo used, and relevant characteristics of bamboo fibers.

4.1.2 Mechanical, chemical, and combined (mechanical and chemical) extraction

Mechanical extraction, chemical extraction, and a combination of both (mechanical and chemical) are the three primary techniques for obtaining bamboo fiber. Chemical retting, alkali/acid retting, and degumming are some methods commonly employed in chemical extraction to lower the lignin concentration in bamboo fibers. Compression molding (CMT) and roller mill technology (RMT) are examples of combined processes [33]. When compared to chemical extraction methods, mechanical extraction of bamboo fibers typically produces better results with fewer fiber breakages [34]. In the mechanical extraction process, chopped fibers are heated for 60 minutes at a temperature of around 175 °C in an autoclave with a pressure of 0.7–0.8 MPa. Steam is evacuated every 5 minutes to ensure cell wall cracking. The fibers are then washed with hot water and dried at 105°C in an oven to ensure fiber softness and prepare them for extraction [33]. In chemical extraction processes, degumming and pectin removal from bamboo fibers are achieved using chemical solutions such as sodium hydroxide or trifluoroacetic acid to dissolve or remove lignin, facilitating extraction. However, alkali treatment is preferred for enhancing interfacial bonding strength in composites [33]. The combined CMT and RMT techniques involve aligning bamboo pieces between two flat plates or within compressed rollers. Bamboo pieces are immersed in a 1% NaOH solution to facilitate fiber extraction, followed by washing and oven drying [33]. A similar approach has been applied by Deshpande et al. [35] for extracting bamboo fibers with three different diameters: 0.15 mm, 0.20 mm, and 0.25 mm.

The chemical composition of bamboo consists of lignin, hemicellulose, cellulose, and pectin, contributing to its high mechanical properties. According to Hu et al., the lignin content in outer bamboo fibers (19%) is higher compared to inner bamboo fibers (14.5%), which have a relatively larger fiber diameter (107.9 μm) [36]. Bamboo fibers treated with alkali, having a diameter of approximately 262 μm, showed a lignin content of 24.4%, whereas fibers extracted using steam explosion had a lignin content of 23.3% with a diameter of 195 μm [37]

4.1.3 Physical Properties

The natural density and moisture content of the fibers are influenced by the fiber extraction technique, plant age, and soil conditions in which the plants are grown [38]. In addition, Porras et al. reported that bamboo fibers have a lower density (0.850 ± 0.076 g/cm^3) than other fibers, including glass fibers (2.5 g/cm^3) and hemp (1.2–1.5 g/cm^3), and a lower moisture content (3.38 ± 0.216%). This enhances the production process' quality by lowering porosity and dimensional variations [38]. The mechanical properties and composite material weight are enhanced through chemical treatments such as alkalinization and acetylation of bamboo fibers, as analyzed by Buson et al. Surface roughness of the fibers increases after chemical treatment, improving their adhesion with the matrix [30]. However, the improvement in composite properties is limited due to bamboo fibers' hygroscopic nature with high water absorption and relatively weak interfacial bonding with the matrix [39]. Bamboo tends to absorb moisture to maintain equilibrium with its surroundings. High moisture content in fibers can negatively impact mechanical properties, as observed by Jakovljevic. The tensile modulus of bamboo fibers typically ranges between 7-20 MPa [32]. Treatment with NaOH is used to reduce the hydrophilic nature of fibers by reacting with hydroxyl groups of hemicellulose, which may cause fiber diameter shrinkage after chemical treatment [30, 39].

4.1.4 Mechanical Properties

It has been demonstrated that fibers extracted through compression molding techniques exhibit maximum tensile strength compared to other methods (645 MPa) [33], consistent with the findings of Deshpande et al. [35], which indicate that larger diameter fibers processed yield higher strength. Kim et al. [40] examined that the properties of bamboo fibers are unaffected by alkali extraction treatments, maintaining a tensile strength of 419 MPa compared to untreated fibers. When bamboo culms are treated with alkali, the mechanical properties of the outer bamboo fibers show significant enhancement: the tensile strength increases by 44.76%, tensile modulus by 124.60%, elongation at break by 20.68%, and fracture work by 123% compared to inner bamboo fibers, with values of 751.70 MPa, 32.50 GPa, 5.31%, and 26.02 J, respectively [37].

Furthermore, fibers extracted through steam explosion and alkaline treatment exhibit lower moisture content compared to untreated fibers at various relative humidity levels (50%, 60%, 70%, 80%, and 90%) at 20°C [37]. Ogawa et al. conducted experiments on fiber extraction using a final grinding machine. To avoid thermal damage to the fibers due to increased grinding speed, the temperature was maintained at 70°C with a cutting speed of 200 mm/minute, ensuring no thermal damage occurred to the fibers [41].

Mechanical extraction, chemical extraction, and a combination of both (mechanical and chemical) are the three primary techniques for obtaining bamboo fiber. Chemical retting, alkali/acid retting, and degumming are some methods commonly employed in chemical extraction to lower the lignin concentration in bamboo fibers. Compression molding (CMT) and roller mill technology (RMT) are examples of combined processes [33]. When compared to chemical extraction methods, mechanical extraction of bamboo fibers typically produces better results with fewer fiber breakage [34]. In the mechanical extraction process, chopped fibers are heated for 60 minutes at a temperature of around 175°C in an autoclave with a pressure of 0.7–0.8 MPa.

4.1.5 Thermal Properties: Thermogravimetric Analysis (TGA)

Thermal-Visual Analysis: A method for figuring out the thermal stability of natural fibers and natural fiber composites is thermogravimetric analysis (TGA). It has been noted that fiber breakdown does not occur above 200 °C. Furthermore, alkaline treatment of fibers raises the temperature at which they degrade thermally by 50 °C [30]. Zhang et al. provide a clear explanation of this, stating that hemicellulose breaks down at 180–280 °C, cellulose at 250–350 °C, and lignin at 356-450 °C. Because natural fibers include highly pure cellulose, alkaline treatment aids in the removal of hemicellulose and lignin from the fibers and improves heat stability [39].

4.1.6 Morphological Properties

Individual fibers are often mechanically oriented lengthwise and joined by lignin and hemicellulose. Weaker bonding is seen in alkali-treated and untreated fibers as compared to steam-exploded fibers [37]. By eliminating waxes and oil content, the use of NaOH improves fiber roughness and improves interface adhesion in composites [30]. The fibers have two different shapes: elliptical and circular. Alkali-treated fiber bundles have a diameter of 90–580 μm, untreated fiber bundles have a diameter of 120–650 μm, and individual fibers have a diameter of 5.20–25 μm. The dissolution of hemicellulose during alkali treatment resulted in a 25% reduction in bamboo fiber diameter after treatment [43]. Figure 1 shows a bamboo field.

Figure.1 A bamboo field (A) with a closer view of the fruit (B) (C) Bamboo Fiber

Table 2 Physical and Mechanical Properties of Bamboo Fiber

Extraction process/Treatment	Density (g/cc)	Moisture (%)	Tensile strength (MPa)	Tensile Modulus (Gpa)	Tensile strain	Diameter (mm)	References
Mechanical	0.91	9.16	503	35.91	1.4	0.24-0.33	[34]
Mechanical	10.16	-	610	46	-	-	[42]
Mechanical	-	-	580,710,750	-	-	0.15, 0.20, 0.25	[35]
Mechanical	0.49	9.3	276.7	2.26	-	-	[30]
Mechanical/alkaline	0.35	10.66	317.3	3.49	-	-	[30]
Mechanical/acetylated	0.48	6.81	198.2	2.55	-	-	[30]
Mechanical	-	-	751.7	32.5	-	-	[36]
Mechanical (steam explosion)	-	-	383	28	2.82	0.19	[40]
Mechanical (steam explosion)	-	-	615-862	35.45	-	0.24-0.33	[33]
Chemical	0.89	10.14	341	19.67	1.73	-	[34]
Chemical	-	-	419	30	2.67	0.27	[40]
Chemical	-	-	329	22	2.35	0.15-0.25	[40]
Chemical	1.3	-	450		-	-	[33]
Combined Chemical + Mechanical	0.8-0.9	-	645	-	-		[33]

4.1.6.1 The Fabrication of Composites Using Bamboo Fibers

Traditionally, the process of creating a composite involves manually organizing fibers, which is followed by mold compression and vacuum-assisted infiltration (VAI). Figure 2 shows the schematic technique for fabricating bamboo composites. Using a spatula or spreader, fibers are progressively submerged in resin to guarantee strong interfacial adhesion. After that, the mixture is poured into the mold, and pressure is used to release any trapped air. A 24-hour room-temperature cure is applied to the manufactured composite to provide a robust adhesion between the thermoset resin and the fibers. The specimens are then further cured at 110 °C for 4 hours [43].

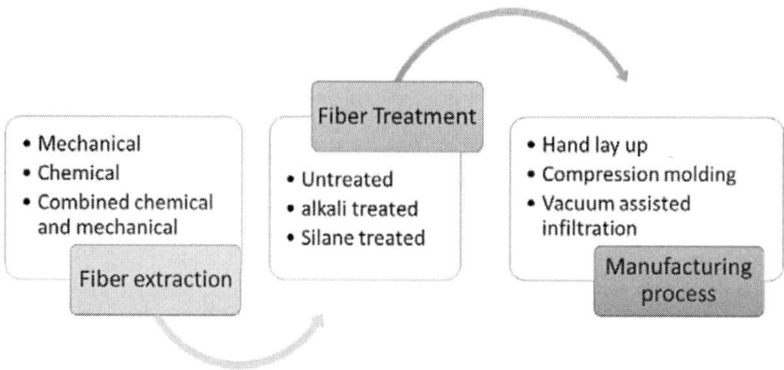

Figure. 2 Schematic flow of fabrication of bamboo composite.

The bamboo fibers are treated with a 0.25 M NaOH solution for 30 minutes at 30 °C in order to improve the bonding of the fibers during manufacture [44]. The natural fibers are dried in a vacuum oven for 24 hours at 60°C, and then they are crushed for 15 minutes at 185°C at 85 bars of pressure to create composites with biodegradable PLA polymer. The composites are annealed in an oven for 30 minutes at 120 °C to enhance their properties [20].

A similar process is discussed by Ochi [45], where metal molds are used for hot pressing composites at 130°C for 5 minutes, followed by pressing at 10 MPa for 10 minutes. In the case of using nanoparticles in composites, the concentration of nano clay is 1% by weight to ensure uniform dispersion in epoxy resin mixtures, stirred at high speed (10000 rpm) using a homogenizer, and then soaked in ice water for 30 minutes to reduce heat generated.

Trapped air is then removed through a 30-minute degassing process before adding the hardener. This preparation process involves manual lay-up in stainless steel molds with fiber reinforcement, followed by curing [46].

Bamboo has been utilized across various industries such as handicrafts, interior furniture, agriculture, paper production, and architecture for many years. Fibers from the bamboo species Neosinocalamus affinis are employed in textiles due to their crystalline structure similar to goni [47]. According to Imadi et al. [48], bamboo is also recognized as a valuable biofuel crop, yielding products like charcoal, firewood, and pellets. Bamboo grass is highly effective in crafting wind instruments such as flutes and string instruments. Its antibacterial properties and hygroscopic nature make bamboo widely used in the pharmaceutical industry for drug preparation, as well as in the manufacture of absorbent pads and medical coats [48]. Most bamboo fibers and their composites are used in structural applications due to their excellent mechanical properties such as tensile strength and flexural strength [49]. Environmental factors are a key consideration in material selection, given the impact of products on the environment [50]. Wind turbine blades are also frequently made from natural fibers. Bamboo composites are extensively used in sports products like surfboards and baseball bats due to their moisture permeability and water-resistant surface properties.

4.2 Pineapple Leaves

4.2.1 Introduction: The Productive System of Pineapple and the Role of Fibers

Pineapple (Ananas comosus) is a perennial plant belonging to the Bromeliaceae family. It is primarily cultivated for its fruit, with global production estimated to reach 31 million tons by 2028, marking an average increase of 1.9% [51]. In the context of a circular economy, there is an increasing need to utilize other products from this plant and its associated waste. Examples include wine production through pineapple fruit fermentation and its waste [52], as well as utilizing residues after food processing as a source of dietary fiber [53]. This has led to various proposals for processing these by-products, such as compacting and formulating them into pellets for animal feed [54]. Another potential application involves phenol extraction from fruit waste and shell husks for medical and chemical purposes [55].

In the food sector, it is known that various parts of the pineapple plant, including its waste, contain potentially valuable polysaccharides, as recently proposed for use as raw materials in the production of biodegradable plastics such as polyhydroxybutyrate (PHB) [56]. However, the primary focus remains on lignocellulosic fiber production and the potential exploitation of biomass for cellulose extraction, particularly if it exhibits high crystalline properties [57]. Fibers from the pineapple plant can be extracted not only from its leaves but also from its stem and roots, each exhibiting distinct characteristics as previously reported [58]. Additionally, the smaller crown leaves atop the pineapple fruit also serve as a food-grade cellulose source [59], recommended for cellulose pulp production [60]. The evaluation of physicochemical characteristics of fibers from leaves and stems of various pineapple varieties provides crucial information for selecting appropriate and accurate fiber materials for various applications, especially when derived from fruit production waste [61]. Despite the large-scale production of pineapple leaf fibers (PALF), their use remains suboptimal due to farmers' limited understanding of alternative potential applications, such as in composites. As a result, pineapple leaves are often underutilized merely as raw materials or for energy, frequently being disposed of through burning or left to decompose unchecked [62]. However, in the polymer composite industry reinforced with plant fibers, the use of PALF continues to grow as it is considered the most promising species in terms of both properties and potential availability [63]. General reviews on pineapple waste also acknowledge this potential, although their primary focus is on the production of chemicals such as natural phenols and antioxidants like bromelain [64]. Further reviews have discussed more specific applications, such as in the textile industry, although the limited length of PALF, which in the best cases reaches only one meter, poses a constraint [65]. On the other hand, pineapple fiber density is relatively high, reaching 1.53 g/cm^2 [66]. Other physical properties have also recently been investigated, such as the wetting contact angle of PALF, which can be altered to become more hydrophobic through appropriate treatments [67]. Comprehensive efforts to address pineapple exploitation issues have been recently undertaken [68].

There are intriguing indications presented up to the semi-finished product level in Figure 14.1 [69], providing a general overview of the system. However, with ongoing research, it is important to present further findings that can summarize and predict relevant and significant trends in the near future. To further exploit this, pineapple leaf fiber (PALF) is typically separated through stripping and decorticating to maintain fiber integrity. Recently, grinding methods have also been considered to enhance the yield of these short fibers, although their utilization remains limited [70]. Following this process, the fibers are soaked in water akin to jute for the retting process. Traditionally, pineapple leaves in the field, as depicted in Figure 3, are often regarded as agricultural waste and seldom utilized for other purposes, although recent characterization has been conducted on various cultivars [71].

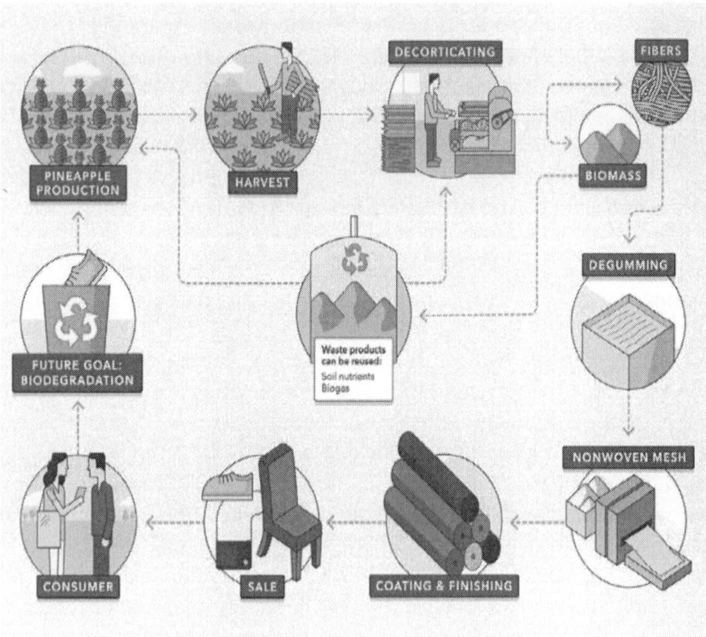

Figure. 3 The productive system of pineapple [19]

The expansion of biocomposite use has suggested increasing mechanical fiber use like PALF to explore the potential of this process [72]. However, it should be noted that due to the limited fiber length and challenges associated with its use in textiles or complex structures, PALF does not directly emerge as the primary candidate to replace glass fibers in composites. On the other hand, within the economically beneficial context, exploring alternative fibers such as hemp, flax, and similar materials, and comparing them with waste from other sectors such as coconut husks and sugarcane bagasse, which have specific uses like heavy metal absorption, is carried out [73].

In recent years, the application of natural fibers in composites has been a focus of research to compare various possibilities and production systems, considering approaches in material selection [74]. One initial possibility is the use of PALF in lightweight concrete in small quantities, which has proven beneficial for enhancing tensile strength, flexural strength, and resistance to degradation [75]. The next section will discuss the characteristics of pineapple leaf fibers and their potential use in composites with various matrices, including thermoplastic polymers, thermosets, oil-based, and bio-based materials.

Figure. 4 A pineapple field with a closer view of the fruit from Aceh, Indonesia [Ahmad Muflih]

4.2.2 General Considerations and Fiber Treatment

In fact, not only PALF, but also shorter leaves from pineapple crowns can be considered as a source of cellulose [59] or even for reinforcing polymer matrices, such as polypropylene [76]. However, in practice, this has been quite rare so far, hence PALF is predominantly obtained from factories. Moreover, pineapple fiber used for composites can originate from agricultural waste, competing with materials used for energy recovery or paper production [77]. The latter application becomes particularly intriguing when considering that the lignin content in pineapple fibers generally does not exceed 4% [78]. However, PALF is often derived from longer fibers intended for textile production, thus requiring chemical treatment for processing, especially to achieve the necessary fineness in cotton weaving machines [79]. Retting is typically done by manual or mechanical stripping, followed by combing. Nonetheless, water retting aided by bacteria has also been proven effective for degumming PALF, although less practical for short fibers with high hydrophilicity [80].

In the application of PALF composites, further reviews can provide interesting perspectives, as concentrated in [81], focusing on relevant tensile and flexural data from applications with polyester, polyethylene, and polypropylene matrices, and [82], comparing them with the most commonly used natural fibers, either from large production systems or wild plants. As previously noted, one obvious drawback of employing PALF in composites with conventional non-biodegradable polymer matrices, such as epoxy, is its high cellulose content, which renders these two materials practically incompatible. This suggests that, when considering PALF application as a reinforcement in polymer composites, it is crucial to consider a wide range of matrices used in the industry, as well as fabrication techniques employed in composite manufacturing: significant efforts have been made with epoxy, as outlined in [83].

PALF needs to be processed to interact effectively with polymer matrices, especially those derived from synthetic oils. One possible solution is to apply specialized treatments aimed at providing adequate protection and enhancing the strength of the fibers. Early studies indicate that classical alkali treatments using sodium hydroxide (up to 18% volume) or bleaching with sodium chlorite (up to 0.7% volume) have improved fiber elongation and reduced hemi-cellulose content, which contributes modestly to their mechanical performance [84]. Further research shows that treatment with silane can reduce the hydrophilic characteristics of the fibers by removing residual alkali and unstable hemi-cellulose [85]. Chemical processing applied to PALF includes methods such as acetylation (esterification), benzoylation, maleation grafting, sodium chlorite bleaching, acrylation and acrylonitrile grafting, calcium hydroxide, isocyanate, triazine, stearic acid, potassium permanganate, and the use of fatty acid derivatives [86]. The process of making PALF (pineapple leaf fiber) into composites is shown in Figure 5. The abundant presence of cellulose in pineapple fibers also makes them highly responsive to cellulase enzymes, which under controlled conditions can depolymerize them into simple sugars. However, well-controlled processes can also produce very fine fibers suitable for textile applications [87].

Figure. 5 The process of making PALF (pineapple leaf fiber) into composites

4.2.3 Characteristics of Pineapple Leaf Fibers

ALF, due to its relatively high density and being primarily composed of cellulose, is highly suitable as a raw material for the extraction of nanocellulose using methods such as bleaching with sodium hypochlorite and steam explosion [74] (see Table 3). Extraction of nanocellulose from pineapple peel has proven to yield high results, which can be used as fillers in various applications, such as 4 wt.% in gellan gum candy films to enhance tensile strength [28], or 0.7 wt.% in starch-chitosan biopolymer, enhancing the storage modulus value [88].

In practice, pineapple leaves possess a multi-cellular fiber structure resembling threads, often containing pectin, with fiber yields typically around 3% by weight of leaf material [30].

Table 3 Composition of PALF Fibers Before and After Nanocellulose Extraction [103]

Material	a-Cellulosa (%)	Hemicellulose (%)	Lignin (%)	Moisture content (%)
Raw PALF	81.27 + 2.45	12.31 + 1.35	3.46 + 0.58	10.52 + 0.48
Steam exploded PALF	93.45 + 2.79	3.72 + 0.73	2.08 + 0.58	10.64 + 0.36
Bleached PALF	98.63 + 0.54	0.53 + 0.03	0.77 + 0.44	10.80 + 0.50

The fiber structure typically obtained resembles a sawtooth pattern, with numerous large voids at its core and several more compressed lumens evenly distributed around it, often in a concave polygonal shape, as shown in Figure 6. Morphologically, the observed variations among different pineapple species can be quite significant. However, most studies have not reported specific cultivar varieties from which these fibers originate. A comparative study of twelve different cultivars noted that the crystallinity index ranged from 61% to 78%, stiffness between 37 and 86 GPa, with a dispersion value reaching about 30%, and strengths varying between approximately 600 and 1300 MPa [89]. Overall, the average performance of PALF from each cultivar tends to be comparable to fibers from other commonly used plants, albeit with variations. Average values for other leaf fibers can be seen in Table 4. Occasionally, differences in values obtained from the same study may be due to different testing parameters, such as the length of samples used in tensile tests, which can influence the recorded results.

Another significant variability relates to the evaluation of different pineapple species originating from various countries like Brazil, Malaysia, or India [90], showing lower values in some studies. Further research explores the potential use of pineapple leaves obtained from local farming for fruit consumption in Ethiopia, where easy extraction processes become

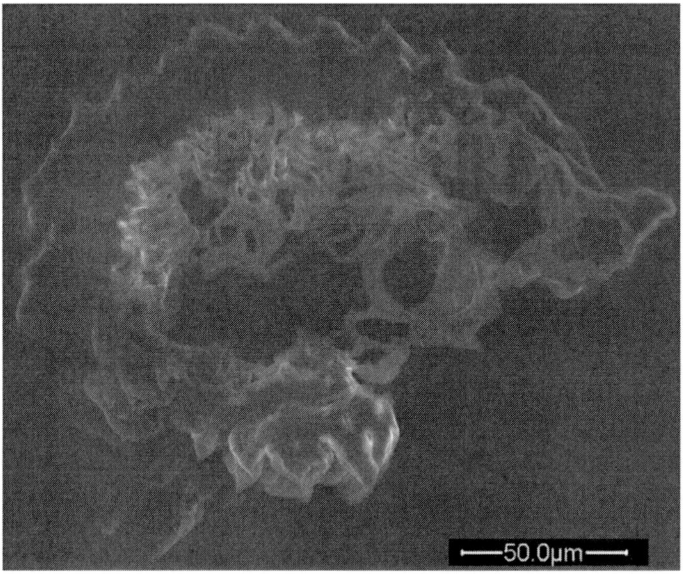

Figure. 6 Multicellular structure of pineapple fiber section (original SEM image by the authors: 933x magnification)

Table 4 Tensile Performance of PALF Compared with Other Natural Fibers

Fiber	Max. strain (%)	Strength (MPa)	Young's modulus (GPa)	Ref
PALF	1.5-3.5	629-1309	37-86	[89]
Curaua	3.9	913	30	[93]
Banana	3.36	600	17.85	[94]
Sisal	5.45	567	10.40	[94]
Palm	13.71	377	2.75	[94]
Phormium	5.04	771	23.89	[95]
Abaca	2.9	756	31.1	[97]
Henequen	4.8	500	13.2	[97]
Fique	5	200	10	[98]
Artisdita hystrix	1.57	440	28	[99]

a primary factor in their utilization decision [91]. However, there are concerns to consider regarding the use of PALF in composites, especially when using thermoset matrices, such as the limited elongation when PALF fibers break, estimated to be only about 0.6% to 2%. This suggests that the benefits of using PALF are more effective when arranged in a unidirectional or bidirectional fiber pattern as reinforcement in composites, rather than in random patterns due to the limited fiber length [92].

Variability in this regard has also been noted in the context where various treatments have been applied to the fiber surfaces to prepare them for use in composites. Evaluating this dispersion requires the use of the Weibull modulus, as described in the findings of a study conducted in [100], which utilized alkali treatment with sodium hydroxide (NaOH) and/or deposition of diglycidyl ether of bisphenol A (DGEBA) to prepare the fibers for their introduction into epoxy resin. Specifically, the distribution of tensile strength values in the fibers appears more uniform, resulting in an increase in the Weibull modulus (WM) values with these treatments and depositions.

4.3 Coconut/Coir Fibers

4.3.1 Coconut/Coir Tree and Fruit, Fiber Extraction and Application

Arecaceae is the plant category, Coir is the common term, and Cocos nucifera is the scientific name for coconut fiber. The word "coconut husk" comes from the Dravidian word "kayiru," which means "rope" and describes the outer layer of the coconut that is coated in coir, a fibrous husk [72]. Over 90 countries throughout the world cultivate coconuts, with India, Indonesia, Sri Lanka, and the Philippines being the top producers. These four nations accounted for 78% of the world's production in 2006–2007, as Table 5 [73] illustrates. Figure 7 depicts the coconut tree, while Figures 8 to 10 illustrate coconuts and the extraction process. The two types of coconuts shown in Figure 8 are green and brown coconuts, from which their respective fibers have been identified and described. Fibers from green coconuts are white due to extraction from immature coconuts, whereas fibers from brown coconuts are brown as they are extracted from mature coconuts. Relatively, white fibers are fine and soft but weaker compared to thick, rough, strong, and highly abrasion-resistant brown fibers.

Table 5 Area of land used for plantation, production, and productivity of coconut in the year 2006-2007 [73].

Country	Area (hectares)	Production (million tonnes)	Productivity (kg/ha)
Brazil	2,200	2,973,700	10,813
China	28,200	289,000	10,248
Dominician	38,000	181,533	4,777
Ghana	55,000	315,000	5,727
India	1,940,000	15,840,000	8,165
Indonesia	22,660,000	16,289,000	6,123
Mozambique	70,000	265,000	3,785
Myanmar	41,000	350,000	8,536
Papua New Guinea	200,000	650,000	3,250
Philippines	3,253,927	14,344,920	4,408
Sri Lanka	447,000	1,950,000	4,362
Tanzania	310,000	370,000	1,193
Thiland	337,000	1,450,000	4,302
Vietnam	132,800	930,600	7,007

Brown coconut coir fibers are typically utilized for technical applications [74]. These fibers are extracted from the outer husk of the coconut fruit, as depicted in Figures 8-10, and exhibit fire resistance as well as durability. The use of scanning electron microscopy (SEM) to scan the morphology of coconut coir fibers has been demonstrated in Figure 11.

Coconut fibers are available in three forms: husk, coir, and decorticated. The utilization of coconut coir fibers in biofiber and bio-composite production has increased their commercial value as agricultural by-products.

Figure. 7 No Permission Required Coconut Tree

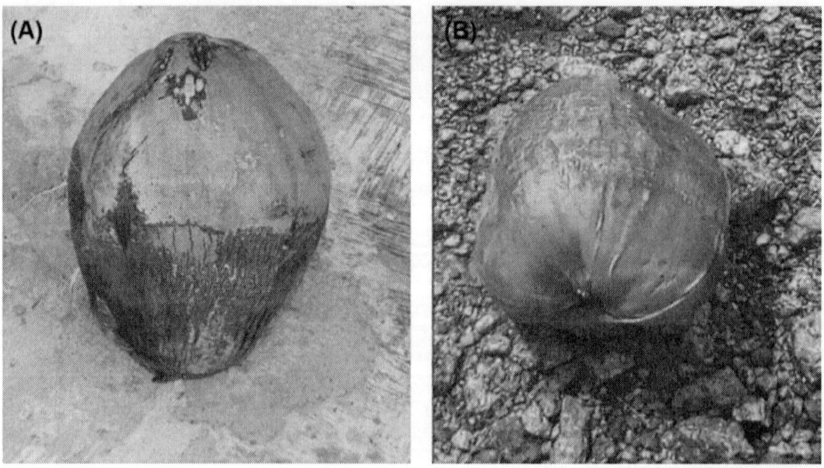

Figure. 8 Coconut fruit (A) premature and (B) mature

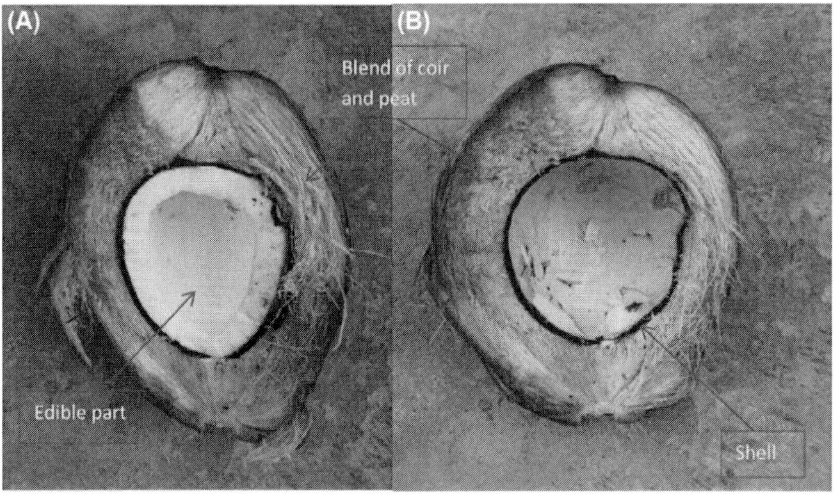

Figure. 9 Cross-section of the fruit (A) with edible part and (B) without the edible part

Figure. 10 Extraction process of coir fiber

Coconut fibers can be extracted using a variety of techniques, including retting and decorticating/defibering procedures. Both mechanical and manual extractions are possible. Retting is a manual procedure that involves soaking the husk in water for three to six months in order to soften it before physically decorticating it. However, mechanical methods require only a five-day retting period, after which the fibers are removed from the husk using a decorticating machine. Following extraction, the fibers are dried, cleaned, and combed.

In addition to these methods, there have been advancements in extraction techniques involving the use of ligninolytic enzymes. This technique reduces the retting time to three to five days and creates environmentally friendly conditions for coconut coir fiber extraction [73]. Several significant properties of coconut fibers that make them versatile in various applications include durability, toughness, fire resistance, moisture resistance, resistance to fungi and decay, as well as temperature resistance.

According to the International Year of Natural Fibres report in 2009, global coconut fiber production reached up to 500,000 tons per year, primarily in India and Sri Lanka, with an estimated value of around $100 million [75].

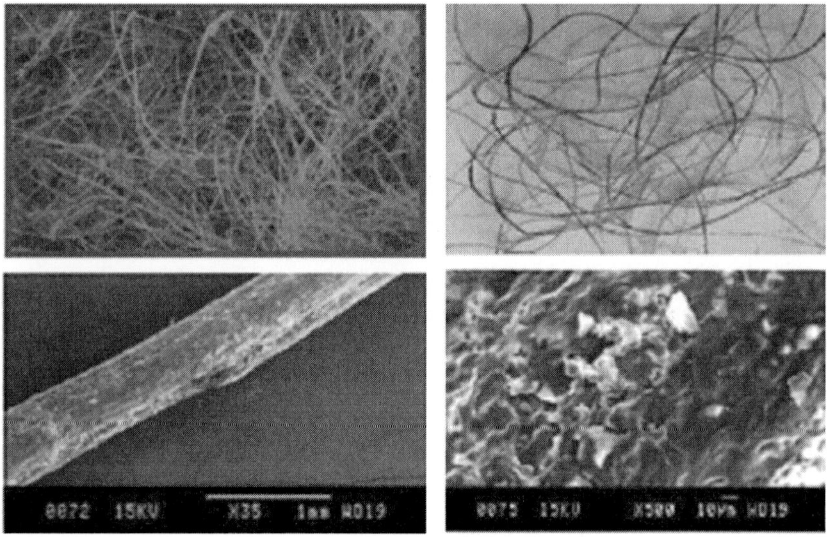

Figure. 11 SEM micrograph of coconut fiber [101]. From Danso, H. (2017). Properties of coconut, oil palm and bagasse fibres: as potential building materials, Procedia Eng. (Vol. 200, pp. 1-9). Elsevier Ltd. https://doi.org/10.1016/j.proeng.2017.07.002 SEM micrograph of coconut fiber [30]

The history of plantations shows that coconuts have a strong ability to adapt to extreme environments and mature faster compared to other lignocellulosic fiber plants [102]. Despite its inherent characteristics that make it effective as a reinforcement in composites, its abundant availability, lightweight nature, and excellent mechanical properties also contribute to the use of coconut fiber or coir in manufacturing composite systems for various applications [103]. However, a major challenge in its use lies in the difficulty of aligning coconut coir fibers into desired orientations due to their naturally curved structure, thus it is recommended to produce them in random orientations [74]. Traditionally, these fibers are used for making ropes, mats, and decorative items.

Many researchers are currently looking into the mechanical characteristics and other elements of ceramic matrix-based composite materials and polymers reinforced with coir fibers for a variety of applications. The constituents of coir fibers consist of lignin, pectin, cellulose, and hemicellulose. According to research findings, the different components and mechanical characteristics of coir fibers are listed in Tables 6 and 7, respectively. Water, fibers, and trace amounts of dissolved substances make up these fibers. Compared to other natural fibers, coir fiber is stronger because of its high lignin content. Variations in chemical constituents and physical and mechanical properties of coir fibers can be influenced by cultivation history, geographical location, and differences in testing conditions [56,103]. The high percentage of cellulose in coconut fibers contributes to why these fibers are tougher, more rigid, and more durable [104].

Table 6 Constituents of Coir Fibers [57,70,73-78]

Hemicellulose (%)	Cellulose (%)	Lignin (%)	Petin (%)	Ash (%)	Moisture (%)	References
31.1	33.2	20.5	-	-	-	[64]
15-28	35-60	20-48	-	-	-	[105]
16.8	68.9	32.1	-	-	-	[106]
0.15-0.25	32-43	40-45	-	-	-	[79]
10-20	36-43	41-45	3-4	-	8	[107]
11.9-15.4	19.9-36.7	32.7-53.3	4.7-7.0	-	0.2-0.5	[102]
0.25	43.44	45.84	3.00	2.22	5.25	[80]
13-20	71	13	-	-	-	[68]

Table 7 Mechanical properties of coir fibers [6,30,31,70,75,78,90,91]

Tensile strength (MPa)	Elongation at failure (%)	Young's Modulus (GPa)	Density (gm/cm^3)	References
95-118	-	8	1.17	[108]
106-175	47	3-6	-	[82]
175	30	4-6	1.2	[19]
150-180	20-40	4-6	1.2	[109]
220	15-25	6	1.25	[79]
131-175	15-40	4-6	1.15	[107]
220	18	6	1.25	[103]
131-220	15-40	4-6	1.15-1.46	[102]

4.4 Characterization Natural Fiber Bamboo/Pineapple Leaf/Coconut Husk As Hybrid Materials Composite

Studies on hybrid composites using bamboo, pineapple leaf, and coconut coir fibers have shown significant improvements in mechanical, thermal, and morphological properties compared to composites using single fiber types. In this paper, we will compare the findings of this research with several previous studies that have explored the use of natural fibers in composite materials.

4.4.1 Mechanical Properties

This study shows that, when compared to composites made of just one type of fiber, hybrid composites made of bamboo fiber, pineapple leaf, and coconut coir have superior tensile strength, flexural strength, and toughness. For example, the hybrid composite has a higher tensile strength than the composite made of only one bamboo fiber. While coconut coir and pineapple leaf fiber improve toughness and flexural strength, bamboo fiber greatly increases tensile strength.

Previous studies by Jawaid et al. (2013) [109] on hybrid composites made from empty oil palm bunches and jute fiber also indicated a substantial improvement in mechanical properties compared to single fiber composites. These findings align with our results, affirming that the integration of various types of natural fibers can produce composites with superior mechanical properties.

In the work by Rihayat et al. (2020) [113], the cracked surface structures were examined using optical and electron microscopy, and the mechanical characteristics were examined using a Universal Testing Machine (UTM). In order to create a hybrid composite material with the best tensile qualities, natural fibers, including coconut, pineapple, and bamboo fiber, were combined with polyester resin. As a renewable material, a blend of polyester and natural fibers (fillers) has been produced. The findings showed that the addition of coconut, pineapple leaf, and bamboo fibers produced a flexural strength of 0.302 MPa and an optimal tensile strength of 366 MPa. Tensile property testing for hybrid composite materials in accordance with the ASTM D 638 standard is shown in Figure 1(a). The cross-head speed, gauge length, and dimensions are chosen in compliance with ASTM D 638 guidelines. The three-point bending technique of measuring flexural strength with specimen samples in accordance with the ASTM D 790 standard is depicted in Figure 1(b). After the specimen samples underwent several tests, the average outcomes were considered to constitute the distinctive qualities of the composite.

Table 8 . Mechanical properties of different composite samples [113]

Sample	Tensile strength (MPa)	Flexural load (kN)
Pineapple leaf + coconut husk	183	0,21
Bamboo + Pineapple leaf	246	0,24
Bamboo + Pineapple leaf + coconut husk	366	0,302

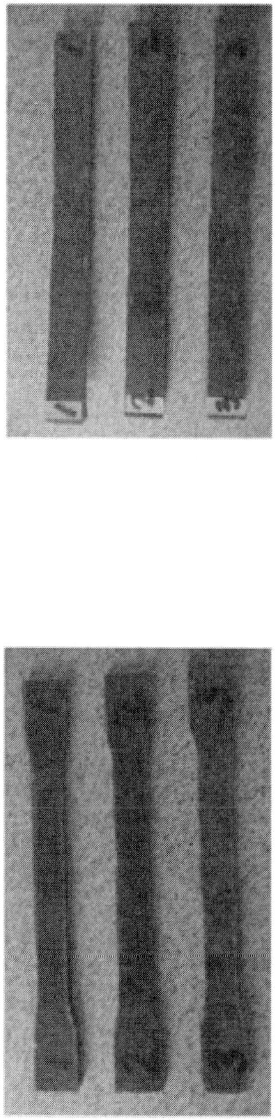

(b)

(a)

Figure 12. (a) Sample specimen ASTM D-638, (b) Sample specimen ASTM D-790

4.4.2 Thermal Properties

The TGA and DSC analyses from this study reveal that hybrid composites exhibit superior thermal stability compared to single fiber composites. Higher decomposition temperatures indicate that hybrid composites are more heat-resistant. This can be attributed to the favorable thermal properties of each fiber type, which synergistically contribute to the hybrid composite. Research by Mishra et al. (2004) [110] on biofiber/glass hybrid composites showed similar results with improved thermal stability. The combination of natural fibers with glass fibers produces composites with enhanced thermal stability, affirming that fiber hybridization can improve the thermal properties of composites.

4.4.3 Morphological Analysis

The SEM images in this study illustrate that the fibers in the hybrid composite are well-dispersed and exhibit strong interfacial bonding with the polymer matrix. Alkali treatment of the fibers enhances surface roughness, thereby improving fiber-matrix adhesion and reducing the likelihood of fiber pull-out under loading. Pothan et al. (2007) [111] also reported that surface treatment of banana fibers enhances fiber-matrix interactions in polyester composites, leading to improvements in mechanical and morphological properties. This is consistent with our findings that fiber treatment plays a crucial role in achieving strong bonding within the composite.

The research by Rihayat et al. (2020) [113] examines the morphology of the hybrid composite as depicted in figures 6(a), (b), and (c) at magnifications of 100x, 500x, and 1000x. These images illustrate improved fiber-matrix adhesion and a uniformly distributed matrix throughout the composite section, indicating an optimal interface attachment between the fibers and matrix. One of the bonding processes found between the fibers and the polymer matrix is interlocking. The diffusion of the polymer matrix into the porous fiber surfaces is responsible for the interlocking features shown in the bamboo-pineapple-coconut husk composite. These surfaces are penetrated by the polymer matrix, which embeds and solidifies to provide a strong link between the fibers and matrix, as indicated by the ideal values reported in earlier flexural test findings [25].

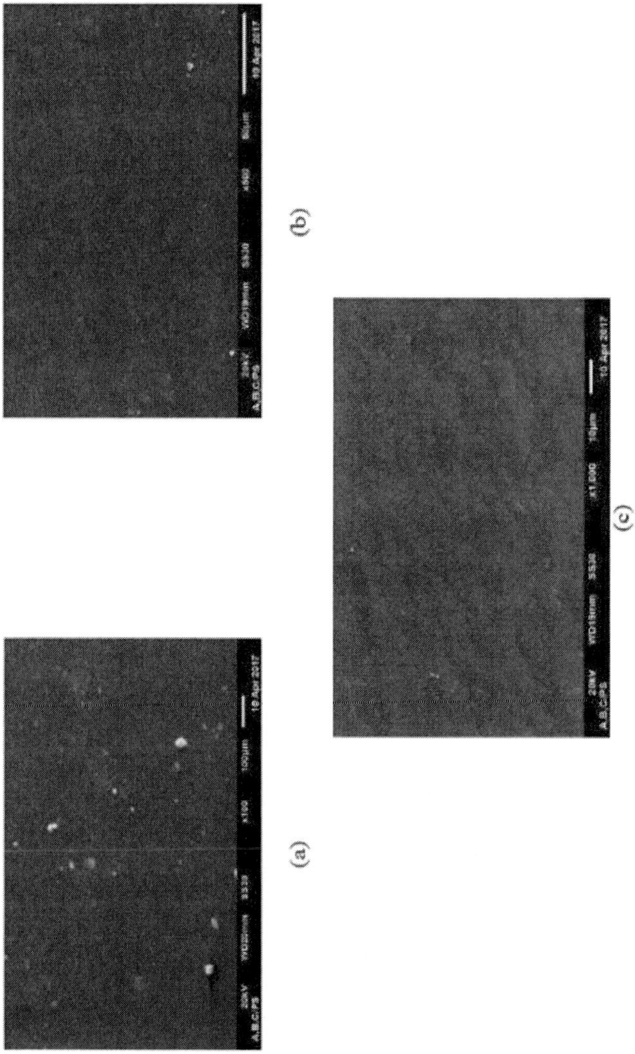

Figure 13. (a) Coconut husk and bamboo leaf 100x, (b) Coconut husk and bamboo leaf 500x, and (c) 1000x coconut husk, pineapple leaf, and bamboo.

4.4.4 Comparison with Single Fiber Composites

When compared to single fiber composites, hybrid composites made from bamboo, pineapple leaf, and coconut fiber demonstrate more pronounced advantages. This combination leverages the unique properties of each fiber, such as the high tensile strength of bamboo, toughness of pineapple leaf fibers, and flexibility of coconut coir. In contrast, single fiber composites tend to exhibit less balanced properties and often excel in only one mechanical or thermal aspect.

Kabir et al. (2012) [112], in their study on chemical treatment of natural fibers, reported that while chemical treatments can enhance the properties of single fiber composites, the improvements are not as significant as those achieved by combining fibers in hybrid composites.

Conclusion

This study confirms that hybrid composites composed of bamboo, pineapple leaf, and coconut coir offer significant advantages in terms of mechanical, thermal, and morphological properties compared to single fiber composites. Comparisons with previous research indicate that combining various types of natural fibers can result in composite materials with superior performance, supporting the potential use of these hybrid composites in applications requiring lightweight, strong, and heat-resistant materials.

References

1. Satyanarayana, K. G., Sukumaran, K., Mukherjee, P. S., et al. (1990). Natural fiber–polymer composite. *Cement Composites, 12*, 117–136.
2. Li, Y., Mai, Y.-W., & Ye, L. (2000). Sisal fiber and its composites: A review of recent developments. *Composites Science and Technology, 60*, 2037–2055.
3. de Andrade Silva, F., Chawla, N., & de Toledo Filho, R. D. (2008). Tensile behavior of high performance natural (sisal) fibers. *Composites Science and Technology, 68*, 3438–3443.
4. Sreenivasan, V. S., Somasundaram, S., Ravindran, D., et al. (2011). Microstructural, physico-chemical and mechanical characterization of Sansevieria cylindrica fibers – an exploratory investigation. *Materials & Design, 32*, 453–461.

5. Athijayamani, A., Thiruchitrambalam, M., Natarajan, U., et al. (2009). Effect of moisture absorption on the mechanical properties of randomly oriented natural fibers/polyester hybrid composite. *Materials Science and Engineering: A, 517*, 344–353.
6. Thamae, T., & Baillie, C. (2007). Influence of fiber extraction method, alkali and silane treatment on the interface of Agave Americana waste HDPE composites as possible roof ceilings in Lesotho. *Composite Interfaces, 14*, 821–836.
7. Davies, P., Morvan, C., Sire, O., et al. (2007). Structure and properties of fibers from sea-grass (Zostera marina). *Journal of Materials Science, 42*, 4850–4857.
8. Gu, H. (2007). Tensile behaviours of the coir fiber and related composites after NaOH treatment. *Materials & Design, 30*, 3931–3934.
9. Jayabal, S., & Natarajan, U. (2011). Influence of fiber parameters on tensile, flexural, and impact properties of nonwoven coir–polyester composites. *International Journal of Advanced Manufacturing Technology, 54*, 639–648.
10. Obi Reddy, K., Sivamohan Reddy, G., Uma Maheswari, C., et al. (2010). Structural characterization of coconut tree leaf sheath fiber reinforcement. *Journal of Forestry Research, 21*, 53–58.
11. Monteiro, S. N., Aquino, R. C. M. P., & Lopes, F. P. D. (2008). Performance of curaua fibers in pullout tests. *Journal of Materials Science, 43*, 489–493.
12. Murali Mohan Rao, K., & Mohana Rao, K. (2007). Extraction and tensile properties of natural fibers: vakka, date and bamboo. *Composite Structures, 77*, 288–295.
13. Alsewailem, F. D., & Binkhder, Y. A. (2010). Preparation and characterization of polymer/date pits composites. *Journal of Reinforced Plastics and Composites, 29*, 1743–1749.
14. De Rosa, I. M., Kenny, J. M., Mohd, M., et al. (2011). Effect of chemical treatments on the mechanical and thermal behaviour of okra (Abelmoschus esculentus) fibers. *Composites Science and Technology, 71*, 246–254.
15. Murali Mohan Rao, K., Ratna Prasad, A. V., Ranga Babu, M. N. V., et al. (2007). Tensile properties of elephant grass fiber reinforced polyester composites. *Journal of Materials Science, 42*, 3266–3272.
16. Mylsamy, K., & Rajendran, I. (2011). Influence of alkaline treatment and fiber length on mechanical properties of short agave fiber reinforced epoxy composites. *Materials & Design, 32*, 4629–4640.

17. Jayabal, S., & Natarajan, U. (2011). Influence of fiber parameters on tensile, flexural, and impact properties of nonwoven coir–polyester composites. *International Journal of Advanced Manufacturing Technology, 54*, 639–648.
18. Le Digabel, F., & Averous, L. (2006). Effect of lignin content on the properties of lignocellulose-based biocomposites. *Carbohydrate Polymers, 66*, 537–345.
19. Munawar, S. S., Umemura, K., & Kawai, S. (2007). Characterization of the morphological, physical, and mechanical properties of seven nonwood plant fiber bundles. *Journal of Wood Science, 53*, 108–113.
20. Murali Mohan Rao, K., Ratna Prasad, A. V., Ranga Babu, M. N. V., et al. (2007). Tensile properties of elephant grass fiber reinforced polyester composites. *Journal of Materials Science, 42*, 3266–3272.
21. Alsewailem, F. D., & Binkhder, Y. A. (2010). Preparation and characterization of polymer/date pits composites. *Journal of Reinforced Plastics and Composites, 29*, 1743–1749.
22. De Rosa, I. M., Kenny, J. M., Mohd, M., et al. (2011). Effect of chemical treatments on the mechanical and thermal behaviour of okra (Abelmoschus esculentus) fibers. *Composites Science and Technology, 71*, 246–254.
23. Murali Mohan Rao, K., Ratna Prasad, A. V., Ranga Babu, M. N. V., et al. (2007). Tensile properties of elephant grass fiber reinforced polyester composites. *Journal of Materials Science, 42*, 3266–3272.
24. Mylsamy, K., & Rajendran, I. (2011). Influence of alkaline treatment and fiber length on mechanical properties of short agave fiber reinforced epoxy composites. *Materials & Design, 32*, 4629–4640.
25. Jayabal, S., & Natarajan, U. (2011). Influence of fiber parameters on tensile, flexural, and impact properties of nonwoven coir–polyester composites. *International Journal of Advanced Manufacturing Technology, 54*, 639–648.
26. Le Digabel, F., & Averous, L. (2006). Effect of lignin content on the properties of lignocellulose-based biocomposites. *Carbohydrate Polymers, 66*, 537–345.
27. Munawar, S. S., Umemura, K., & Kawai, S. (2007). Characterization of the morphological, physical, and mechanical properties of seven nonwood plant fiber bundles. *Journal of Wood Science, 53*, 108–113.
28. Murali Mohan Rao, K., Ratna Prasad, A. V., Ranga Babu, M. N. V., et al. (2007). Tensile properties of elephant grass fiber reinforced polyester composites. *Journal of Materials Science, 42*, 3266–3272.

29. Venkateshwaran, N., & Elayaperumal, A. (2010). Banana fiber reinforced polymer composites - a review. *Journal of Reinforced Plastics and Composites, 29*, 2387–2396.
30. John, M. J., & Anandjiwala, R. D. (2008). Recent developments in chemical modification and characterization of natural fiber-reinforced composites. *Polymer Composites, 29*, 187–207.
31. Joseph, K., Toledo, R. D., James, B., et al. (1999). A review on sisal fiber reinforced polymer composites. *Revista Brasileira de Engenharia Agricola e Ambiental, 3*, 367–379.
32. Venkateshwaran, N., & Elayaperumal, A. (2010). Banana fiber reinforced polymer composites - a review. *Journal of Reinforced Plastics and Composites, 29*, 2387–2396.
33. Li, X., Tabil, L. G., & Panigrahi, S. (2007). Chemical treatments of natural fiber for use in natural fiber-reinforced composites: A review. *Journal of Polymers and the Environment, 15*, 25–33.
34. Obi Reddy, K., Sivamohan Reddy, G., Uma Maheswari, C., et al. (2010). Structural characterization of coconut tree leaf sheath fiber reinforcement. *Journal of Forestry Research, 21*, 53–58.
35. Alsewailem, F. D., & Binkhder, Y. A. (2010). Preparation and characterization of polymer/date pits composites. *Journal of Reinforced Plastics and Composites, 29*, 1743–1749.
36. Mylsamy, K., & Rajendran, I. (2010). Investigation on physio-chemical and mechanical properties of raw and alkali treated agave Americana fiber. *Journal of Reinforced Plastics and Composites, 29*, 2925–2935.
37. Depuydt, D. E. C., Billington, L., Fuentes, C., et al. (2018). Physical and mechanical characterization of surface treated bamboo fibers. *Science and Technology of Materials, 1–7.*
38. Buson, R. F., Melo, L. F. L., Oliveira, M. N., Rangel, G. A. V. P., & Deus, E. P. (2018). Physical and mechanical characterization of surface treated bamboo fibers. *Science and Technology of Materials, 1-7.*
39. Lo, T. Y., Cui, H. Z., & Leung, H. C. (2004). The effect of fiber density on strength capacity of bamboo. *Materials Letters, 58(21)*, 2595–2598.
40. Jakovljević, S., & Lisjak, D. (2019). Investigation into the effects of humidity on the mechanical and physical properties of bamboo. *Construction and Building Materials, 194*, 386–396.
41. Zakikhani, P., Zahari, R., Sultan, M. T. H., & Majid, D. L. (2014). Extraction and preparation of bamboo fiber reinforced composites. *Materials Design, 63*, 820–828.

42. Rao, K. M. M., & Rao, K. M. (2007). Extraction and tensile properties of natural fibers: Vakka, date and bamboo. *Composite Structures, 77*(3), 288–295.
43. Hu, M., Wang, C., Lu, C., et al. (2020). Investigation on the classified extraction of the bamboo fiber and its properties. *Journal of Natural Fibers, 17*(12), 1798–1808.
44. Deshpande, A. P., Bhaskar Rao, M., & Lakshmana Rao, C. (2000). Extraction of bamboo fibers and their use as reinforcement in polymeric composites. *Journal of Applied Polymer Science, 76*(1), 83–92.
45. Phong, N. T., Fujii, T., Chuong, B., & Okubo, K. (2011). Study on how to effectively extract bamboo fibers from raw bamboo and wastewater treatment. *Journal of Materials Science Research, 1*(1), 144–155.
46. Porras, A., & Maranon, A. (2012). Development and characterization of a laminate composite material from polylactic acid (PLA) and woven bamboo fabric. *Composites Part B: Engineering, 43*(7), 2782–2788.
47. Zhang, X., Wang, F., & Keer, L. M. (2015). Influence of surface modification on the microstructure and thermo-mechanical properties of bamboo fibers. *Materials (Basel), 8*(10), 6597–6608.
48. Kim, H., Okubo, K., Fujii, T., & Takemura, K. (2013). Influence of fiber extraction and surface modification on mechanical properties of green composites with bamboo fiber. *Journal of Adhesion Science and Technology, 27*(12), 1348–1358.
49. Ogawa, K., Hirogaki, T., Aoyama, E., & Imamura, H. (2008). Bamboo fiber extraction method using a machining center. *Journal of Advanced Mechanical Design, Systems, and Manufacturing, 2*(4), 550–559.
50. Amada, S., & Untao, S. (2001). Fracture properties of bamboo. *Composites Part B: Engineering, 32*(5), 451–459.
51. Chin, S. C., Tee, K. F., Tong, F. S., Ong, H. R., & Gimbun, J. (2020). Thermal and mechanical properties of bamboo fiber reinforced composites. *Materials, 100876.*
52. Sukmawan, R., Takagi, H., & Nakagaito, A. N. (2016). Strength evaluation of cross-ply green composite laminates reinforced by bamboo fiber. *Composites Part B: Engineering, 84*, 9–16.
53. Zhang, J., Khatibi, A. A., Castanet, E., Baum, T., Komeily-Nia, Z., Vroman, P., & Wang, X. (2019). Effect of natural fibre reinforcement on the sound and vibration damping properties of bio-composites

compression moulded by nonwoven mats. *Composites Communications, 13*, 12–17.

54. Chee, S. S., Jawaid, M., Alothman, O. Y., & Fouad, H. (2021). Effects of nanoclay on mechanical and dynamic mechanical properties of bamboo/kenaf reinforced epoxy hybrid composites. *Polymers, 13*(3), 1–17.
55. Yueping, W., Ge, W., Haitao, C., Genlin, T., Zheng, L., Feng, X. Q., Xiangqi, Z., & Xiaojun, H. (2010). Structures of bamboo fiber for textiles. *Textile Research Journal, 80*(4), 334–343.
56. Imadi, S. R., & Mahmood, I. (2014). Bamboo fiber processing, properties, and applications. *Biomass and Bioenergy, 27–46.*
57. Yu, Y., Wang, H., Lu, F., Tian, G., & Lin, J. (2014). Bamboo fibers for composite applications: A mechanical and morphological investigation. *Journal of Materials Science, 49*(6), 2559–2566.
58. Sharma, B., Gatoo, A., Bock, M., Mulligan, H., & Ramage, M. (2015). Engineered bamboo: State of the art. *Proceedings of the Institution of Civil Engineers - Construction Materials, 168*(2), 57–67.
59. Food and Agriculture Organisation (FAO). (2020). *Medium-term outlook: Prospects for global production and trade in bananas and tropical fruits 2019 to 2028*. Rome.
60. Isitua, C. C., & Ibeh, I. N. (2010). Novel method of wine production from banana (Musa acuminata) and pineapple (Ananas comosus) wastes. *African Journal of Biotechnology, 9*(44), 7521–7524.
61. Hu, H., & Zhao, Q. (2018). Optimization extraction and functional properties of soluble dietary fiber from pineapple pomace obtained by shear homogenization-assisted extraction. *RSC Advances, 8*(72), 41117-41130.
62. Zainuddin, M. F., Rosnah, S., Noriznan, M. M., & Dahlan, I. (2014). Effect of moisture content on physical properties of animal feed pellets from pineapple plant waste. *Agriculture and Agricultural Science Procedia, 2*, 224-230.
63. Alothman, M., Bhat, R., & Karim, A. A. (2009). Antioxidant capacity and phenolic content of selected tropical fruits from Malaysia, extracted with different solvents. *Food Chemistry, 115*(3), 785-788.
64. Sukruansuwan, V., & Napathorn, S. C. (2018). Use of agro-industrial residue from the canned pineapple industry for polyhydroxybutyrate production by Cupriavidus necator strain A-04. *Biotechnology for Biofuels, 11*(1), 1-15.

65. Dos Santos, R. M., Neto, W. P. F., Silvério, H. A., Martins, D. F., Dantas, N. O., & Pasquini, D. (2013). Cellulose nanocrystals from pineapple leaf, a new approach for the reuse of this agro-waste. *Industrial Crops and Products, 50*, 707-714.
66. Yves, O. R., Christian, F. B., Akum, O. B., Theodore, T., & Bienvenu, K. (2020). Physical and mechanical properties of pineapple fibers (leaves, stems and roots) from Awae Cameroon for the improvement of composite materials. *Journal of Fiber Science and Technology, 76*(12), 378-386.
67. Choquecahua Mamani, D., Otero Nole, K. S., Chaparro Montoya, E. E., Mayta Huiza, D. A., Pastrana Alta, R. Y., & Aguilar Vitorino, H. (2020). Minimizing organic waste generated by pineapple crown: A simple process to obtain cellulose for the preparation of recyclable containers. *Recycling, 5*(4), 24.
68. Van Tran, A. (2006). Chemical analysis and pulping study of pineapple crown leaves. *Industrial Crops and Products, 24*(1), 66-74.
69. Zainuddin, M. F., Shamsudin, R., Mokhtar, M. N., & Ismail, D. (2014). Physicochemical properties of pineapple plant waste fibers from the leaves and stems of different varieties. *BioResources, 9*(3), 5311-5324.
70. Uddin, N., Miah, S., Abdul, M., Jalil, M. A., Islam, M., & Ayesha, S. (2017). A review on extraction, characterization and application of pineapple leaf fiber (PALF) in textiles and other fields. *International Journal of Advanced Research, 5*, 112-116.
71. Thyavihalli Girijappa, Y. G., Mavinkere Rangappa, S., Parameswaranpillai, J., & Siengchin, S. (2019). Natural fibers as sustainable and renewable resource for development of eco-friendly composites: A comprehensive review. *Frontiers in Materials, 6*, 226.
72. Upadhyay, A., Lama, J. P., & Tawata, S. (2010). Utilization of pineapple waste: A review. *Journal of Food Science and Technology Nepal, 6*, 10-18.
73. Yusof, Y., Yahya, S. A., & Adam, A. (2015). Novel technology for sustainable pineapple leaf fibers productions. *Procedia CIRP, 26*, 756-760.
74. Naidu, A. L., Jagadeesh, V., & Bahubalendruni, M. R. (2017). A review on chemical and physical properties of natural fiber reinforced composites. *Journal of Advanced Research in Engineering and Technology, 8*(1), 56-68.

75. Atmakuri, A., Janušas, G., Siddabathula, M., & Palevicius, A. (2020). Wettability and moisture analysis on natural fiber reinforced epoxy resin hybrid composites. In *2020 International Conference on Mechatronic Systems and Materials (MSM)* (pp. 1-6). IEEE.
76. Jawaid, M., Asim, M., Tahir, P. M., & Nasir, M. (2020). *Pineapple Leaf Fibers: Processing, Properties and Applications*. Springer Nature.
77. Todkar, S. S., & Patil, S. A. (2019). Review on mechanical properties evaluation of pineapple leaf fibre (PALF) reinforced polymer composites. *Composites Part B: Engineering, 174*, 106927.
78. Kengkhetkit, N., & Amornsakchai, T. (2012). Utilisation of pineapple leaf waste for plastic reinforcement: A novel extraction method for short pineapple leaf fiber. *Industrial Crops and Products, 40*, 55-61.
79. Mohamed, A. R., Sapuan, S. M., Shahjahan, M., & Khalina, A. (2009). Characterization of pineapple leaf fibers from selected Malaysian cultivars. *Journal of Food, Agriculture and Environment, 7*(1), 235-240.
80. Mishra, S., Mohanty, A. K., Drzal, L. T., Misra, M., & Hinrichsen, G. (2003). A review on pineapple leaf fibers, sisal fibers and their biocomposites. *Macromolecular Materials and Engineering, 289*(11), 955-974.
81. Rosmi, M. S., Azhari, S., & Ahmad, R. (2014). Adsorption of cadmium from aqueous solution by biomass: Comparison of solid pineapple waste, sugarcane bagasse, and activated carbon. *Advanced Materials Research, 832*, 810-815.
82. Al-Oqla, F. M., & Sapuan, S. M. (2017). *Materials Selection for Natural Fiber Composites*. Woodhead Publishing.
83. Patrick, K., Ocharo Onchiri, R., & Manguriu, G. (2019). Developing suitable proportions for the production of pineapple leaf fibers reinforced normal strength concrete. *Open Journal of Civil Engineering, 9*, 185-194.
84. Sipiao, B., Paiva, R., Goulart, S., & Mulinari, D. (2011). Effect of chemical modification on mechanical behaviour of polypropylene reinforced pineapple crown fibers composites. *Procedia Engineering, 10*, 2028-2033.
85. Daud, Z., Hatta, M. Z. M., Kassim, A. S. M., Awang, H., & Aripin, A. M. (2014). Exploring of agro waste (pineapple leaf, corn stalk, and napier grass) by chemical composition and morphological study. *BioResources, 9*(1), 872-880.

86. Zawawi, D., Mohd, Z., Angzzas, S., & Ashuvila, M. A. (2014). Analysis of the chemical compositions and fiber morphology of pineapple (Ananas comosus) leaves in Malaysia. *Journal of Applied Sciences, 14*(12), 1355-1358.
87. Debnath, S. (2016). Pineapple leaf fibre—a sustainable luxury and industrial textiles. In *Handbook of Sustainable Luxury Textiles and Fashion* (pp. 35-49). Springer, Singapore.
88. Jose, S., Salim, R., & Ammayappan, L. (2016). An overview on production, properties, and value addition of pineapple leaf fibers (PALF). *Journal of Natural Fibers, 13*(3), 362-373.
89. Begum, K., & Islam, M. A. (2013). Natural fiber as a substitute to synthetic fiber in polymer composites: A review. *Research Journal of Engineering Sciences, 2*(3), 46-53.
90. Thyavihalli Girijappa, Y. G., Mavinkere Rangappa, S., Parameswaranpillai, J., & Siengchin, S. (2019). Natural fibers as sustainable and renewable resource for development of eco-friendly composites: A comprehensive review. *Frontiers in Materials, 6*, 226.
91. Shirke, A., Narayankar, J., Shaikh, M. U., Satpute, T., & Sreejith, S. (2018). Fabrication and characterization of pineapple fiber reinforced epoxy composite. *International Journal of Scientific Engineering and Research, 9*(5), 430-434.
92. Saha, S. C., Das, B. K., Ray, P. K., Pandey, S. N., & Goswami, K. (1993). Some physical properties of pineapple leaf fiber (PALF) influencing its textile behavior. *Journal of Applied Polymer Science, 50*(3), 555-556.
93. Asim, M., Jawaid, M., Abdan, K., & Ishak, M. R. (2016). Effect of alkali and silane treatments on mechanical and fibre-matrix bond strength of kenaf and pineapple leaf fibres. *Journal of Bionic Engineering, 13*, 426-435.
94. Siakeng, R., Jawaid, M., Tahir, P. M., Siengchin, S., & Asim, M. (2020). Improving the properties of pineapple leaf fibres by chemical treatments. In *Pineapple Leaf Fibers* (pp. 55-71). Springer, Singapore.
95. Maneeintr, K., Leewisuttikul, T., Kerdsuk, S., & Charinpanitkul, T. (2018). Hydrothermal and enzymatic treatments of pineapple waste for energy production. *Energy Procedia, 152*, 1260-1265.
96. Almendárez-Camarillo, A., Flores-Hernandez, C. G., Balcázar-Enríquez, V., Aguirre-García, M. S., Martínez-Hernandez, A. L., & Velasco-Santos, C. (2020). Nanocellulose extraction of pineapple

leaves for chitosan-starch nanocomposites. *Journal of Natural Fibers*. In press.

97. Neto, A. R. S., Araujo, M. A., Barboza, R. M., Fonseca, A. S., Tonoli, G. H., Souza, F. V., & Marconcini, J. M. (2015). Comparative study of 12 pineapple leaf fiber varieties for use as mechanical reinforcement in polymer composites. *Industrial Crops and Products, 64*, 68-78.
98. Baldoino Da Paixão, M. F. (2017). Variability assessment of mechanical properties of pineapple leaf fibers. *International Journal of Engineering Research and Technology, 6*(1), 27-30.
99. Gebino, G., & Muhammed, N. (2018). Extraction and characterization of Ethiopian pineapple leaf fiber. *Current Trends in Fashion Technology & Textile Engineering, 4*(4), 77-83.
100. Teles, M. C. A., Glória, G. O., Altoé, G. R., Amoy Netto, P., Margem, F. M., Braga, F. O., & Monteiro, S. N. (2015). Evaluation of the diameter influence on the tensile strength of pineapple leaf fibers (PALF) by Weibull method. *Materials Research, 18*, 185-192.
101. Gomes, A., Matsuo, T., Goda, K., & Ohgi, J. (2007). Development and effect of alkali treatment on tensile properties of curaua fiber green composites. *Composites Part A, 38*, 1811-1820.
102. Rao, K. M. M., Rao, K. M., & Prasad, A. R. (2010). Fabrication and testing of natural fibre composites: Vakka, sisal, bamboo and banana. *Materials and Design, 31*(1), 508-513.
103. De Rosa, I. M., Kenny, J. M., Puglia, D., Santulli, C., & Sarasini, F. (2010). Tensile behavior of New Zealand flax (Phormium tenax) fibers. *Journal of Reinforced Plastics and Composites, 29*(23), 3450-3454.
104. Shibata, M., Takachiyo, K. I., Ozawa, K., Yosomiya, R., & Takeishi, H. (2002). Biodegradable polyester composites reinforced with short abaca fiber. *Journal of Applied Polymer Science, 85*, 129-138.
105. Valadez-Gonzalez, A., Cervantes-Uc, J. M., Olayo, R., & Herrera-Franco, P. J. (1999). Effect of fiber surface treatment on the fiber–matrix bond strength of natural fiber reinforced composites. *Composites Part B, 30*, 309-320.
106. Gañán, P., & Mondragon, I. (2004). Fique fiber-reinforced polyester composites: Effects of fiber surface treatments on mechanical behaviour. *Journal of Materials Science, 39*, 3121-3128.
107. Kathiresan, P. P., & Senthamaraikannan, S. S. S. (2016). Physicochemical properties of new cellulosic Artisdita hystrix leaf fiber. *International Journal of Polymer Analysis and Characterization, 21*, 663-668.

108. Lopattananon, N., Payae, Y., & Seadan, M. (2008). Influence of fiber modification on interfacial adhesion and mechanical properties of pineapple leaf fiber-epoxy composites. *Journal of Applied Polymer Science, 110*(1), 433-443.
109. Dai, H., Ou, S., Huang, Y., & Huang, H. (2018). Utilization of pineapple peel for production of nanocellulose and film application. *Cellulose, 25*(3), 1743–1756.
110. Sibaly, S., & Jeetah, P. (2017). Production of paper from pineapple leaves. *Journal of Environmental Chemical Engineering, 5*(6), 5978–5986.
111. Cherian, N. M., Leão, A. L., de Souza, S. F., Thomas, S., Pothan, L. A., & Kottaisamy, M. (2010). Isolation of nanocellulose from pineapple leaf fibres by steam explosion. *Carbohydrate Polymers, 81*(3), 720–725.
112. Almendárez-Camarillo, A., Flores-Hernandez, C. G., Balcázar-Enríquez, V., Aguirre-García, M. S., Martínez-Hernandez, A. L., & Velasco-Santos, C. (2020). Nanocellulose extraction of pineapple leaves for chitosan-starch nanocomposites. *Journal of Natural Fibers*. In press.
113. Reddy, M. I., Kumar, M. A., & Raju, C. R. B. (2018). Tensile and flexural properties of jute, pineapple leaf and glass fiber reinforced polymer matrix hybrid composites. *Materials Today: Proceedings, 5*(1), 458–462.
114. Chollakup, R., Tantatherdtam, R., Ujjin, S., & Sriroth, K. (2011). Pineapple leaf fiber reinforced thermoplastic composites: Effects of fiber length and fiber content on their characteristics. *Journal of Applied Polymer Science, 119*(4), 1952–1960.
115. Arib, R. M. N., Sapuan, S. M., Ahmad, M. H. M., Paridah, M. T., & Khairul Zaman, H. M. D. (2006). Mechanical properties of pineapple leaf fiber reinforced polypropylene composites. *Materials & Design, 27*(5), 391–396.
116. Atmakuri, A., Palevicius, A., Siddabathula, M., Vilkauskas, A., & Janusas, G. (2020). Analysis of mechanical and wettability properties of natural fiber-reinforced epoxy hybrid composites. *Polymers, 12*(12), 2827.
117. Jawaid, M. et al. (2013). Mechanical performance of oil palm empty fruit bunches/jute fibres reinforced epoxy hybrid composites. *Materials Science and Engineering: A, 527*(29-30), 7944-7949.
118. Mishra, S., et al. (2004). Studies on mechanical performance of biofiber/glass reinforced polyester hybrid composites. *Composites Science and Technology, 64*(6), 955-965.

119. Pothan, L. A. et al. (2007). Effect of fiber surface treatments on the fiber-matrix interaction in banana fiber reinforced polyester composites. *Composites Part A: Applied Science and Manufacturing, 38*(5), 1116-1121.

120. Kabir, M. M., et al. (2012). Chemical treatments on plant-based Natural Fiber Reinforced Polymer composites: An overview. *Composites Part B: Engineering, 43*(7), 2883-2892.

121. Rihayat, T., Siregar, J. P., Jaafar, J., & Cionita, T. (2020, April). Green composites of natural fiber bamboo/pineapple leaf/coconut husk as hybrid materials. In *IOP Conference Series: Materials Science and Engineering* (Vol. 788, No. 1, p. 012038). IOP Publishing.

CHAPTER VII

DESIGN AND FABRICATION ROSELLE FIBER-REINFORCED GREEN EPOXY COMPOSITES FOR DINNERWARE

J. YUSUF, MOHD SAPUAN SALIT

Abstract

The potential of Roselle (Hibiscus sabdariffa) fibers for developing sustainable composite materials for cutlery and dinnerware is examined in this chapter. An annual tropical plant called Roselle produces fibers that are perfect for eco-friendly composites because of their exceptional mechanical qualities and biodegradability. The chapter highlights how using Roselle fibers can help the environment and contribute to global sustainability goals by lowering the use of synthetic materials. Highlighted applications include rice spoons, chopping boards, plates, bowls, trays, and other items that show off the adaptability and usefulness of Roselle fiber composites. The significance of creative, sustainable material selections in promoting a future with more environmental responsibility is emphasized.

1. Introduction

Hibiscus species thrive in over 300 tropical and subtropical regions. One uncommon variety, known as roselle or Jamaica sorrel (Hibiscus sabdariffa), is cultivated in different tropical climates for its leaves, seeds, stems, and calyces. The dried calyces are utilized in the production of tea, syrup, jams, jellies, and various other beverages (Mohamed 2021). Roselle, an annual plant, matures in about six months. As illustrated in Figure 1, its leaves consist of three to five lobes arranged alternately along the stem. Each Roselle flower features two peripheral ribs and a prominent central rib on its calyx lobes, classifying it within the Furcaria section. The flowers possess soft, fleshy calyces and come in colors ranging from white to pale yellow, with petals that can be white, pink, red, orange, purple, or yellow.

When fully mature, the fruits exhibit a vivid red color(Ilyas et al. 2021). On the Indian subcontinent, Roselle is commonly referred to as 'Mesta' or 'Meshta.' It has garnered significant interest from food, beverage, and pharmaceutical manufacturers due to its commercial potential as a natural food ingredient and coloring agent, offering an alternative to some synthetic products (Riaz et al. 2021).

Fig 1. Roselle plant (Sapian et al. 2023).

2. Composite Fabrication Techniques

The production of composites using Roselle fibers involves various methods. The simplest technique is the hand layup method, which involves manually placing layers, or plies, of either dry fabrics or prepreg (fabric pre-impregnated with resin) onto a tool to create a stack of laminates. After the layup process is complete, resin is typically applied to the dry plies using the resin infusion method. Another approach involves using machinery that combines glue and sprays it through a gun along with short, chopped strands of Roselle fibers. This technique, known as "chopper guns" or "spray-up" equipment, allows for rapid lamination, with the quality depending on careful rolling of the laminate. Additionally, advanced methods such as hot compression molding are used to produce laminate composites. Experimental evaluation and comparison of mono polypropylene and its hybrid composites include analyzing their density, X-ray diffraction (XRD) patterns, and mechanical properties like tensile/flexural strength and fracture toughness.

3. Conceptual design

Conceptual design is employed to select the most cost-effective material. Although the initial cost of adjustments is low, significant potential savings can be achieved using these strategies, as illustrated in Figure 2. Conceptual design comprises three main stages: "pre-concept generation," "concept generation," and "concept evaluation." The goal of conceptual design is to minimize waste during disposal and positively impact the environment. Therefore, recycling should be considered when designing both products and materials (Jameel, Leman, and Sapuan 2023).

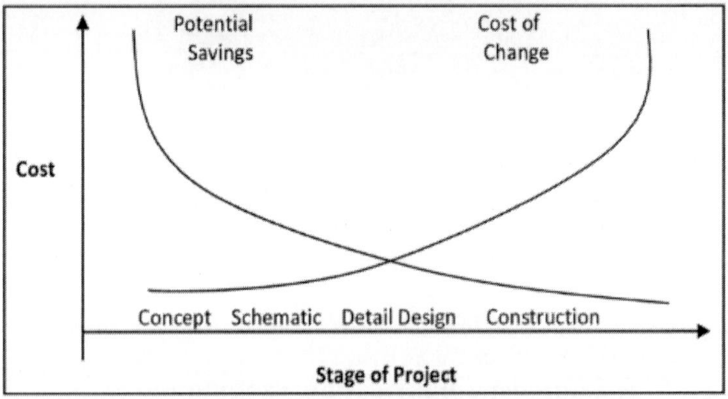

Fig 2. Stages of project and saving potential ((Mishra, Kumar, and Shrestha 2019)

4. Applications of Roselle Fibers in Composite Materials

Roselle fibers, due to their excellent properties, have found applications in various industries. The use of these fibers in composite materials has been explored extensively, and they have shown promising results in several applications.

5. Textiles and Consumer Goods

In the textile industry, Roselle fibers can be incorporated into various consumer goods. They serve as potential reinforcement fillers in polymer composites for a range of applications. These fibers can aid in developing and producing new products, promoting more sustainable and circular textile materials. Utilizing Roselle fibers supports the broader global industry trends toward a circular economy and striving for a closed-loop production cycle (Sigaard and Laitala 2023).

6. Roselle Fiber Composite Tray

Begin harvesting the roselle plants when the calyx, the part enclosing the seedpod, is juicy, plump, and still green. Use a knife or sharp pruning shears to cut the stems close to the plant's base. Then, remove any leaves or small branches from the stems, leaving only the main stem for fiber extraction. Prepare the stems for retting, a process that breaks down the non-fibrous components to aid fiber extraction. Retting can be done either in the sun or in water. Water retting involves submerging the stems in water, preferably flowing water from a river or stream, for a few days to a few weeks. For this process, roselle stems were submerged in a pail of water for two weeks to achieve high-quality fibers (Osei-Kwarteng, Gweyi-Onyango, and Mahunu 2021).

The next step involves removing the shattered outer layer using a technique called scutching to extract the fiber from the woody core. After scutching, the fibers are further separated and straightened using several combs or hackles in a process known as hackling or combing. To prevent mold and decay, thoroughly dry the cleaned and separated fibers in a well-ventilated area away from direct sunlight.

The roselle fiber composite tray undergoes a 6-hour curing process at room temperature to initiate resin hardening. It is then left to cool at room temperature for 2 days to avoid thermal stress. On the third day, the tray is gently demolded after fully curing, facilitated by a previously applied releasing agent. Any excess material is trimmed, and the tray's surface is finished for the desired aesthetic look. Figure 3 shows examples of the final products (Landrock 1995).

Fig 3. Roselle fiber composite tray

7. Eco-Friendly Plates using Roselle Fiber-reinforced Epoxy Composites

Roselle fibers offer significant potential as an eco-friendly and durable alternative to conventional synthetic fibers in composite materials. Incorporating roselle fibers into epoxy composites provides a practical solution for reducing the environmental impact of industrial materials. The fibers' high cellulose content, excellent mechanical properties, and biodegradability align with global sustainability objectives, including reducing reliance on non-renewable resources and minimizing ecological footprints. The water retting method used to extract these fibers has proven effective in maintaining their quality and properties. This approach enhances the mechanical performance of the composites while promoting environmentally friendly manufacturing practices. Future research should aim to refine processing techniques and investigate other natural fibers with similar attributes to further enhance composite performance and applicability.

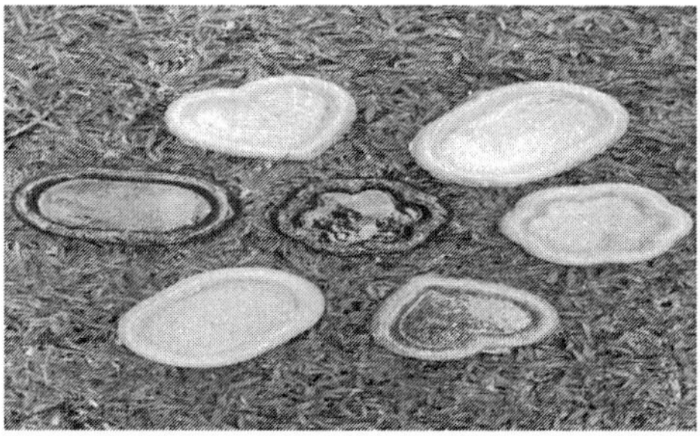

Fig. 4. Plates from roselle fiber-reinforced epoxy composites

Additionally, conducting life cycle assessments and durability studies under various environmental conditions will offer valuable insights into the effectiveness and sustainability of roselle fiber composites in real-world applications (refer Figure 4).

8. Roselle Fiber-reinforced Epoxy Composites for Making Chopping Board

Utilizing a Design for Sustainability (DfS) approach in the creation of roselle composite chopping boards provides a method for producing eco-friendly products without compromising functionality. This approach emphasizes using roselle fiber-reinforced composites as the primary material, offering a sustainable alternative to traditional materials like plastic and ceramics. The project progresses through a conceptual design phase, exploring various iterations to integrate the unique properties of roselle fibers effectively. These fibers are naturally biodegradable, lightweight, and strong enough for everyday use. By implementing the DfS approach, the development of roselle composite chopping boards sets a benchmark for sustainable product design, reflecting a commitment to environmental protection and innovation.

The roselle fiber-reinforced composite chopping board was designed and produced as a high-quality alternative to existing options on the market. The development process involved several stages: planning, selection, and execution. Initially, the product background was researched, and market analysis was conducted to understand current chopping board options. In the selection stage, various design concepts were sketched and evaluated to finalize the design. Calculations were performed to determine material requirements before proceeding to the execution phase, where the mold was created, and the composite was fabricated (refer Figure 5). This DfS approach is anticipated to capture the interest of global manufacturers.

Fig 5. The final product of chopping board

9. Bowl Made by Roselle Fiber-reinforced Epoxy Composites

Creating a powdered roselle fiber biocomposite reinforced epoxy bowl using hand layup involves a detailed process to ensure both durability and visual appeal. First, the mold is thoroughly cleaned to remove dust and residues, which is essential for proper epoxy resin adhesion. Next, Meguiar's Mirror Glaze 8 Release Wax is applied to create a non-stick surface, making it easier to remove the bowl without damaging the mold. A gel coat is then applied to provide a smooth finish and enhance durability, ensuring an even application to avoid imperfections that could impact the final product's quality.

The roselle fibers are treated to improve their bonding with the epoxy resin (Zwawi 2021). The treatment generally involves chemical processes that alter the surface of the fibers, enhancing their roughness and improving adhesion to the resin. This step is crucial for establishing a strong bond between the fibers and the epoxy matrix. The epoxy resin and methyl ethyl ketone peroxide (MEKP) are mixed according to the specified ratio, ensuring a uniform and consistent blend. Accurate mixing is essential to prevent incomplete curing or weak areas in the final product (refer Figure 6) (Tiwari et al. 2020).

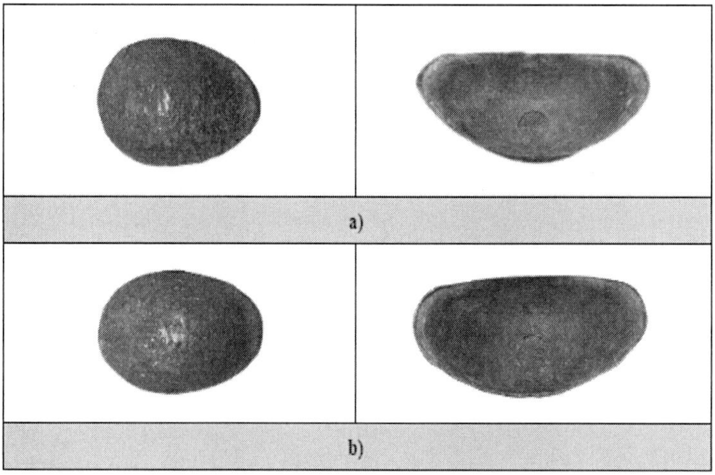

Fig 6. a) Composite bowl fabrication b) Composite bowl fabrication

10. Sustainable Roselle Epoxy Composite Rice Spoon

The research effectively highlights the potential of using roselle stems as a sustainable material for making rice spoons. Through detailed characterization, innovative treatment, and advanced processing, the natural properties of roselle stems have been enhanced to produce a durable biocomposite suitable for rice spoon production. Comparative analysis underscores the environmental and competitive benefits of roselle fibers compared to traditional materials such as plastic and wood. By introducing roselle stems into utensil manufacturing, this project provides an eco-friendly alternative and promotes sustainable practices in daily use. The roselle composite rice spoon exemplifies the effectiveness of renewable and biodegradable materials in minimizing environmental impact.

Additionally, the development of a sustainable ladle made from roselle fiber and epoxy composite demonstrates that combining natural fibers with advanced resin technology can produce durable, lightweight, and environmentally responsible products. This approach leverages the strength and versatility of epoxy resin along with the biodegradable qualities of roselle fibers. The resulting composite material meets the performance needs of everyday tools and offers a sustainable replacement for conventional plastics and metals (refer Figure 7). The method highlights the importance of sustainable manufacturing techniques and the role of natural fiber composites in fostering a more eco-friendly future.

Fig. 7. The roselle composite rice spoon

Conclusions

The chapter emphasizes the fibers from the Roselle plant (Hibiscus sabdariffa), which have a promising future in the creation of sustainable and green composite materials. Excellent mechanical capabilities, biodegradability, and adaptability for a variety of industrial uses are the characteristics of Roselle fibers, which find use in everything from dinnerware like as trays, plates, and bowls to tools like chopping boards and rice spoons. Because Roselle fibers have a lower environmental effect and a decreased dependency on synthetic materials, they support global sustainability goals. It is recommended that future studies concentrate on refining processing methods and performing life cycle analyses to improve Roselle fiber composite's performance and practicality. This creative method emphasizes the significance of making sustainable material choices and their part in creating a more environmentally friendly future. Prospective research avenues encompass streamlining procedures and carrying out life cycle assessments.

Acknowledgements

The authors are thankful to Universiti Pura Malaysia (UPM) for providing financial support through Putra IPS grant vote number 9742900. The authors are also thankful to Fillip Sdn Bhd, Putrajaya Malaysiaand Sekolah Kebangsaan Bukit Kemuning, Shah Alam, Selangor, Malaysia for providing raw material. The authors also want to thank Zhou Tiancheng, Norman Amirul Bin Jasni, Muhammad Irfan In Suffian, Ahmad Danish, Syafwan Bin Azeiz Badlei Syah, Azzam Luqman Bin Rosdi, Faiz Daniel Bin Mohd Nasser, Kiroshini A/P Muniandy, Muhammad Arif Bin Saifulazmi, Muhammad Nabil Hakim Bin Abdul Rahim, Karveish A/L Yeesvaran, Ikmal Bin Zainuddin, Ahmad Ameer Bin Ahmad, and Shuhairil from Universiti Putra Malaysia for their support.

References

R. A. Ilyas, S.M. Sapuan, W. Kirubaanand, Z.M. Zahfiq, M.S.N. Atikah, R. Ibrahim, A.M. Radzi, R. Nadlene, M.R.M. Asyraf and M.D. Hazrol (2021), Roselle: production, product development, and composites, in Roselle: Production, Processing, Products and Biocomposites (editors: S.M. Sapuan, R. Nadlene, A.M. Radzi and R.A. Ilyas), Academic Press, London, pp. 1–23

Jameel, Yusuf, Z Leman, and S M Sapuan. 2023. "Conceptual Design of the Car Gear Console Panel Made by Glass / Natural Fiber Conceptual Design of the Car Gear Console Panel Made by Glass / Natural Fiber reinforced Hybrid Polymer Composite," no. March 2022.

Landrock, Arthur H. 1995. *Handbook of Plastic Foams: Types, Properties, Manufacture and Applications*. Elsevier.

Mishra, Anjay Kumar, Chandra Kumar, and Pan Shrestha. 2019. "Status Assessment of Life Cycle Costing in Baitadi Town Water Supply and Sanitation Project , Nepal," no. January.

Mohamed, Bahaeldeen Babiker. 2021. "Roselle (Hibiscus Sabdariffa L.) in Sudan: Production and Uses." In *Roselle*, 121–27. Elsevier.

Osei-Kwarteng, Mildred, Joseph Patrick Gweyi-Onyango, and Gustav Komla Mahunu. 2021. "Harvesting, Storage, Postharvest Management, and Marketing of Hibiscus Sabdariffa." In *Roselle (Hibiscus Sabdariffa)*, 15–31. Elsevier.

Riaz, Ghazala, Satya Narayan Naik, Meenakshi Garg, and Rajni Chopra. 2021. "Phytochemical Composition of an Underutilized Plant Sorrel/Roselle (Hibiscus Sabdariffa L.) Cultivated in India." *Lett. Appl. NanoBioScience* 10: 2138–47.

Sapian, Syaifuzah, Asma Ali Ibrahim Mze, Fatin Farhana Jubaidi, Nor Anizah Mohd Nor, Izatus Shima Taib, Zariyantey Abd Hamid, Satirah Zainalabidin, Nur Najmi Mohamad Anuar, Haliza Katas, Jalifah Latip, Juriyati Jalil, Nur Faizah Abu Bakar, and Siti Balkis Budin. 2023. "Therapeutic Potential of Hibiscus Sabdariffa Linn. in Attenuating Cardiovascular Risk Factors." *Pharmaceuticals* 16 (6): 1–25. doi:10.3390/ph16060807.

Sigaard, Anna Schytte, and Kirsi Laitala. 2023. "Natural and Sustainable? Consumers' Textile Fiber Preferences." *Fibers* 11 (2). doi:10.3390/fib11020012.

Tiwari, Shilpi, Kavita Srivastava, C. L. Gehlot, and Deepak Srivastava. 2020. "Studies on Mechanical and Thermal Properties of Epoxy/Fly Ash/Nanofiller Nanocomposite: A Review." *International Journal of Civil Engineering and Technology (IJCIET)* 11 (2): 120–39. doi:10.34218/IJCIET.11.2.2020.013.

Zwawi, Mohammed. 2021. "A Review on Natural Fiber Bio-Composites, Surface Modifications and Applications." *Molecules (Basel, Switzerland)* 26 (2). doi:10.3390/molecules26020404.

CHAPTER VIII

BACTERIAL CELLULOSE IN BIOCOMPOSITES

RAZAN MUHAMMAD RAILIS, MUHAMMAD ADLAN AZKA, ABDUL HABIB, HAIRUL ABRAL, MOHD SAPUAN SALIT, MARIA YOSEFINA MEINADIA SEKAR KINANTI ASWIRAWAN

Abstract

Researchers are now focusing on using cellulose as a reinforced biopolymer matrix to create biodegradable and environmentally friendly materials, also known as biocomposites, in an effort to reduce the consumption of high-consumption synthetic polymers. This material consists of a matrix and reinforcement made from bioresources, particularly plant extracts. However, the extraction of this cellulose results in a high pollution of chemical waste from the extraction process, which could also impact the environment. To overcome this problem, several researchers introduced cellulose from bacteria (BC) as a potential alternative.

This chapter gives an overview of bacterial cellulose (BC), its production method, and the properties of pure BC material. Furthermore, BC's biocompatibility enables its use as a biobased reinforcement in the production of biocomposites. To enhance the use of BC, this chapter delves into its potential applications in biocomposite production, as well as the characteristics of the biocomposite reinforced with BC. This chapter offers a comprehensive overview of the BC material, its properties, and the development of bacterial cellulose biocomposite, with the aim of enhancing its utilization and minimizing the use of synthetic polymers.

1. Introduction

The demand for renewable and sustainable bio-based materials has considerably increased in recent years due to environmental concerns. Bio-based materials with high purity, mechanical stability, and barrier properties have become a widespread choice for the research community to develop environmentally friendly systems (Galiano et al., 2018). This allows for the manufacturing of several pieces of equipment, such as packaging, home appliances, electronic devices, construction, and textiles (Andrady, 2017). Due to this benefit, the utilization of synthetic polymers has increased greatly, with Yusuf et al. (2022) reporting that the synthetic polymer products have reached over 400 million tonnes worldwide. However, besides their benefits, synthetic polymers possess a significant drawback: they exhibit a formidable resistance to natural degradation (Yusuf et al., 2023). As per the findings of H. Yang et al. (2022), it takes a long time for a single piece of synthetic polymer to completely decompose into soil, which poses a risk to the ecosystem's balance and has the potential to result in contamination. Biopolymers, also referred to as natural polymers, offer a potential solution to address this issue due to their inherent biodegradability. Cellulose is a captivating biopolymer that holds great interest for scientists.

Cellulose is currently one of the most abundant materials on Earth (McNamara et al., 2015; Tang et al., 2024; Zennifer et al., 2021). Cellulose is extensively employed in the production of ecologically friendly products, including extracts from ginger (Babaei-Ghazvini et al., 2024; Dang et al., 2024; Abral et al., 2020), jute (Baigh et al., 2023; Bledzki et al., 2015), pineapple leaf(Armynah et al., 2022; Aurelio et al., 2023), bengkuang skin (Ilyas et al., 2023), pandan duri leaf (Syafri et al., 2022), and sugar palm leaf (Ilyas et al., 2019; Mohammed et al., 2023), as a consequence of these advantages. Numerous studies have demonstrated that the incorporation of cellulose enhances the biocomposite's thermal and antibacterial properties (Abral, Basri et al., 2019; Cho & Park, 2011; Kargarzadeh et al., 2015; Nazrin et al., 2020; W. Yang et al., 2021). The production process is still impeded by restrictions, despite the advantages and impact of the cellulose blending process using plant-derived cellulose fibers (Abolore et al., 2024). The residual byproduct of the extraction process is a potential environmental hazard, as it still necessitates a considerable amount of chemicals and a protracted chemical procedure to extract cellulose from the plants (Abolore et al., 2024). To resolve this issue, it is imperative to identify an alternative solution to replace plant-derived cellulose materials, which necessitate a time-consuming procedure and substantial expenditures. Cellulose, which is derived from microorganisms called bacterial cellulose (BC), has significant potential to address this issue.

Finally, this chapter provides an overview of bacterial cellulose, the history of bacterial cellulose, and also explains how this material can be developed and utilized as an alternative substitute for plant-based cellulose. In addition, this chapter also explains the producing method for BC material and potential application of BC as a bioreinforcement on biodegradable biocomposite.

2. Overview of Bacterial Cellulose

Bacterial cellulose (BC) is a biopolymer and polysaccharide composed of glucose units that is located outside of cells (Moniri et al., 2017; Saleh et al., 2022). This cellulose type was discovered by Adrian Brown in 1886 while conducting research with bacteria xylinum (Choi et al., 2022). The BC is distinct from plant cellulose in that it is not adhered to hemicellulose, lignin, or pectin (Muiruri et al., 2023). This distinction provides BC with numerous advantages over plant-based cellulose. One of the main advantages of BC is a refined cellulose that eliminates the necessity for the energy-intensive process of purifying cellulose from plants (Abolore et al., 2024). Moreover, the purification process of BC is environmentally sustainable, as it employs alkali to eliminate any remaining pathogens (Girard et al., 2024).

BC has a crystalline structure and solid material that exhibits extraordinary mechanical properties due to its high purity (Carvalho et al., 2019). Furthermore, BC has excellent compatibility with both natural and synthetic materials (Popa et al., 2022), rendering it very suitable for application in the realm of composites, particularly in the development of biocomposite materials such as packaging, electronics, and environmental applications (Halib et al., 2012; Ilyas et al., 2018). Bacterial nanocellulose (BNC) possesses remarkable hydrogel bodies that have a high water content of 99% and are resistant to mechanical and thermal stress. Additionally, BNC exhibits exceptional purity since it consists solely of cellulose, and it showcases a network structure composed of nanofibers (Klemm et al., 2018).

A biosynthetic process produces bacterial cellulose, unlike other types of cellulose. A network of much smaller fibers forms them into a ribbon-like fibril. The lack of additional byproducts or impure molecules like hemicellulose, lignin, or pectin makes this nanocellulose the purest (Ribeiro et al., 2019). Bacterial cellulose manufacturing involves two crucial procedures: fermentation and purification (Z. Zhu et al., 2021). Methods such as enhanced fermentation techniques, genetic engineering to alter behavior, manipulation of strain types, and utilization of agricultural by-products were employed to enhance the yield of bacterial nanocellulose (Goswami et al., 2023). BNC does not possess essential features such as antibacterial activity, antioxidant property, magnetic property, or electroactivity (Ullah et al., 2023). The bioprocessing sector, biomedical applications, wastewater treatment, electroconductive materials, and food packaging are just a few of the expanding application fields for BNC (Perera et al., 2023). There are an abundance of hydroxyl groups (OH) within its chemical structure, as can be seen from Figure 1. BC serves as an enabling environment for the absorption and corporation of other hydrophilic substances and nanoparticles (Mensah et al., 2022).

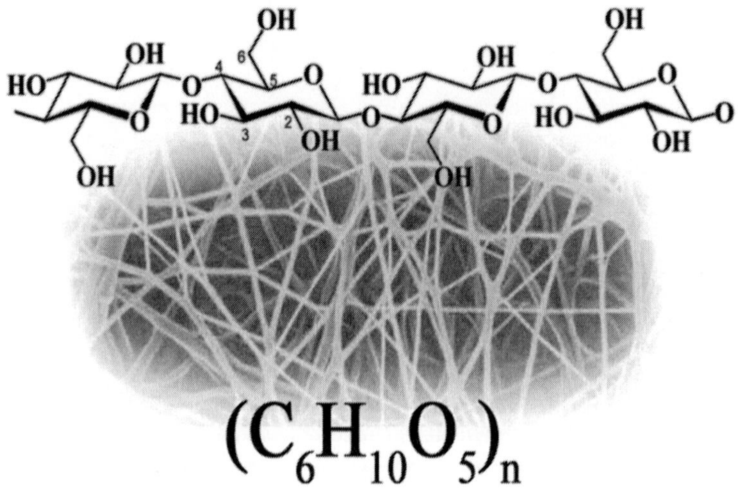

Figure 1. Chemical structure of bacterial cellulose (Mensah et al., 2022).

3. Bacterial cellulose production

As a fundamental structure of fibrils (Andriani et al., 2020), BC is composed of a β-1→4 glucan chain with the molecular formula $(C_6H_{10}O_5)n$ (Esa et al., 2014). The primary method of producing BC typically entails the oxidative fermentation of various types of bacteria in a culture medium. BC can be produced by a variety of bacterial types, including agrobacterium, achromobacter, sarcina, acetobacter, rhizobium, and pseudomonas, as identified through research (Lin et al., 2020). There are two methods for processing this material, which are the static method and the agitated method (Cazón & Vázquez, 2021). A gel-like pellicle is formed at the air-liquid interface during static fermentation (Lima et al., 2024), whereas small, irregularly shaped granules are suspended in the culture fluid during agitated fermentation (Lima et al., 2024; Saleh et al., 2024).

Air/liquid interface area is the determining factor in BC production's straightforward and widely used static culture cultivation method. The BC membrane can adjust to the shape of the material, and the culture medium can be deposited in containers of varying sizes and shapes. Thus, this process allows for the creation of materials with pre-established forms. The BC pellicle captures the CO_2 molecules generated by the bacteria during incubation, forcing them to rise and remain on the surface of the culture medium. The cultivation procedure typically takes between 5 and 20 days. The static approach produces bacterial cellulose (BC) as a cohesive film on the surface of the medium, making it appropriate for various applications (Navya et al., 2022).

An agitated method differs from a static method in that it utilizes motion and turbulence to join the fibril. The medium is continuously aerated during turbulent cultivation, leading to increased productivity and reduced production expenses. The size and form of the grown BC depend on the spinning speed employed. An agitated culture has a notable drawback: it increases the probability of cellulose-negative mutants emerging from cellulose-producing cells when there is substantial turbulence and shear stress (Navya et al., 2022).

The main difference between the static and agitated procedures is that the BC produced under static settings displayed more excellent crystallinity than that produced under agitated conditions (Gao et al., 2020). In addition, the BCs produced under static conditions displayed a compact network, enhanced tensile strength (Young's modulus), and increased resilience to temperature (Torres et al., 2019). Because of the excellent properties of BC, this material has potential as a reinforcement on biocomposite for several applications in our entire lives and also can be a candidate for making a bio-based composite with superior properties.

Fatima et al. (2023) produced bacterial cellulose using saccharides and ethanol as growth media. The growth media was poured into 60mm plates under static conditions. Moreover, the bacterial was introduced into the plate and incubated for 7 days. After the bacterial cellulose was obtained, the bacterial cellulose was washed using NaOH with deionized water for 12 hours. Following this, the pre-treated gel-like membranes were soaked overnight in deionized water at room temperature to remove the remaining growth media embedded inside. The bacterial cellulose production is illustrated in Figure 2.

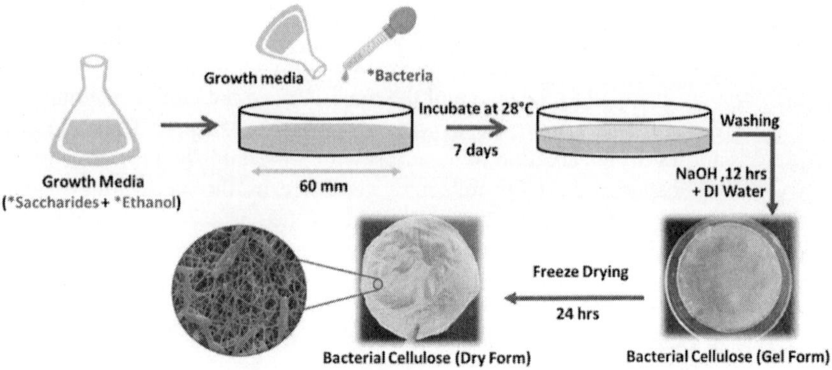

Figure 2. Procedure for Bacterial Cellulose membrane fabrication (Fatima et al., 2023).

4. Bacterial Cellulose in Biocomposites

According to the latest study, BC is well-suited for producing biodegradable materials, specifically composite material. C. Zhu et al. (2021) created a superior BC/polyphenylene sulfide (PPS) composite separator. The composite separator containing 20% BC particle demonstrated a good thermal stability, higher rate of electrolyte wettability, and ionic conductivity. This enhancement could increase consumption rates and improve cycling performance, resulting in faster and more consistent performance compared to cells with commercial separators (Song et al., 2024).

An investigation conducted by Abral et al. (2021) revealed that the incorporation of bacterial cellulose into starch/chitosan biofilm resulted in enhanced mechanical and physical characteristics, rendering it more suitable for bio packaging applications. The characteristics of the biofilm increased with the addition of 15% of the BC combination. The effective dispersion of bacterial cellulose within the mixture, which is essential for the formation of additional hydrogen bonds in the biofilm, was the cause of this observed increase. Also, another study from Abral, Kadriadi, et al. (2019) showed that the incorporation of BC nanoparticles into pure polyvinyl alcohol (PVA) film resulted in enhanced characteristics. BC nanoparticles decrease the free O-H chain in pure PVA films by 7.5%, leading to the formation of strong intermolecular chain in biocomposite structure. This results in the development of a robust and water-repellent PVA/BC biocomposite, which is well-suited for biodegradable food packaging.

Another study by J. Zhu et al. (2024) investigated bacterial cellulose-reinforced carbon nanotubes (CNTs) as flexible electroconductive film. They found that the presence of CNTs with only 0.36 wt% on BC composite successfully enhanced the mechanical properties of the films. Incorporating CNTs also reduced the mass loss in the composites because of the superior thermal stability of the BC/CNTs films. Furthermore, BC/CNTs composite film found practical application as the bottom electrode of flexible electroluminescent, offering a promising avenue for the development of flexible electronic devices. Figure 3 shows the flexibility and electroconductivity of the BC/CNTs composites.

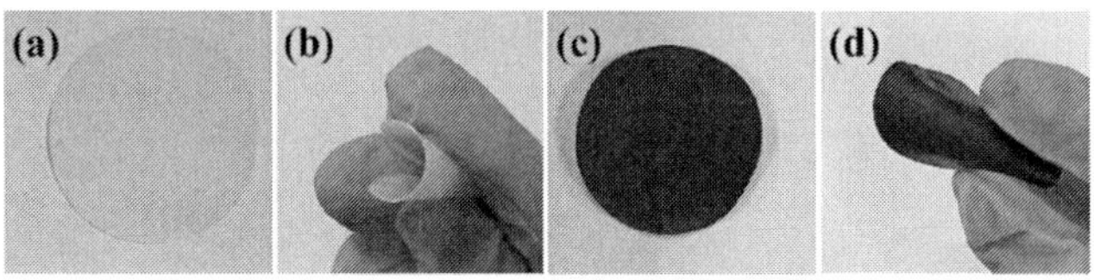

Figure 3 Macroscopic pictures of BC film (a), a bent BC film (b), BC/CNTs composite film (c), and a bent BC/CNTs film (J. Zhu et al., 2024).

5. Advanced application of bacterial cellulose

Bacterial cellulose possesses exceptional thermal stability, minimal chronic inflammatory response, and notable mechanical qualities, rendering it a viable material for diverse applications (Kadier et al., 2021). Figure 3 depicts the extensive utilization of bacterial cellulose across several applications. BC's chemical composition contains a large number of hydroxyl groups (OH), which allows it to effectively absorb and incorporate other hydrophilic chemicals and nanoparticles as reinforcement (Vasconcelos et al., 2017). BC's exceptional characteristics, including its ability to biodegrade, its controllability during production, and its net-like morphology resembling human collagen, make it highly desirable in the medical field. It has been extensively used in various applications such as controlled drug delivery, medical pads, artificial skin, cartilage and bone regeneration, bone tissue engineering scaffolds, hormone delivery, and nerve guides for treating spinal cord injuries. The mentioned items include vascular grafts, tubes used for blood vessels, dental implants, proteins, and hormones, as well as nerve guides used for treating spinal cord injuries. (Kadier et al., 2021).

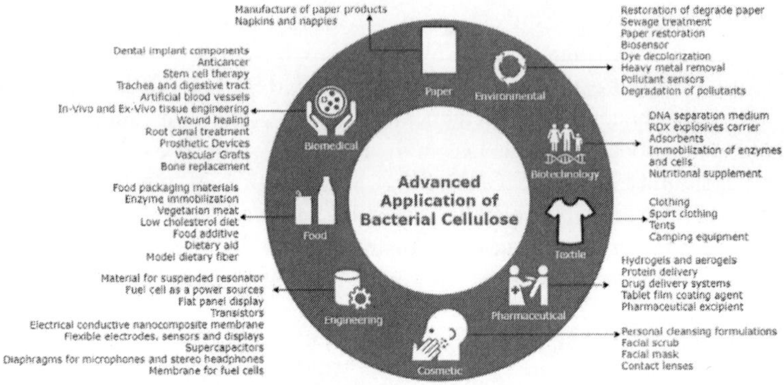

Figure 4. Advanced application of bacterial cellulose (Kadier et al., 2021).

a. Membrane filtration

The bacterial cellulose membranes possess a three-dimensional porous structure composed of cellulose nano and micro fibrils with diameters ranging from 10 to 100 nm. However, the cellulose membranes created by various bacterial strains have unique physical and chemical characteristics that are significantly impacted by the bacterial growing methods (Iguchi et al., 2000; Wang et al., 2019). Fatima et al. (2023) examined the attributes of bacterial cellulose as a membrane filtration. Various cultivation processes and characterization studies were conducted to determine the suitability of Komagataeibacter sp. FXV3, Komagataeibacter sp. NFXK3, and K. intermedius LMG 18909 bacterial strains for producing cellulose membranes with different properties. The resulting bacterial cellulose (BC) membranes were freeze-dried to achieve stability. Upon examination of the bacterial cellulose bottom surface, it was seen that each bacterial strain resulted in a distinct interwoven pattern, fiber size, pore size, and surface porosity. These findings are depicted in Figure 5. They infused the gas permeation with nitrogen (N2), oxygen (O2), and carbon dioxide (CO2) gas. The findings indicate that NFXK3 membranes exhibited the lowest gas permeability as a result of their compact and less porous nature.

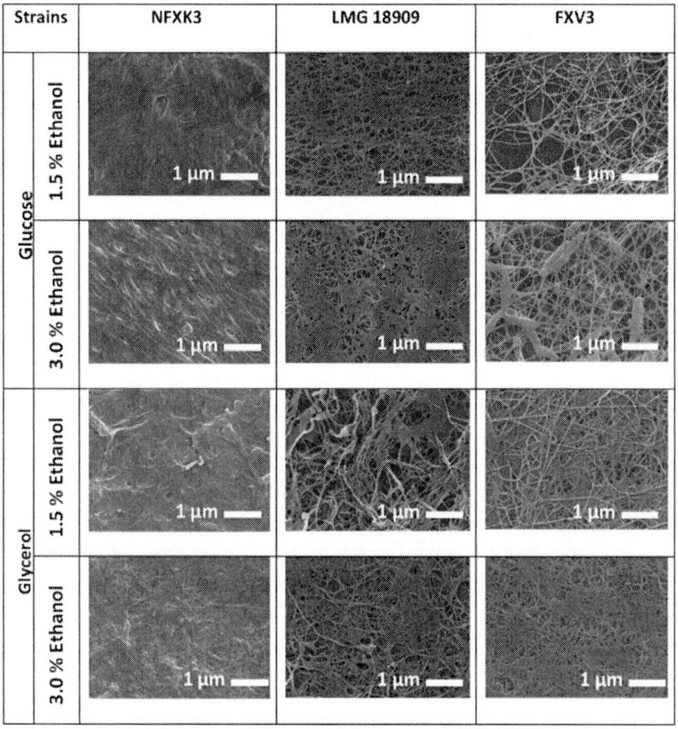

Figure 5. SEM bottom surface morphology of bacterial cellulose membranes produced by NFXK3, LMG 18909, and FXV3 strains under different studied conditions (Fatima et al., 2023).

b. Medical application

Bacterial cellulose possesses an exceptionally high level of purity and a structure that resembles the net-like morphology of human collagen. This biomimetic characteristic enables its use in various applications, including artificial skin, vascular grafts, tissue-engineering scaffolds, dental implants, medical pads, artificial bone, and cartilage, as well as the delivery of drugs, proteins, and hormones (Kadier et al., 2021). The biocompatibility assessment of the BC implant indicates that there are no visible signs of inflammation, such as chronic inflammation, foreign body responses, or angiogenesis. Additionally, there is no presence of fibrotic capsule or giant cells, and the infiltration of fibroblasts does not result in a chronic inflammatory reaction (Helenius et al., 2006).

c. Electro conductive films

J. Zhu et al. (2024) investigated the utilization of carbon nanotube composite to reinforce bacterial cellulose, resulting in the creation of conductive and flexible film composites. Vacuum filtration was utilized to separate water from a mixed dispersion liquid containing bacterial cellulose (BC) and single-walled carbon nanotubes (SWCNTs) in order to manufacture the composites. The final outcome was a flexible composite film (BC/CNTs) that exhibited conductivity. The conductive composite films exhibited remarkable mechanical characteristics, and the inclusion of single-walled carbon nanotubes (SWCNTs) enhanced the conductivity of the composites. The luminous qualities of the material were highly noteworthy, as the brightness remained rather stable even after undergoing 500 bending cycles. This can be observed in Figure 6.

d. Sensor application

Flexible sensors want a specific level of pliability and steadiness, and can be categorized into materials such as metals and cellulose. Bacterial cellulose possesses abundant natural resources, exceptional characteristics, and a distinctive structure, making it a highly promising polymer resource (Pirsa et al., 2018). Bai et al. (2024) employed bacterial cellulose as a reinforcement material along with zinc nitrate hexahydrate (ZnO) and pyrrole and ferric chloride (PPy). The results demonstrated that piezo-resistive sensors made of BC/ZnO/PPy composites exhibit the potential for use in sound sensor applications. The researchers conducted experiments involving many types of sounds, including those produced by animals, the environment, and musical instruments. The diagram illustrates the outcome of the experimental configuration.

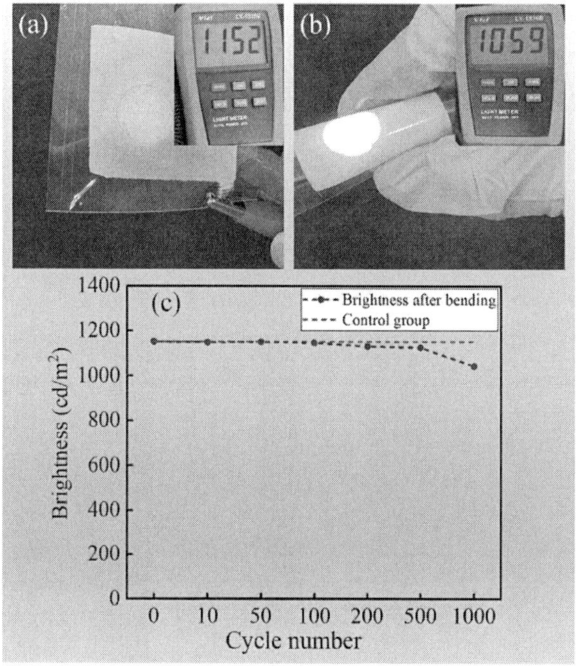

Figure 6. Luminescence of the device in two states: without bending (a) and under bending conditions (b);
Luminescence under the bending cycle test (c)
(J. Zhu et al., 2024).

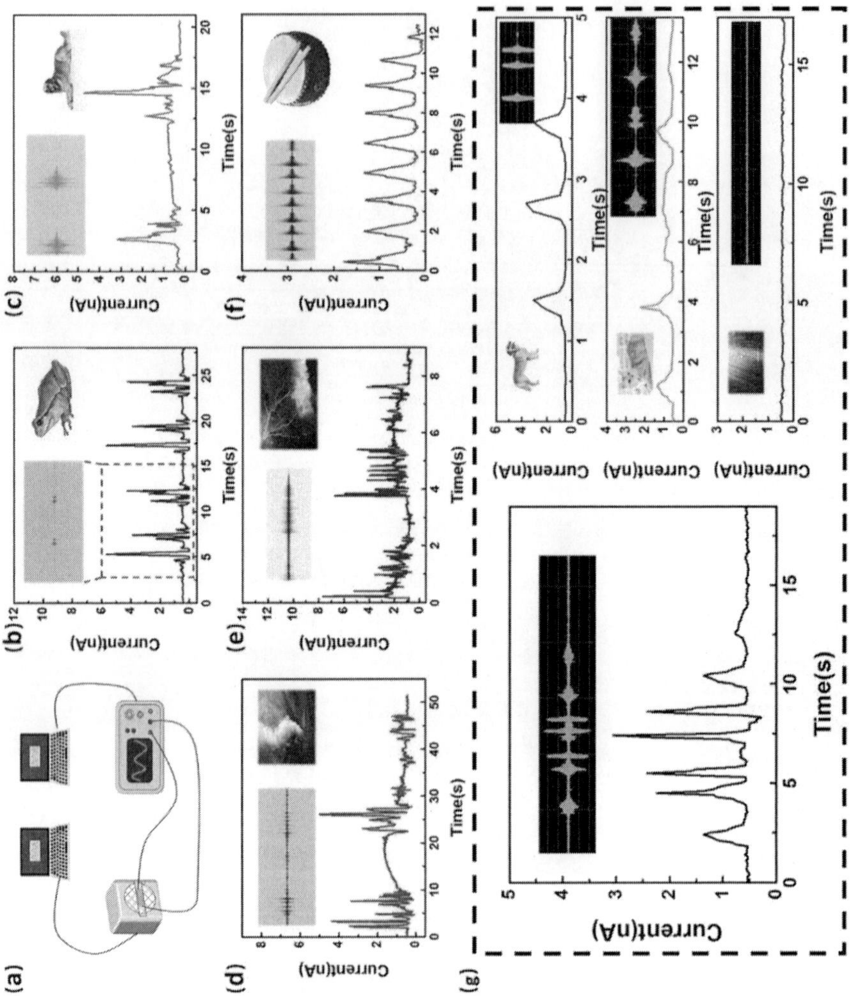

Figure 7. (a) Schematic diagram of experimental setup. (b-c) Recognition of animal sound signals (frog and dog). (d-f) Recognition of non-biological sound signals (Bai et al., 2024).

5. Conclusion

Bacterial cellulose is considered the optimal choice for manufacturing biocomposites due to its unique properties and advantages. One of the key reasons is its remarkable ability to establish hydrogen bonds by binding with the unoccupied O-H bonds present in the matrix. This bonding significantly enhances the strength and durability of the biocomposite, contributing to its superior mechanical, thermal, and physical properties. These enhanced properties make bacterial cellulose-based biocomposites not only more robust but also more sustainable compared to traditional materials.

Extensive research has been conducted on the utilization of bacterial cellulose as the primary constituent of biocomposites. These studies have consistently demonstrated the material's exceptional characteristics, providing valuable insights into its potential applications. The findings from this body of research offer inspiration and a solid foundation for exploring more sophisticated and innovative uses of bacterial cellulose in various fields.

As a result, bacterial cellulose has garnered significant attention in the scientific community and industry alike, driving ongoing efforts to develop advanced biocomposites that leverage its unique properties. This continuous exploration promises to unlock new possibilities and applications, further cementing bacterial cellulose's status as a crucial material for sustainable development.

References

Abolore, R. S., Jaiswal, S., & Jaiswal, A. K. (2024). Green and sustainable pretreatment methods for cellulose extraction from lignocellulosic biomass and its applications: A review. *Carbohydrate Polymer Technologies and Applications*, *7*, 100396. https://doi.org/https://doi.org/10.1016/j.carpta.2023.100396

Abral, H., Ariksa, J., Mahardika, M., Handayani, D., Aminah, I., Sandrawati, N., Sapuan, S. M., & Ilyas, R. A. (2020). Highly transparent and antimicrobial PVA based bionanocomposites reinforced by ginger nanofiber. *Polymer Testing*, *81*(July 2019), 106186. https://doi.org/10.1016/j.polymertesting.2019.106186

Abral, H., Basri, A., Muhammad, F., Fernando, Y., Hafizulhaq, F., Mahardika, M., Sugiarti, E., Sapuan, S. M., Ilyas, R. A., & Stephane, I. (2019). A simple method for improving the properties of the sago starch films prepared by using ultrasonication treatment. *Food Hydrocolloids*,

93(January), 276–283. https://doi.org/10.1016/j.foodhyd.2019.02.012

Abral, H., Kadriadi, Mahardika, M., Handayani, D., Sugiarti, E., & Muslimin, A. N. (2019). Characterization of disintegrated bacterial cellulose nanofibers/PVA bionanocomposites prepared via ultrasonication. *International Journal of Biological Macromolecules*, *135*, 591–599. https://doi.org/https://doi.org/10.1016/j.ijbiomac.2019.05.178

Abral, H., Pratama, A. B., Handayani, D., Mahardika, M., Aminah, I., Sandrawati, N., Sugiarti, E., Muslimin, A. N., Sapuan, S. M., & Ilyas, R. A. (2021). Antimicrobial edible film prepared from bacterial cellulose nanofibers/starch/chitosan for a food packaging alternative. *International Journal of Polymer Science*, *2021*. https://doi.org/10.1155/2021/6641284

Andrady, A. L. (2017). The plastic in microplastics: A review. *Marine Pollution Bulletin*, *119*(1), 12–22. https://doi.org/https://doi.org/10.1016/j.marpolbul.2017.01.082

Andriani, D., Apriyana, A. Y., & Karina, M. (2020). The optimization of bacterial cellulose production and its applications: a review. *Cellulose*, *27*(12), 6747–6766. https://doi.org/10.1007/s10570-020-03273-9

Armynah, B., Anugrahwidya, R., & Tahir, D. (2022). Composite cassava starch/chitosan/Pineapple Leaf Fiber (PALF)/Zinc Oxide (ZnO): Bioplastics with high mechanical properties and faster degradation in soil and seawater. *International Journal of Biological Macromolecules*, *213*, 814–823. https://doi.org/https://doi.org/10.1016/j.ijbiomac.2022.06.038

Aurelio, R.-H., Mextlisol, C.-V. S., Páramo-Calderón, D. E., Acevedo-Gómez, R., Gerardo, G.-G., Nolasco-Hipolito, C., Eduardo, B.-G. J., Carlos, C.-A. J., & Alejandro, A.-S. (2023). Functionality and characterization of modified starch films with pineapple leaf fibers. *International Journal of Biological Macromolecules*, *246*, 125611. https://doi.org/https://doi.org/10.1016/j.ijbiomac.2023.125611

Babaei-Ghazvini, A., Vafakish, B., Patel, R., Falua, K. J., Dunlop, M. J., & Acharya, B. (2024). Cellulose nanocrystals in the development of biodegradable materials: A review on CNC resources, modification, and their hybridization. *International Journal of Biological Macromolecules*, *258*, 128834. https://doi.org/https://doi.org/10.1016/j.ijbiomac.2023.128834

Bai, Z., Zeng, W., Deng, J., Zhou, S., Yang, C., Jin, T., & Wei, H. (2024). Piezoresistive sensors based on bacterial cellulose/ZnO/PPy composite materials for human signal monitoring and sound detection. *Sensors and Actuators A: Physical*, *377*(July), 115714. https://doi.org/10.1016/j.sna.2024.115714

Baigh, T. A., Nanzeeba, F., Hamim, H. R., & Habib, M. A. (2023). A comprehensive study on the effect of hybridization and stacking sequence in fabricating cotton-blended jute and pineapple leaf fibre biocomposites. *Heliyon*, *9*(9), e19792. https://doi.org/https://doi.org/10.1016/j.heliyon.2023.e19792

Bledzki, A. K., Franciszczak, P., Osman, Z., & Elbadawi, M. (2015). Polypropylene biocomposites reinforced with softwood, abaca, jute, and kenaf fibers. *Industrial Crops and Products*, *70*, 91–99. https://doi.org/https://doi.org/10.1016/j.indcrop.2015.03.013

Carvalho, T., Guedes, G., Sousa, F. L., Freire, C. S. R., & Santos, H. A. (2019). Latest Advances on Bacterial Cellulose-Based Materials for Wound Healing, Delivery Systems, and Tissue Engineering. *Biotechnology Journal*, *14*(12), 1900059. https://doi.org/https://doi.org/10.1002/biot.201900059

Cazón, P., & Vázquez, M. (2021). Improving bacterial cellulose films by ex-situ and in-situ modifications: A review. *Food Hydrocolloids*, *113*, 106514. https://doi.org/https://doi.org/10.1016/j.foodhyd.2020.106514

Cho, M. J., & Park, B. D. (2011). Tensile and thermal properties of nanocellulose-reinforced poly(vinyl alcohol) nanocomposites. *Journal of Industrial and Engineering Chemistry*, *17*(1), 36–40. https://doi.org/10.1016/j.jiec.2010.10.006

Choi, S. M., Rao, K. M., Zo, S. M., Shin, E. J., & Han, S. S. (2022). Bacterial Cellulose and Its Applications. In *Polymers* (Vol. 14, Issue 6). https://doi.org/10.3390/polym14061080

Dang, X., Li, N., Yu, Z., Ji, X., Yang, M., & Wang, X. (2024). Advances in the preparation and application of cellulose-based antimicrobial materials: A review. *Carbohydrate Polymers*, *342*, 122385. https://doi.org/https://doi.org/10.1016/j.carbpol.2024.122385

Esa, F., Tasirin, S. M., & Rahman, N. A. (2014). Overview of Bacterial Cellulose Production and Application. *Agriculture and Agricultural Science Procedia*, *2*, 113–119. https://doi.org/10.1016/j.aaspro.2014.11.017

Fatima, A., Ortiz-Albo, P., Neves, L. A., Nascimento, F. X., & Crespo, J. G. (2023). Biosynthesis and characterization of bacterial cellulose membranes presenting relevant characteristics for air/gas filtration. *Journal of Membrane Science*, *674*(March). https://doi.org/10.1016/j.memsci.2023.121509

Galiano, F., Briceño, K., Marino, T., Molino, A., Christensen, K. V., & Figoli, A. (2018). Advances in biopolymer-based membrane preparation and applications. *Journal of Membrane Science*, *564*, 562–586. https://doi.org/https://doi.org/10.1016/j.memsci.2018.07.059

Gao, H., Sun, Q., Han, Z., Li, J., Liao, B., Hu, L., Huang, J., Zou, C., Jia, C., Huang, J., Chang, Z., Jiang, D., & Jin, M. (2020). Comparison of bacterial nanocellulose produced by different strains under static and agitated culture conditions. *Carbohydrate Polymers*, *227*, 115323. https://doi.org/https://doi.org/10.1016/j.carbpol.2019.115323

Girard, V.-D., Chaussé, J., & Vermette, P. (2024). Bacterial cellulose: A comprehensive review. *Journal of Applied Polymer Science*, *141*(15), e55163. https://doi.org/https://doi.org/10.1002/app.55163

Goswami, R., Singh, S., Pavithra, N., Ramamurthy, P. C., Mishra, A., Mishra, P. K., Joshi, H. C., Pant, G., Singh, J., Kumar, G., Khan, N. A., & Yousefi, M. (2023). Nanocellulose: A comprehensive review investigating its potential as an innovative material for water remediation. *International Journal of Biological Macromolecules*, *254*(P3), 127465. https://doi.org/10.1016/j.ijbiomac.2023.127465

Halib, N., Amin, M. C. I. M., & Ahmad, I. (2012). Physicochemical Properties and Characterization of Nata de Coco from Local Food Industries as a Source of Cellulose. *Sains Malaysiana*, *41*(2), 205–211.

Helenius, G., Bäckdahl, H., Bodin, A., Nannmark, U., Gatenholm, P., & Risberg, B. (2006). In vivo biocompatibility of bacterial cellulose. *Journal of Biomedical Materials Research Part A*, *76A*(2), 431–438. https://doi.org/https://doi.org/10.1002/jbm.a.30570

Iguchi, M., Yamanaka, S., & Budhiono, A. (2000). Bacterial cellulose—a masterpiece of nature's arts. *Journal of Materials Science*, *35*(2), 261–270. https://doi.org/10.1023/A:1004775229149

Ilyas, R. A., Hamid, N. H. A., Ishak, K. A., Norrrahim, M. N. F., Thiagamani, S. M. K., Rangappa, S. M., Siengchin, S., Bangar, S. P., & Nurazzi, N. M. (2023). 16 - Advanced applications of biomass nanocellulose-reinforced polymer composites. In N. M. Nurazzi, R. A. Ilyas, S. M. Sapuan, & A. B. T.-S. and N. N. in P. C. Khalina (Eds.), *Woodhead Publishing Series in Composites Science and Engineering* (pp. 349–385). Woodhead Publishing. https://doi.org/https://doi.org/10.1016/B978-0-443-19053-7.00013-5

Ilyas, R. A., Sapuan, S. M., Ishak, M. R., & Zainudin, E. S. (2019). Sugar palm nanofibrillated cellulose (Arenga pinnata (Wurmb.) Merr): Effect of cycles on their yield, physic-chemical, morphological and thermal behavior. *International Journal of Biological Macromolecules*, *123*, 379–388. https://doi.org/10.1016/j.ijbiomac.2018.11.124

Ilyas, R. A., Sapuan, S. M., Ishak, M. R., Zainudin, E. S., & Atikah, M. S. N. (2018). Nanocellulose reinforced starch polymer composites : a review of preparation, properties and application. *5th International Conference on Applied Sciences and Engineering (ICASEA, 2018),*

April, 325–341.

Kadier, A., Ilyas, R. A., Huzaifah, M. R. M., Harihastuti, N., Sapuan, S. M., Harussani, M. M., Azlin, M. N. M., Yuliasni, R., Ibrahim, R., Atikah, M. S. N., Wang, J., Chandrasekhar, K., Amirulislam, M., Sharma, S., Punia, S., Rajasekar, A., Asyraf, M. R. M., & Ishak, M. R. (2021). Use of industrial wastes as sustainable nutrient sources for bacterial cellulose (BC) production: Mechanism, advances, and future perspectives. *Polymers*, *13*(19). https://doi.org/10.3390/polym13193365

Kargarzadeh, H., Sheltami, R. M., Ahmad, I., Abdullah, I., & Dufresne, A. (2015). Cellulose nanocrystal: A promising toughening agent for unsaturated polyester nanocomposite. *Polymer*, *56*, 346–357. https://doi.org/10.1016/j.polymer.2014.11.054

Klemm, D., Cranston, E. D., Fischer, D., Gama, M., Kedzior, S. A., Kralisch, D., Kramer, F., Kondo, T., Lindström, T., Nietzsche, S., Petzold-Welcke, K., & Rauchfuß, F. (2018). Nanocellulose as a natural source for groundbreaking applications in materials science: Today's state. *Materials Today*, *21*(7), 720–748. https://doi.org/10.1016/j.mattod.2018.02.001

Lima, N. F., Maciel, G. M., Lima, N. P., Fernandes, I. de A. A., & Haminiuk, C. W. I. (2024). Bacterial cellulose in cosmetic innovation: A review. *International Journal of Biological Macromolecules*, *xxxx*. https://doi.org/10.1016/j.ijbiomac.2024.133396

Lin, D., Liu, Z., Shen, R., Chen, S., & Yang, X. (2020). Bacterial cellulose in food industry: Current research and future prospects. *International Journal of Biological Macromolecules*, *158*, 1007–1019. https://doi.org/10.1016/j.ijbiomac.2020.04.230

McNamara, J. T., Morgan, J. L. W., & Zimmer, J. (2015). A Molecular Description of Cellulose Biosynthesis. *Annual Review of Biochemistry*, *84*(Volume 84, 2015), 895–921. https://doi.org/https://doi.org/10.1146/annurev-biochem-060614-033930

Mensah, A., Chen, Y., Christopher, N., & Wei, Q. (2022). Membrane Technological Pathways and Inherent Structure of Bacterial Cellulose Composites for Drug Delivery. In *Bioengineering* (Vol. 9, Issue 1). https://doi.org/10.3390/bioengineering9010003

Mohammed, A. A. B. A., Hasan, Z., Borhana, A. A., Elfaghi, A. M., Hassan, Y., Akeel, N. A. A., Ilyas, R. A., & Sapuan, S. M. (2023). Effect of sugar palm fibers on the properties of blended wheat starch / polyvinyl alcohol (PVA) -based biocomposite films. *Journal of Materials Research and Technology*, *24*, 1043–1055. https://doi.org/10.1016/j.jmrt.2023.02.027

Moniri, M., Boroumand Moghaddam, A., Azizi, S., Abdul Rahim, R., Bin

Ariff, A., Zuhainis Saad, W., Navaderi, M., & Mohamad, R. (2017). Production and Status of Bacterial Cellulose in Biomedical Engineering. In *Nanomaterials* (Vol. 7, Issue 9). https://doi.org/10.3390/nano7090257

Muiruri, J. K., Yeo, J. C. C., Zhu, Q., Ye, E., Loh, X. J., & Li, Z. (2023). Bacterial cellulose: Recent advances in biosynthesis, functionalization strategies and emerging applications. *European Polymer Journal*, *199*(September). https://doi.org/10.1016/j.eurpolymj.2023.112446

Navya, P. V., Gayathri, V., Samanta, D., & Sampath, S. (2022). Bacterial cellulose: A promising biopolymer with interesting properties and applications. *International Journal of Biological Macromolecules*, *220*(July), 435–461. https://doi.org/10.1016/j.ijbiomac.2022.08.056

Nazrin, A., Sapuan, S. M., & Zuhri, M. Y. M. (2020). Mechanical, physical and thermal properties of sugar palm nanocellulose reinforced thermoplastic starch (Tps)/poly (lactic acid) (pla) blend bionanocomposites. *Polymers*, *12*(10), 1–18. https://doi.org/10.3390/polym12102216

Perera, K. Y., Pradhan, D., Rafferty, A., Jaiswal, A. K., & Jaiswal, S. (2023). A comprehensive review on metal oxide-nanocellulose composites in sustainable active and intelligent food packaging. *Food Chemistry Advances*, *3*(June), 100436. https://doi.org/10.1016/j.focha.2023.100436

Pirsa, S., Shamusi, T., & Kia, E. M. (2018). Smart films based on bacterial cellulose nanofibers modified by conductive polypyrrole and zinc oxide nanoparticles. *Journal of Applied Polymer Science*, *135*(34), 46617. https://doi.org/https://doi.org/10.1002/app.46617

Popa, L., Ghica, M. V., Tudoroiu, E.-E., Ionescu, D.-G., & Dinu-Pîrvu, C.-E. (2022). Bacterial Cellulose-A Remarkable Polymer as a Source for Biomaterials Tailoring. *Materials (Basel, Switzerland)*, *15*(3). https://doi.org/10.3390/ma15031054

Rana, A. K., Guleria, S., Gupta, V. K., & Thakur, V. K. (2023). Cellulosic pine needles-based biorefinery for a circular bioeconomy. *Bioresource Technology*, *367*, 128255. https://doi.org/https://doi.org/10.1016/j.biortech.2022.128255

Ribeiro, R. S. A., Pohlmann, B. C., Calado, V., Bojorge, N., & Pereira Jr, N. (2019). Production of nanocellulose by enzymatic hydrolysis: Trends and challenges. *Engineering in Life Sciences*, *19*(4), 279–291. https://doi.org/https://doi.org/10.1002/elsc.201800158

Riseh, R. S., Vazvani, M. G., Hassanisaadi, M., & Thakur, V. K. (2024). Agricultural wastes: A practical and potential source for the isolation and preparation of cellulose and application in agriculture and different industries. *Industrial Crops and Products*, *208*, 117904. https://doi.org/https://doi.org/10.1016/j.indcrop.2023.117904

Saleh, A. K., El-Gendi, H., Soliman, N. A., El-Zawawy, W. K., & Abdel-Fattah, Y. R. (2022). Bioprocess development for bacterial cellulose biosynthesis by novel Lactiplantibacillus plantarum isolate along with characterization and antimicrobial assessment of fabricated membrane. *Scientific Reports*, *12*(1), 2181. https://doi.org/10.1038/s41598-022-06117-7

Saleh, A. K., Ray, J. B., El-Sayed, M. H., Alalawy, A. I., Omer, N., Abdelaziz, M. A., & Abouzeid, R. (2024). Functionalization of bacterial cellulose: Exploring diverse applications and biomedical innovations: A review. *International Journal of Biological Macromolecules*, *264*(January). https://doi.org/10.1016/j.ijbiomac.2024.130454

Song, D., Liu, W., Liu, C., & Li, H. (2024). Recent progress of bacterial cellulose-based separator platform for lithium-ion and lithium-sulfur batteries. *International Journal of Biological Macromolecules*, *274*, 133419. https://doi.org/https://doi.org/10.1016/j.ijbiomac.2024.133419

Syafri, E., Jamaluddin, Harmailis, Umar, S., Mahardika, M., Amelia, D., Mayerni, R., Mavinkere Rangappa, S., Siengchin, S., Sobahi, T. R., Khan, A., & Asiri, A. M. (2022). Isolation and Characterization of New Cellulosic Microfibers from Pandan Duri (Pandanus Tectorius) for Sustainable Environment. *Journal of Natural Fibers*, *19*(16), 12924–12934. https://doi.org/10.1080/15440478.2022.2079582

Tang, Z., Lin, X., Yu, M., Mondal, A. K., & Wu, H. (2024). Recent advances in TEMPO-oxidized cellulose nanofibers: Oxidation mechanism, characterization, properties and applications. *International Journal of Biological Macromolecules*, *259*, 129081. https://doi.org/https://doi.org/10.1016/j.ijbiomac.2023.129081

Torres, F. G., Arroyo, J. J., & Troncoso, O. P. (2019). Bacterial cellulose nanocomposites: An all-nano type of material. *Materials Science and Engineering C*, *98*(May 2018), 1277–1293. https://doi.org/10.1016/j.msec.2019.01.064

Ullah, M. W., Alabbosh, K. F., Fatima, A., Islam, S. U., Manan, S., Ul-Islam, M., & Yang, G. (2023). Advanced biotechnological applications of bacterial nanocellulose-based biopolymer nanohybrids: A review. *Advanced Industrial and Engineering Polymer Research*, *xxxx*. https://doi.org/10.1016/j.aiepr.2023.07.004

Vasconcelos, N. F., Feitosa, J. P. A., da Gama, F. M. P., Morais, J. P. S., Andrade, F. K., de Souza Filho, M. de S. M., & Rosa, M. de F. (2017). Bacterial cellulose nanocrystals produced under different hydrolysis conditions: Properties and morphological features. *Carbohydrate*

Polymers, *155*, 425–431. https://doi.org/https://doi.org/10.1016/j.carbpol.2016.08.090

Wang, J., Tavakoli, J., & Tang, Y. (2019). Bacterial cellulose production, properties and applications with different culture methods – A review. *Carbohydrate Polymers*, *219*, 63–76. https://doi.org/https://doi.org/10.1016/j.carbpol.2019.05.008

Yang, H., Dong, H., Huang, Y., Chen, G., & Wang, J. (2022). Interactions of microplastics and main pollutants and environmental behavior in soils. *Science of The Total Environment*, *821*, 153511. https://doi.org/https://doi.org/10.1016/j.scitotenv.2022.153511

Yang, W., Ding, H., Qi, G., Li, C., Xu, P., Zheng, T., Zhu, X., Kenny, J. M., Puglia, D., & Ma, P. (2021). Highly transparent PVA/nanolignin composite films with excellent UV shielding, antibacterial and antioxidant performance. *Reactive and Functional Polymers*, *162* (November 2020), 104873. https://doi.org/10.1016/j.reactfunctpolym.2021.104873

Yusuf, A. A., Ampah, J. D., Veza, I., Atabani, A. E., Hoang, A. T., Nippae, A., Powoe, M. T., Afrane, S., Yusuf, D. A., & Yahuza, I. (2023). Investigating the influence of plastic waste oils and acetone blends on diesel engine combustion, pollutants, morphological and size particles: Dehalogenation and catalytic pyrolysis of plastic waste. *Energy Conversion and Management*, *291*, 117312. https://doi.org/https://doi.org/10.1016/j.enconman.2023.117312

Yusuf, A. A., Dankwa Ampah, J., Soudagar, M. E. M., Veza, I., Kingsley, U., Afrane, S., Jin, C., Liu, H., Elfasakhany, A., & Buyondo, K. A. (2022). Effects of hybrid nanoparticle additives in n-butanol/waste plastic oil/diesel blends on combustion, particulate and gaseous emissions from diesel engine evaluated with entropy-weighted PROMETHEE II and TOPSIS: Environmental and health risks of plastic wa. *Energy Conversion and Management*, *264*(March). https://doi.org/10.1016/j.enconman.2022.115758

Zennifer, A., Senthilvelan, P., Sethuraman, S., & Sundaramurthi, D. (2021). Key advances of carboxymethyl cellulose in tissue engineering & 3D bioprinting applications. *Carbohydrate Polymers*, *256*, 117561. https://doi.org/https://doi.org/10.1016/j.carbpol.2020.117561

Zhu, C., Zhang, J., Qiu, S., Jia, Y., Wang, L., & Wang, H. (2021). Tailoring the pore size of polyphenylene sulfide nonwoven with bacterial cellulose (BC) for heat-resistant and high-wettability separator in lithium-ion battery. *Composites Communications*, *24*, 100659. https://doi.org/https://doi.org/10.1016/j.coco.2021.100659

Zhu, J., Yang, Q., Tao, S., Zhou, W., Su, J., Zhang, N., Li, Y., Xu, L., Pan, H., Zhang, H., & Wang, J. (2024). Fabrication and characterization of bacterial cellulose/carbon nanotube composite conductive film and its impact on the luminance of flexible electroluminescent devices as the bottom electrode. *Materials Science and Engineering: B*, *308*(May), 117572. https://doi.org/10.1016/j.mseb.2024.117572

Zhu, Z., Wang, W., Wang, X., Zhao, X., Xia, N., Kong, F., & Wang, S. (2021). *Study on the Preparation of Nanocellulose Powder and Its Formation Mechanism*. 1–18.

CHAPTER IX

GINGER NANOCELLULOSE AS A PROMISING BIO REINFORCEMENT IN BIOCOMPOSITES

MUHAMMAD ADLAN AZKA, HAIRUL ABRAL, MOHD SAPUAN SALIT, DIAN HANDAYANI

Abstract

Researchers are now prioritizing the development of biocomposite materials due to increasing environmental concerns regarding waste management and plastic pollution. This is because biocomposites are renewable and biodegradable. Biocomposites consist of two natural-derived materials, which are bioreinforcement and biomatrix. Plant fiber is a naturally derived reinforcement material with abundant availability that is renewable, sustainable, and biodegradable. There are many types of natural fiber, and one of them is ginger. This chapter provides an overview of ginger, including its bioactive compounds and its various applications. Furthermore, plant fiber can be produced into nanocellulose, which is intriguing due to the remarkable properties of the nanocellulose. To promote the use of ginger, this chapter explores the process of producing nanocellulose from ginger and its potential application as a bioreinforcement in biocomposites. This chapter provides researchers with information about ginger, including ginger nanocellulose production and ginger nanocellulose biocomposites development, in order to overcome waste disposal management and plastic pollution.

1. Introduction

The scientific and industrial communities are currently focusing on renewable and biodegradable materials due to the ecological risks posed by conventional petroleum-based polymer products, including global warming and plastic pollution (Kargarzadeh et al., 2017). According to Horton (2022), in 2016 more than 23 million tons of plastic entered aquatic ecosystems. This plastic pollution affects not only the environment, but also the economy. Lau et al. (2020) reported that the minimum estimate of the economic consequences of plastic pollution on fishing, tourism, and shipping expenses amounts to $13 billion per year. This problem can be solved by using biodegradable materials (Azkaet al., 2024). Biocomposites are an example of a biodegradable material. Biocomposites offer numerous environmental benefits. Biocomposites have the ability to decrease the environmental consequences of extracting and processing fossil fuels by reducing the dependence on non-renewable resources like polymers derived from petroleum (Azka et al., 2024).

Biocomposites can be classified as biodegradable materials because biocomposites contain natural-derived reinforcement and renewable polymers. Researchers have shown interest in natural fibers as a material for bioreinforcement because of their outstanding properties, including their low density, affordability, widespread availability, ability to decompose naturally, and ease of handling. Additionally, they exhibit significant mechanical, thermal, and exceptional acoustic characteristics, along with a high level of resistance to fractures (Vinod et al., 2020). Natural fibers are divided into three different types, which are animal fibers, plant fibers, and mineral fibers (Jino et al., 2023; Rajeshkumar et al., 2021), as shown in Figure 1. Plant fibers are mainly composed of cellulose (Gholampour & Ozbakkaloglu, 2020). Cellulose can be considered one of the world's most illimitable renewable green materials (Gan et al., 2020). According to Pandey et al. (2010), approximately 2 x 1011 tons of lignocellulosics biomass, including agricultural crops, animal, forest resources, and municipal wastes, are produced annually. The cellulose content in natural fibers can influence their properties due to the presence of hydrogen bonding in cellulose, microfibril angles, and the degree of cellulose polymerization (Siakeng et al., 2019). Furthermore, cellulose can be extracted or reduced into nanocellulose.

Nanocellulose can be extracted from plant fiber with several methods which are the chemical method (acid hydrolysis), and mechanical method (Phanthong et al., 2018). The main reason for the increasing interest in nanocellulose is its highly uniform nature, enhanced mechanical properties, and its biodegradable nature, which makes it a sustainable material (Nasir et al., 2017). Furthermore, this material is very useful as a bioreinforcement in biocomposites. Biocomposite materials fabricated utilizing nanocellulose consistently exhibit distinctive behavior, including exceptional mechanical behavior, elevated thermal properties, and a combination of transparency and lightweight (Phanthong et al., 2018). According to Dufresne (2013), the use of nanocellulose as bioreinforcement have attracted significant interest over the past 20 years. Hence, the high population of ginger as the source of natural fiber has attracted research to develop ginger into nanocellulose and reinforce it with biopolymer to produce bio composites, which are biodegradable, renewable, and environmentally friendly.

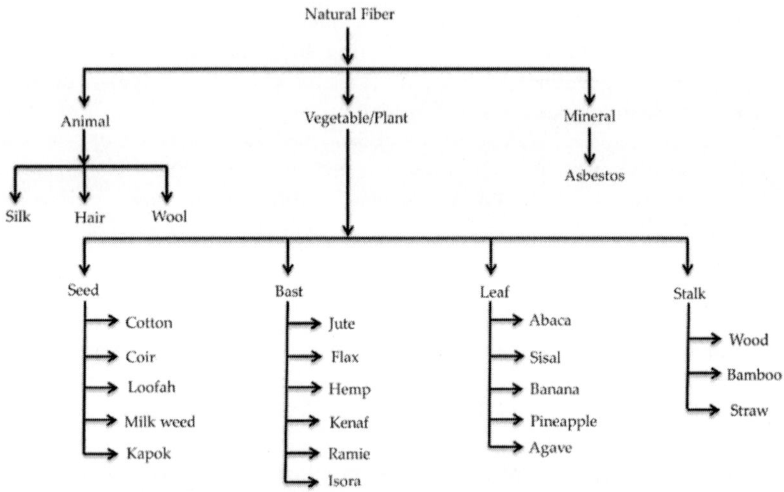

Figure 1. Natural fiber classification (Hasan et al., 2020).

Therefore, this chapter focuses on an overview of ginger, including its bioactive compounds, its applications, and how this plant can contribute to human life. Additionally, this chapter explores the potential of producing ginger nanocellulose and combining it with a biopolymer matrix to create renewable and biodegradable ginger nanocellulose-reinforced biocomposites.

2. Overview of Ginger

Ginger, scientifically known as Zingiber officinale Roscoe, belongs to the Zingiberaceae family of plants (Ballester et al., 2022). According to Wang et al. (2020), approximately 500,000 hectares of ginger are growing around the world.In a study by Ersedo et al. (2023), it was stated that the leading ginger-producing countries were Thailand (1.61 × 105 tons), Nepal (2.76 × 105 tons), Indonesia (2.66 × 105 tons), China (4.51 × 105 tons), and India (6.55 × 105 tons). This high population of ginger enables its use in various areas. Kiyama (2020) stated that ginger has been utilized in various countries for over 3000 years for its culinary, aromatic, nutritional, and medicinal properties, owing to its advantageous attributes such as aroma, pungency, nutrients, and pharmacological activity.

2.1. Ginger bioactive compound

The ginger rhizome possesses numerous bioactive compounds present in its extracts, which possess antibacterial, antioxidant, and anti-aging properties (Asoka et al., 2022). A multitude of bioactive compounds found in ginger have been identified, including lipids, carbohydrates, phenolics, and terpene compounds (Ballester et al., 2022; Mao et al., 2019). The primary phenolic compounds found in ginger are gingerol, shogaol, zingerone, and paradol. These compounds are responsible for the diverse range of biological activities exhibited by ginger, as depicted in Figure 2 (Ballester et al., 2022). Ginger contains various terpene compounds such as geranium, zingiberene, farnesene, bisabolene, sesquifellandrene, linalool, limonene, cineol, borneol, and cucumene (Utami & Abdullah Qarrah, 2022). Ginger is not only composed of this compound but also contains amino acids, ash, raw fiber, protein, vitamins such as vitamin A and nicotinic acid, phytosterols, and minerals (Prasad & Tyagi, 2015).

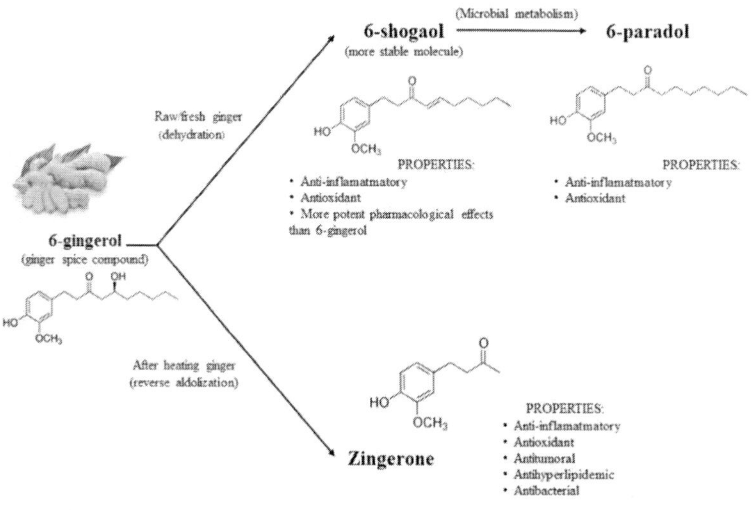

Figure 2. Properties and structure of the 4 main compound of ginger (Ballester et al., 2022).

2.2. Application of ginger

Ginger is extremely useful and valuable in human life nowadays, as shown in Figure 3. In terms of medication, ginger can be used to cure many diseases such as stomach ache, heart conditions, rheumatic issues, diarrhoea, nausea, respiratory disorders, cholera, tooth ache, asthma, motion sickness, and migraines (Vasala, 2012).

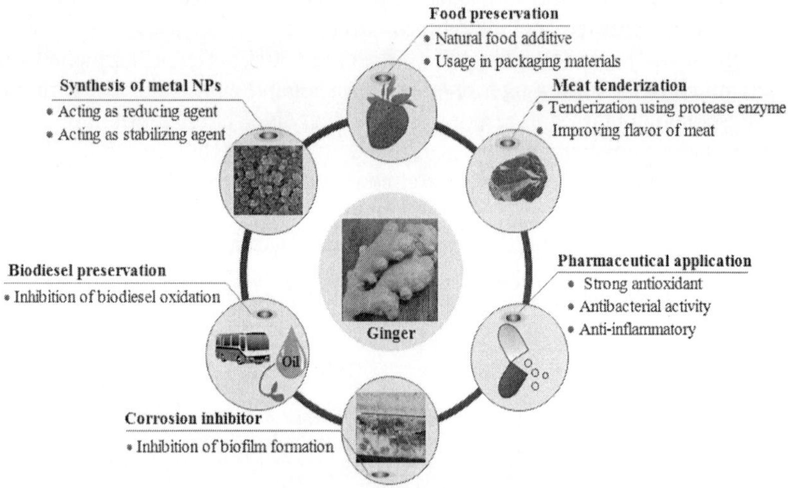

Figure 3. Main overall applications of ginger rhizomes (Akhlaghi & Najafpour-Darzi, 2023).

Ginger's medical benefits come from its terpene compounds, which have many different pharmacological properties, such as fighting cancer, reducing inflammation, fighting viruses and bacteria, lowering blood sugar, relieving pain, protecting the digestive system, and neuroprotective effects (Utami & Abdullah Qarrah, 2022).

Not only in the medical area, ginger has also benefited the food industry. According to Yang et al. (2024) ginger possesses numerous nutritional and functional characteristics that show potential for application in the food industry. Ginger can be utilized as a fortifying agent in the development of functional foods, enhancing their functional characteristics and nutrition. Ginger can be used in food preservation as an antimicrobial and antioxidant agent to prolong the durability of food products. In addition, ginger can serve as both an active ingredient and film matrix in food packaging and edible coatings.

Besides these applications, ginger can also be used as a promising green reinforcement in biocomposites. Ginger can be extracted into nanocellulose and reinforced with biopolymer, as explained in the next section. These biocomposites have various applications in the areas of water treatment, biomedical science, drug delivery systems, and tissue engineering (Jacob et al., 2019b).

3. Production of ginger nanocellulose

There are several studies that have extracted ginger into nanocellulose and one of these is a study by Ariksa et al. (2022), who conducted research on ginger nanocellulose. In this research, ginger fiber was immersed into 2:1 toluene and ethanol mixture then heated with hot plate stirrer at 500 rpm and 50 C for 48 h and then rinsed and filtered until it reached pH 7 and was then dried in an oven. After the fiber was dried, it was smoothed with a blender and then alkalinize was put into it with 5% sodium hydroxide in a hot plate magnetic stirrer at 50 C for 4 h and neutralized pH and then dried in an oven. After this alkalization process, the fiber underwent bleaching treatment with 4:1 ratio of sodium chloride and acetic acid at 500 rpm and 60 C for 2 h and then rinsed until pH 7. Next, the fiber underwent acid hydrolysis with 5M HCl at 50 C for 12 h and was rinsed until pH neutral. Lastly, the fiber was ultrasonicated at 600 W to obtain ginger nanocellulose, as shown in Figure 4.

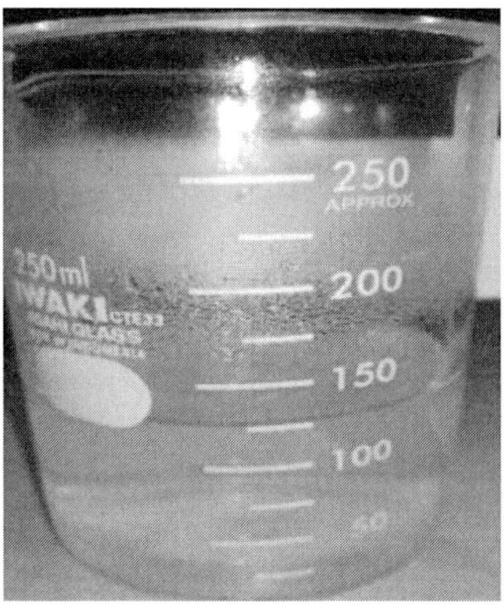

Figure 4. Ginger nanofiber suspension (Ariksa et al., 2022).

Developing ginger nanocellulose is intriguing since the nanocellulose itself can be produced into transparent film. A study conducted by Abral et al. (2020) investigated the utilization of ginger nanocellulose for the production of a transparent to light film. This is illustrated in Figure 5. The process consisted of alkalization, bleaching, acid hydrolysis to extract ginger nanocellulose, and ultrasonication to create a transparent film. Furthermore, the study not only created a transparent film, but also incorporated anti-microbial properties. According to the study, the ginger nanocellulose film can transmit 83.3% of light at a wavelength of 650 nm after being subjected to 6 minutes of ultrasonication and has good antibacterial performance.

4. Ginger nanocellulose in biocomposites

Ginger offers numerous advantages in daily life, especially in the field of medicine. On the other hand, ginger also offers multitude advantages when it is extracted as a biomaterial, such as nanocellulose and then developed into biocomposites. Rahmadiawan et al. (2022) developed nanocellulose derived from white ginger as a reinforcement in polyvinyl alcohol (PVA) for food packaging applications. This study highlights the numerous benefits of using ginger nanocellulose to reinforce PVA. In terms of mechanical strength, using ginger nanocellulose can increase the composites' tensile strength. Not only does the use of this nanocellulose enhance mechanical strength, but it also significantly inhibits the growth of bacteria in the composite samples, as shown in Figure 6.

Another study by Abral et al. (2020) looked at PVA reinforced with ginger nanocellulose composites. In this study, different loadings of ginger nanocellulose were used to reinforce the PVA composites. According to the study, the increased ginger nanocellulose loading in the composites increased the sample's tensile strength. Furthermore, the composites' moisture absorption decreased as the amount of nanocellulose increased. This phenomenon of increased tensile strength and moisture absorption occurred due to the formation of strong hydrogen bonds between the nanocellulose and the PVA matrix. Additionally, increasing the amount of ginger nanocellulose in the composites enhanced their antimicrobial performance. This is due to the presence of monoterpenoid, sesquiterpenoid, and phenolic compounds in ginger, as well as derivatives such as aldehydes, ketones, alcohol, and esters. These compounds demonstrate a broad spectrum of antimicrobial activity against various microorganisms. This behavior of ginger nanocellulose-reinforced PVA composites makes them potential biocomposites that can replace synthetic food packaging.

Figure 5. Drying of the nano-sized fiber suspension resulted in an optically transparent film showing antimicrobial activity (Abral et al., 2020).

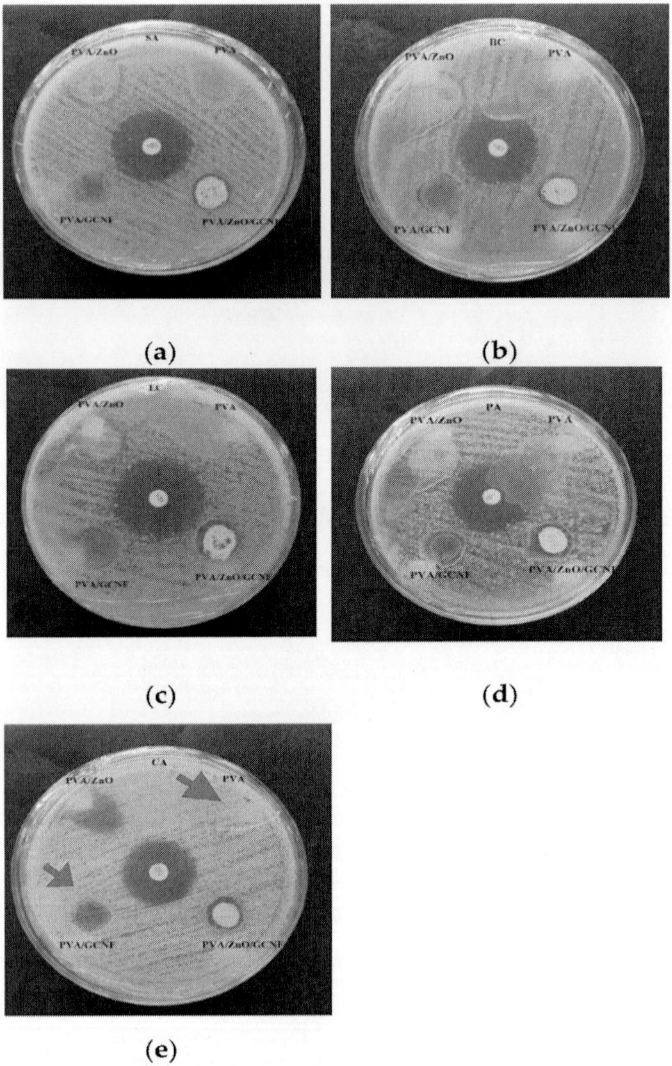

Figure 6. Antimicrobial activity of PVA and biocomposite film against gram-positive SA (a), BS (b), PA (c), EC (d), and CA (e). The more transparent area (blue and red arrows) is the inhibition zone diameter (Rahmadiawan et al., 2022).

Jacob et al. (2018) developed ginger nanocellulose as a reinforcement for potato and tapioca starch biocomposites. To make ginger nanocellulose in this study, the ginger was extracted by alkalizing, bleaching, acid hydrolysis, and homogenizing it. Once the nanocellulose was made, potato and tapioca starch were used to make biocomposites. The study stated that the use of this nanocellulose increased the tensile strength of the biocomposites due to decreases in molecular mobility in the biocomposites. Additionally, the use of ginger nanocellulose also increases the anti-bacterial performance of the biocomposites, as shown in Table 1.

Table 1. Antibacterial activity of the GNF and its PS and TS bionanocomposites, with 5% GNF, in comparison with the control (Jacob et al., 2018).

Samples	Minimum Inhibitory Concentration (µg/mL)			
	Bacillus cereus	***Escherichia coli***	***Staphylococcus aureus***	***Salmonella typhimurium***
GNF	14 ± 2	13 ± 1	18 ± 0	31 ± 0
GNF–PS	9.1 ± 1	8.1 ± 1	15 ± 1	24 ± 2
GNF–TS	8.3 ± 1	3.1 ± 0	12 ± 2	19 ± 1
Positive Control	1.6 ± 0	0.4 ± 0	6 ± 0	12 ± 1

Jacob et al. (2019a) developed an environmentally friendly composite that was ginger nanocellulose reinforced with both PVA and chitosan composites. The study stated that using 5% loading of ginger nanocellulose to reinforce the composites increased the tensile strength and modulus. Moreover, the incorporation of this nanocellulose enhances the composites' antibacterial performance. The composites' beneficial behavior makes them applicable in many areas, such as food packaging or medical areas.

5. Conclusion

This chapter covers an overview of ginger, including its bioactive compounds and its various applications. Furthermore, this chapter also delved into the development of biocomposites to overcome waste disposal and plastic pollution issues, with an especially strong focus on the production of ginger nanocelluse and ginger nanocellulose-reinforced biocomposites.

Ginger contains many bioactive compounds that possess antibacterial, antioxidant, and anti-aging properties. Because of its remarkable characteristics, this bioactive compound makes ginger highly desirable in medicine. Moreover, ginger finds applications in various other sectors, including the food industry, and it becomes particularly intriguing when it is extracted into nanocellulose and used as a bioreinforcement in biocomposites. Ginger nanocellulose can be produced by immersing it in a toluene-ethanol mixture, followed by alkalization, bleaching, and acid hydrolysis. The ultrasonication process can also be done in nanocellulose to make a highly transparent film, which is intriguing. When ginger nanocellulose is used as bioreinforcement, several studies show that adding nanocellulose to biocomposites could make them more effective against microbes because ginger contains a bioactive compound that has antibacterial properties. In addition to its anti-microbial properties, the incorporation of ginger nanocellulose as a reinforcement in biocomposites also enhances their tensile strength by reducing molecular mobility and forming strong hydrogen bonds with the polymer matrix. Furthermore, these robust hydrogen bonds significantly impacted the biocomposites' ability to absorb moisture, leading to a reduction in the composites' moisture absorption rate. This beneficial characteristic makes these biocomposites possible as renewable and biodegradable biocomposites, which can overcome waste disposal and plastic pollution issues.

References

Abral, H., Ariksa, J., Mahardika, M., Handayani, D., Aminah, I., Sandrawati, N., Pratama, A. B., Fajri, N., Sapuan, S. M., & Ilyas, R. A. (2020). Transparent and antimicrobial cellulose film from ginger nanofiber. *Food Hydrocolloids*, *98*, 105266. https://doi.org/10.1016/j.foodhyd.2019.105266

Abral, H., Ariksa, J., Mahardika, M., Handayani, D., Aminah, I., Sandrawati, N., Sapuan, S. M., & Ilyas, R. A. (2020). Highly transparent and antimicrobial PVA based bionanocomposites reinforced by ginger nanofiber. *Polymer Testing*, *81*, 106186. https://doi.org/10.1016/j.polymertesting.2019.106186

Akhlaghi, N., & Najafpour-Darzi, G. (2023). Potential Applications of Ginger Rhizomes as a Green Biomaterial: A Review. *International Journal of Engineering*, *36*(2), 372–383. https://doi.org/10.5829/IJE.2023.36.02B.16

Ariksa, J., Suhdi, Setiawan, Y., Saparin, Wijianti, E. S., Wibowo, B. S., & Yandi, W. (2022). Characterization biodegradable film of ginger nanofiber renvorced PVA bionanocomposites. *IOP Conference Series: Earth and Environmental Science*, *1108*(1), 012063.

https://doi.org/10.1088/1755-1315/1108/1/012063

Asoka, S. F., Batubara, I., Lestari, A. R., Wahyuni, W. T., & Wahyudi, S. T. (2022). Compounds in Indonesian Ginger Rhizome Extracts and Their Potential for Anti-Skin Aging Based on Molecular Docking. *Cosmetics*, *9*(6), 128. https://doi.org/10.3390/cosmetics9060128

Azka, M. A., Adam, A., Ridzuan, S. M., Sapuan, S. M., & Habib, A. (2024). A review on the enhancement of circular economy aspects focusing on nanocellulose composites. *International Journal of Biological Macromolecules*, *269*, 132052. https://doi.org/10.1016/j.ijbiomac.2024.132052

Azka, M. A., Sapuan, S. M., Abral, H., Zainudin, E. S., & Aziz, F. A. (2024). An examination of recent research of water absorption behavior of natural fiber reinforced polylactic acid (PLA) composites: A review. *International Journal of Biological Macromolecules*, 131845. https://doi.org/10.1016/j.ijbiomac.2024.131845

Ballester, P., Cerdá, B., Arcusa, R., Marhuenda, J., Yamedjeu, K., & Zafrilla, P. (2022). Effect of Ginger on Inflammatory Diseases. *Molecules*, *27*(21), 7223. https://doi.org/10.3390/molecules27217223

Dufresne, A. (2013). Nanocellulose: a new ageless bionanomaterial. *Materials Today*, *16*(6), 220–227. https://doi.org/10.1016/j.mattod.2013.06.004

Gan, P. G., Sam, S. T., Abdullah, M. F. bin, & Omar, M. F. (2020). Thermal properties of nanocellulose-reinforced composites: A review. *Journal of Applied Polymer Science*, *137*(11). https://doi.org/10.1002/app.48544

Gholampour, A., & Ozbakkaloglu, T. (2020). A review of natural fiber composites: properties, modification and processing techniques, characterization, applications. *Journal of Materials Science*, *55*(3), 829–892. https://doi.org/10.1007/s10853-019-03990-y

Hasan, K. M. F., Horváth, P. G., & Alpár, T. (2020). Potential Natural Fiber Polymeric Nanobiocomposites: A Review. *Polymers*, *12*(5), 1072. https://doi.org/10.3390/polym12051072

Horton, A. A. (2022). Plastic pollution: When do we know enough? *Journal of Hazardous Materials*, *422*, 126885. https://doi.org/10.1016/j.jhazmat.2021.126885

Jacob, J., Haponiuk, J., Thomas, S., Peter, G., & Gopi, S. (2018). Use of Ginger Nanofibers for the Preparation of Cellulose Nanocomposites and Their Antimicrobial Activities. *Fibers*, *6*(4), 79. https://doi.org/10.3390/fib6040079

Jacob, J., Peter, G., Thomas, S., Haponiuk, J. T., & Gopi, S. (2019a). Chitosan and polyvinyl alcohol nanocomposites with cellulose nanofibers from ginger rhizomes and its antimicrobial activities. *International*

Journal of Biological Macromolecules, *129*, 370–376. https://doi.org/10.1016/j.ijbiomac.2019.02.052

Jacob, J., Peter, G., Thomas, S., Haponiuk, J. T., & Gopi, S. (2019b). In-situ synthesis and characterization of biocompatible magnetic ginger nanofiber composites for copper (II) removal from water. *Materials Today Communications*, *21*, 100690. https://doi.org/10.1016/j.mtcomm.2019.100690

Jino, L., Dev Prasad, V., Ajay Eswar, M., Manoj, E., Jacob, A., Arockia Suthan, S., Jayaganthan, A., & Anderson, A. (2023). Review on natural fibre composites reinforced with nanoparticles. *Materials Today: Proceedings*, *xxxx*, 1–6. https://doi.org/10.1016/j.matpr.2023.01.126

Kargarzadeh, H., Mariano, M., Huang, J., Lin, N., Ahmad, I., Dufresne, A., & Thomas, S. (2017). Recent developments on nanocellulose reinforced polymer nanocomposites: A review. *Polymer*, *132*, 368–393. https://doi.org/10.1016/j.polymer.2017.09.043

Kiyama, R. (2020). Nutritional implications of ginger: chemistry, biological activities and signaling pathways. *The Journal of Nutritional Biochemistry*, *86*, 108486. https://doi.org/10.1016/j.jnutbio.2020.108486

Laelago Ersedo, T., Teka, T. A., Fikreyesus Forsido, S., Dessalegn, E., Adebo, J. A., Tamiru, M., & Astatkie, T. (2023). Food flavor enhancement, preservation, and bio-functionality of ginger (Zingiber officinale): a review. *International Journal of Food Properties*, *26*(1), 928–951. https://doi.org/10.1080/10942912.2023.2194576

Lau, W. W. Y., Shiran, Y., Bailey, R. M., Cook, E., Stuchtey, M. R., Koskella, J., Velis, C. A., Godfrey, L., Boucher, J., Murphy, M. B., Thompson, R. C., Jankowska, E., Castillo Castillo, A., Pilditch, T. D., Dixon, B., Koerselman, L., Kosior, E., Favoino, E., Gutberlet, J., … Palardy, J. E. (2020). Evaluating scenarios toward zero plastic pollution. *Science*, *369*(6510), 1455–1461. https://doi.org/10.1126/science.aba9475

Mao, Q.-Q., Xu, X.-Y., Cao, S.-Y., Gan, R.-Y., Corke, H., Beta, T., & Li, H.-B. (2019). Bioactive Compounds and Bioactivities of Ginger (Zingiber officinale Roscoe). *Foods*, *8*(6), 185. https://doi.org/10.3390/foods8060185

Nasir, M., Hashim, R., Sulaiman, O., & Asim, M. (2017). Nanocellulose. In *Cellulose-Reinforced Nanofibre Composites* (pp. 261–276). Elsevier. https://doi.org/10.1016/B978-0-08-100957-4.00011-5

Pandey, J. K., Ahn, S. H., Lee, C. S., Mohanty, A. K., & Misra, M. (2010). Recent Advances in the Application of Natural Fiber Based Composites. *Macromolecular Materials and Engineering*, *295*(11), 975–989.

https://doi.org/10.1002/mame.201000095

Phanthong, P., Reubroycharoen, P., Hao, X., Xu, G., Abudula, A., & Guan, G. (2018). Nanocellulose: Extraction and application. *Carbon Resources Conversion*, *1*(1), 32–43. https://doi.org/10.1016/j.crcon.2018.05.004

Prasad, S., & Tyagi, A. K. (2015). Ginger and Its Constituents: Role in Prevention and Treatment of Gastrointestinal Cancer. *Gastroenterology Research and Practice*, *2015*, 1–11. https://doi.org/10.1155/2015/142979

Rahmadiawan, D., Abral, H., Yesa, W. H., Handayani, D., Sandrawati, N., Sugiarti, E., Muslimin, A. N., Sapuan, S. M., & Ilyas, R. A. (2022). White Ginger Nanocellulose as Effective Reinforcement and Antimicrobial Polyvinyl Alcohol/ZnO Hybrid Biocomposite Films Additive for Food Packaging Applications. *Journal of Composites Science*, *6*(10), 316. https://doi.org/10.3390/jcs6100316

Rajeshkumar, G., Arvindh Seshadri, S., Devnani, G. L., Sanjay, M. R., Siengchin, S., Prakash Maran, J., Al-Dhabi, N. A., Karuppiah, P., Mariadhas, V. A., Sivarajasekar, N., & Ronaldo Anuf, A. (2021). Environment friendly, renewable and sustainable poly lactic acid (PLA) based natural fiber reinforced composites – A comprehensive review. *Journal of Cleaner Production*, *310*, 127483. https://doi.org/10.1016/j.jclepro.2021.127483

Siakeng, R., Jawaid, M., Ariffin, H., Sapuan, S. M., Asim, M., & Saba, N. (2019). Natural fiber reinforced polylactic acid composites: A review. *Polymer Composites*, *40*(2), 446–463. https://doi.org/10.1002/pc.24747

Utami, A. T., & Abdullah Qarrah, A. G. (2022). Benefits of Ginger as Medicine for COVID-19: Literature Study. *Journal of Biomedical Research & Environmental Sciences*, *3*(10), 1208–1223. https://doi.org/10.37871/jbres1580

Vasala, P. A. (2012). Ginger. In *Handbook of Herbs and Spices* (pp. 319–335). Elsevier. https://doi.org/10.1533/9780857095671.319

Vinod, A., Sanjay, M. R., Suchart, S., & Jyotishkumar, P. (2020). Renewable and sustainable biobased materials: An assessment on biofibers, biofilms, biopolymers and biocomposites. *Journal of Cleaner Production*, *258*, 120978. https://doi.org/10.1016/j.jclepro.2020.120978

Wang, C., Song, R., Wei, S., Wang, W., Li, F., Tang, X., & Li, N. (2020). Modification of insoluble dietary fiber from ginger residue through enzymatic treatments to improve its bioactive properties. *LWT*, *125*, 109220. https://doi.org/10.1016/j.lwt.2020.109220

Yang, Z., Guo, Z., Yan, J., & Xie, J. (2024). Nutritional components, phytochemical compositions, biological properties, and potential food applications of ginger (Zingiber officinale): A comprehensive review. *Journal of Food Composition and Analysis*, *128*, 106057. https://doi.org/10.1016/j.jfca.2024.106057

CHAPTER X

STATE-OF-THE-ART REVIEW AND FUTURE CHALLENGES OF POLYMERS AND NANOCELLULOSE-BASED BIO COMPOSITES FOR BIOMEDICAL AND HEALTH APPLICATIONS

J. YUSUF, MOHD SAPUAN SALIT, R. A. ILYAS, MOKLIS MUHAMMAD HARUSSANI, NISA ATHIRAH ROSLAN, NUR DINI NABIHAH ROSIDI, NUR MUNIRAH MOHD SHUKOR, NURLIYANA SYAHIRAH NORAZLI

Abstract

Thermoplastic, natural polymer, elastomer, and synthetic polymer are part of the polymer's group that is being studied in this paper. These polymers are chemical substances that are able to offer functions to the biomedical and health sector on their own. All these polymers are further discussed to a few other poly-groups stationed under each of them. Some of the applications of these polymers in the biomedical and health sector are discussed such as genomic mapping and DNA sequencing, cancer and respiratory injury treatment, medical gowns material, breast implant, contact lens, and many others. From the single polymer chains, it could be modified or recycled and still able to offer valuable usages to the sector. Other than that, the properties of polymeric materials are also elaborated in terms of its thermal and mechanical aspects. Lastly, despite offering many applications in the biomedical and health sector, there are consequences faced within the fabrication and application process. Due to a number of applications of these polymers to the daily biomedical and health sector, this paper only elaborates on the challenges upon applying the trend of additive manufacturing and biodegradable polymers.

1. Introduction

The term polymer originated from the Greek phrases poly (many) and meros (pieces or units of high molecular mass). Every molecule is made up of a huge number of distinct structural components that are held together in a uniform pattern by covalent bonds. Polymers are huge molecules with a large molecular mass known as macromolecules that are created by joining together an enormous number of tiny molecules known as monomers. Polymerization is the process through which monomers bond to produce polymers. Polymerization is described as a chemical process where two or many chemicals mix with or without the emergence of heat, water, or some liquids to produce a large molecular mass molecule. The resulting product is known as a polymer, and the basic substance from which polymers are created is known as a monomer (Kaushik et al., 2016).

In other words, polymers are a diverse category of materials that we encounter in our daily lives, and they are defined as molecules or macromolecules made up of subunits. The order, orientation, disposition, and type of the subunits along the polymer chain ensure that this wide class of materials has different features, such as a wide range of molecular weights and qualities. For example, while polypropylene, a synthetic polymer consisting of repeating propylene units, is presently used for the synthesizing of substances like tables with great mechanical qualities and hardness, elastomers just do not have these, but they do have flexibility and plasticity suited to produce tyres as well as other substances (H. Iván Meléndez-Ortiz et al., 2016). From a biomedical standpoint, the creation of smart polymers and materials provided effective and appropriate materials capable of responding specifically to the condition, illness to be treated, or a specific requirement. While cellulose is readily available, it is increasingly becoming a focus for researchers. Novel possibilities for the application of nanocellulose in high-tech science have been revealed by recent scientific breakthroughs.

Nanocellulose-based biomaterials are employed in wound healing as injury bandages and regeneration structures. The capability of hydrogels to store and expel water is highly important in managing excessive injury exudates whilst preserving a wet wound environment to facilitate recovery (ter Horst et al., 2019).

2. Biomedical applications of different types of polymers from 2012-2022

2.1 Thermoplastic

According to Biron (2012), thermoplastics have the most basic molecular arrangement, consisting of chemically free macromolecules. Thermoplastics are one type of plastic that can be softened or melted by the heating process. Then, by cooling, it can be shaped, formed, welded, and solidified. Various cooling and heating processes can be done indefinitely without causing serious disturbance, offering reprocessing and recycling. According to Ferreira et al. (2012), depending on the nature of resin (polymer in melted liquid form), thermoplastics give excellent strength, flexibility, and shrinkage resistance. For molecular structure, thermoplastics are linear polymers with straight-chain constructions that have poor molecular bonds. However, according to Weerakoon-Ratnayake et al. (2017), when compared to elastomers, thermoplastics are high-molecular-weight polymers that are straight or branched, have a higher Young's modulus, and have a wider range of physico-chemical properties.

Next, Ferreira et al. (2012) also stated that the melting point of thermoplastics is lower than the degradation temperature. The mechanical properties of thermoplastics are malleable, bendable, and have high impact resistance; the strength comes from crystallinity. It is also highly chemical resistant and can dissolve in organic solvents. Engineering thermoplastics are usually flame retardant and can resist temperatures as high as 100°C. With the inclusion of carbon or metal fibers, thermoplastic composites can be rendered electrically conductive. They are commonly employed as thermal and electrical insulators. However, in this paper, the applications of this particular polymer to biomedical needs are to be elaborated.

Application of Thermoplastics

Nanofluidic Devices

Thermoplastics were used in nanofluidic devices. According to Weerakoon-Ratnayake et al. (2017), thermoplastic substrates have a lot of potential in areas like genomic mapping and DNA sequencing. Electrochemical detection of tiny compounds, research of kinetics of enzyme reactions, and single nucleotide polymorphisms identification are also part of the usage of thermoplastic nanofluidic devices. These are possible because the biocompatibility of thermoplastics makes them suitable for microfluidic channel creation by casting, thermal forming, hot embossing, injection moulding, and compression moulding. As glass transition temperatures (Tg) of popular thermoplastics, such as polymethyl methacrylate (PMMA), polycarbonate (PC), cyclic olefin copolymer (COC), and polyethylene terephthalate (PET), are much lesser than those of glass, it enables the manufacture of nanostructures at rapid production rates, high fidelity, and low prices. However, this application still confronts issues. For example, in order to guide the creation of ideal nanoscale devices, it is still required to understand the physical phenomena happening in nanochannels using theory and simulations, complemented by experimentation.

2.2 Natural polymer

According to Khan et al. (2019), a natural polymer, commonly known as a biopolymer, can be defined as an organic compound that has been generated from the life process of green plants, animals, bacteria, and fungi. The most common types of biopolymers can be categorized into polysaccharides, polypeptides, and polynucleotides. According to John et al. (2012), there are two different types of natural polymers which are from living things such as carbohydrates and proteins and polymerized renewable resources such as lactic acid and triglycerides.

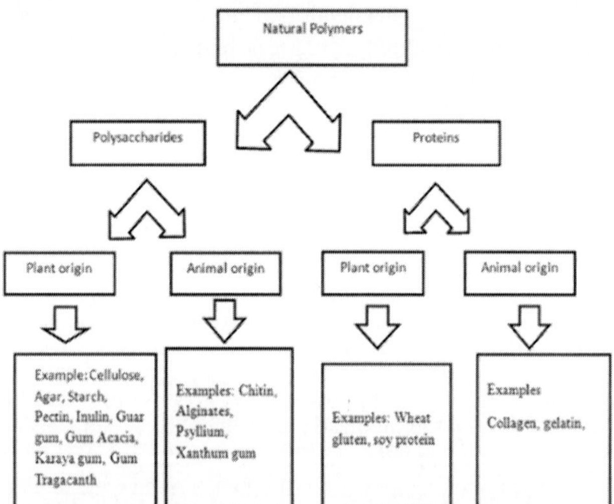

Figure 2.2.1: The classification of natural polymers (Kaushik et al., 2016).

Referring to Rana (2021), the natural polymer has characteristics of hydrophilicity and hydrophobicity where proteins and polysaccharides are normally used for nanomedicine. There are a few examples of proteins which are albumin, gelatine, lectin, and polysaccharides which are cellulose, dextran, chitosan, and alginates. Both types are natural hydrophilic polymers. Figure 2.2.1 below shows the detailed classification of natural polymers.

Biomedical Application of Natural Polymer

Cancer Treatment

Polysaccharides and polypeptides are employed as medication carriers in cancer treatment. There are several polysaccharides that are used in the production of nanocarriers for cancer therapy such as starch, dextran, pullulan, and many more. Polysaccharides – based polymers are usually used for cancer treatment as they are biodegradable, biocompatible, accessible, stable, non-toxic, and inexpensive. There are various strategies utilized for the arrangement of polysaccharide-based nanoparticles that incorporate polyelectrolyte complexation, self-assembly of hydrophobically changed polymers, and covalent or ionic cross-connecting. Natural polymers are available nowadays, where they are being applied as an establishment to which less solvent medications can be joined and transported to tumour cells (Khan et al., 2019).

Respiratory injury treatment

As reported by Rana (2021), Covid-19 related to respiratory injury treatments, polymeric nanomedicine techniques could be a viable option for delivering a wide range of active molecules such as antiviral biologics and nucleic acids to the specific region. Polymeric nanomedicine technique is based on the stimuli-responsive materials that can effectively prevent the biomolecule (drug) breakdown before arriving at the target site of actions to reduce a general adverse effect and increase the solubility of the supplied therapeutic components. The combination of stimuli-responsive polymer material and nanocarrier fabrication techniques allows for the construction of biomolecule-loaded nanocarriers with enhanced antiviral effects. For the treatment of new viruses, roughly 90 antiviral medication approaches have been authorized.

2.3 Elastomer

Elastomers can be described as polymers that are flexibly cross-linked. In regards to plasticity and versatility, elastomers are similar to rubber. The lengthy arbitrarily looped, flexibly cross-linked materials are easily expanded, however, when the tension or load is released, they revert to their former forms. A high number of cross-links will cause the structure to be stiff, tough, and more similar to a thermoset (Figure 2.3.1) (Sastri, 2010).

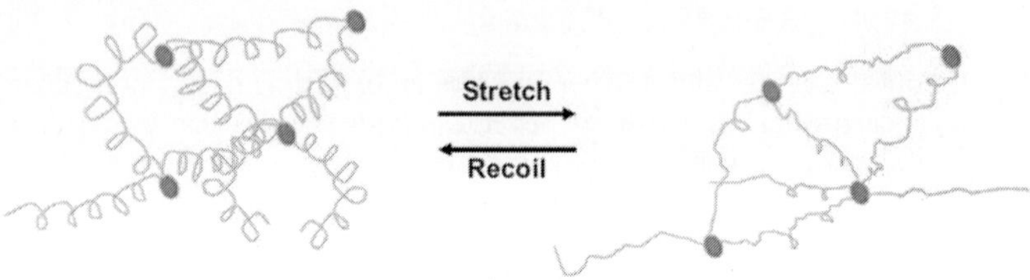

Figure 2.3.1: Elastomers (Sastri, 2010).

Silicones, thermoplastic elastomers, polyolefin and polydiene elastomers, poly (vinyl chloride), natural rubber, heparinized polymers, hydrogels, and polypeptide elastomers are examples of elastomeric biomaterials. Besides, cardiovascular equipment, prosthetic equipment, and basic health care tools are among the biomedical applications of elastomers.

Elastomers have been employed in pharmaceutical products since the beginning of the latex business. As for vulcanized natural rubber, it has been used in pharmaceutical products shortly after the advent of the rubber-vulcanization technique.

In the biomedical field, three kinds of elastomers have been utilized. First, there are commercial elastomers that are employed in the biomedical sector. Second, clinical grades of elastomers are authorized for brief physiological interaction and lastly, a low number of elastomers are suited for relatively long physiological interaction or installation.

Excellent quality, desirable physical, chemical and mechanical qualities, ease of fabrication, and high stability and sterilizability are all essential for biomedical elastomers. When implemented for probable interaction with living tissue and fluids, they must never trigger thrombosis, damage biological components, modify plasma protein, kill enzymes, decrease electrolytes, create immunological reactions and cancer, and should not generate mild allergic reactions (Yoda, 1998).

Elastic Polymers for Biomedical Practice

Table 2.3.1 below illustrates the characteristics and biomedical applications of silicones, polyurethanes, and thermoplastic elastomers (TPE) lacking polyurethane structures.

Table 2.3.1: The Characteristics of Polymers and Their Biomedical Applications (Yoda, 1998).

Polymer	Characteristics	Biomedical Application
Silicones	• Excellent elastomeric • Insensitivity to physiology • High-temperature stability • High biodegradability • High tolerance to rust and Ultraviolet (UV) light • Medium biocompatibility • Excellent age tolerance • Great absorption of vapors and gases	• Metal implantable encapsulants • Extrusions used to transport blood, medicine, or minerals. • Method assists in drug production • Unique production: rods, rings, gaskets • Blood oxygenator membranes in heart-lung flow • Medication transfer systems
Polyurethanes	• Pyrolysis • Permeability of gas • Biostability • Extractability • Compatibility of Blood	• Bandages and straps • Breast implants • Fracture adhesive • Incision tools • Medical gowns
Thermoplastic elastomers (TPE) lacking polyurethane structures	• Low toxic • Exceptional bend transparency, physiologically safe Longer-bending endurance • Blood and medication tolerance • Tolerance to biting and tearing • Low extractables • Thermal sealant	• Syringe blockers • Tubing • Tubing for Dialysis • Catheter • Newborn teats • Gasket • Seal

Biomedical Application of Elastomer

Cardiac support devices

Cardiac shock occurs when the heart is unable to supply enough blood and oxygen to the brain and other essential organs due to ventricular hypokinesis. The intra-aortic balloon (IAB) pump and artificial hearts are the two treatment technologies that are able to maintain the malfunctioning ventricle when cardiac shock occurs. The IAB pump could supply the output of the cardiac without requiring an emergency operation. The IAB pump is a polyurethane-based catheter which is placed inside the thoracic aorta through a skin incision via the femoral artery as depicted in Figure 2.3.2. The polyurethane balloon at the catheter's tip is continuously expanded and compressed in time with the heart, assisting in the stabilization of the failing left ventricle.

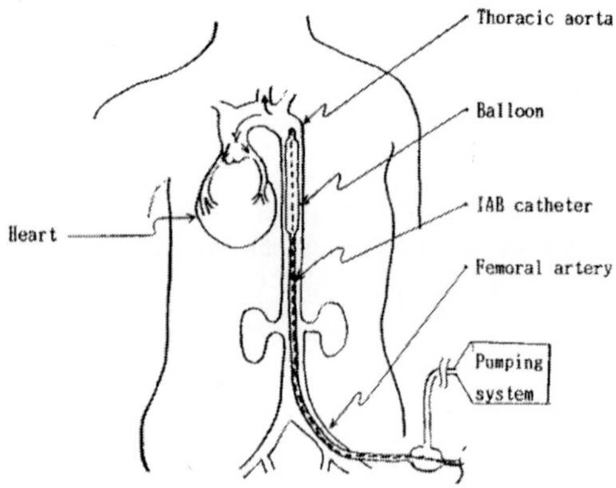

Figure 2.3.2: Simplified drawing of IAB construction (Yoda, 1998).

Artificial hearts in medical use are classified into two categories: temporary technologies created to help the pounding action of the injured heart and indefinitely implantable ventricular aid technologies if the injury is incurable. Figure 2.3.3 depicts the schematic design of a sac-type, air-driven artificial heart.

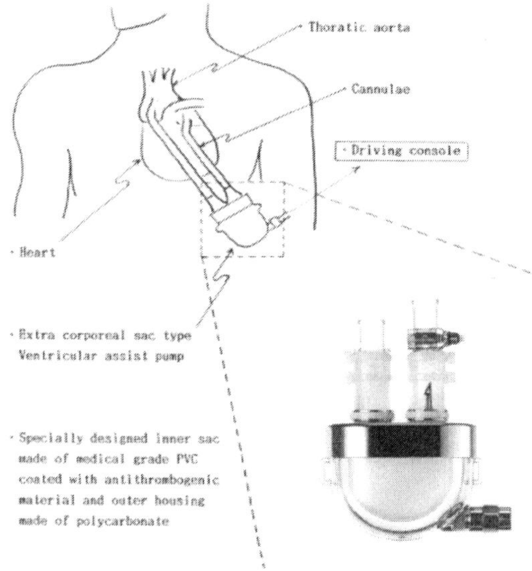

Figure 2.3.3: Simplified drawing of a sac-type, air-driven artificial heart construction (Yoda, 1998).

One of the most challenging practices for polyurethane elastomers is in artificial hearts due to the demand for intrinsically excellent blood tolerance, flexural tolerance, and wear tolerance. Currently, five commonly produced polyurethane elastomers are labelled as biomedical standard: Biomer™, TM-3™, Pellethane™, Tecoflex™, and Cardiothane™, as illustrated in Table 2.3.2 having the appropriate molecular structure. Table 2.3.2 displays a variety of elastomers utilized by various groups to create prosthetic hearts.

Table 2.3.2: Elastomers utilized to make artificial hearts (Yoda, 1998).

Investigator	**Blood Pump's Type**	**Elastomer**
Novacor's LVAD	Pusher-plate	Biomer™
Symbion	Diaphragm	Biomer™
Aviomed	Tuberous	Angioflex™
Nimbus	Pusher-plate	Hexsyn™/gelatine
Thermedics LVAD	Tuberous	Tecoflex™
Nippon Zeon VAD	Sac	Cardiothane™
Totobo VAD	Diaphragm	TM-3™, TM-5™
Berlin University	Diaphragm	Pellethane™

Nanocellulose-based composites

Nanocellulose has a significant impact on biomedical applications (Shi et al., 2022). Nanocellulose bio composites are used in the fabrication of several products, such as medical implants, drug delivery vectors, and vascular grafts. These materials are also utilized in tissue engineering due to their sustainability, biodegradability, and biostability. Because of its exceptional mechanical strength, enhanced flexibility, and great water-holding capacity, nanocellulose is an ideal material for use in wound care dressings. Moreover, it possesses antibacterial action, which is an added advantage to fulfil the goal. Since BCNC has a microstructure that closely resembles human tissue and is incredibly biocompatible, it is the perfect material to replace damaged or injured tissues (Czaja et al., 2007; Aditya et al., 2022).

Biomedical Application of Nanocellulose-based composites

Tissue engineering

Tissue engineering uses nanocellulose-based scaffolds because of their optical clarity, water retention, and other important features. The polyvinyl alcohol (PVA)/CNC scaffolds that were freeze-dried showed interconnected pore architectures. Researchers have discovered an increase in pore size when the concentration of CNC increases (Lam et al., 2017). The mechanical, thermal, and swelling properties of scaffolds were greatly improved using CNC. Compared to pure PVA, PVA/CNC with 4 weight percent and 6 weight percent CNC had higher equilibrium swelling ratios of 85.3% and 85.0%. For efficient cell adhesion, development, and metabolism, the homogeneous porosity structures created by freeze-drying and the hydrophilicity of CNC combined to make the perfect microenvironment. When utilized as scaffold materials, several naturally occurring macromolecules, such as collagen, show good biocompatibility but low mechanical strength (Seifu et al., 2018; Tavşanlı and Okay, 2018).

Wound dressing

Film or injectable hydrogel materials for wound dressings can be produced directly with a CNC machine. To introduce self-healing behavior, it can be modified by incorporating active molecules into the wound dressing materials, such as vitamins, enzymes, antioxidants, and antibacterial medications. Researchers created an injectable self-healing nanocomposite hydrogel that can be used to treat severe burn injuries (Huang et al., 2018). It has both carboxymethyl chitosan (CMC) and modified CNC. It absorbs and retains wound exudate while preserving the perfect moist environment. They used periodate oxidation to make dialdehyde CNC (DACNC), and they used dynamic Schiff-base couplings between the amines in carboxymethyl chitosan and DACNC to cross-link the hydrogels. The results indicated more self-healing behavior than chitosan-based ones, which could be attributed to the chemical cross-linking function of DACAC and its reinforcing impact.

Artificial blood vessels

Medical implants featuring different matrices are equipped with CNC to facilitate the creation of vascular prosthesis. Hydrophilic polyurethane and CNC are utilized to create new nanocomposites with characteristics similar to human tissues, which can be used for vascular transplants. Hydrophilic PVA and CNC-produced nanocomposites have demonstrated efficacy in the replacement of soft tissue within the heart. PVA was used in one of the earliest studies in this field, which involved PVA-CNC nanocomposites based on CNC manufactured from tunicates (Potter et al., 2014). Bhat et al. employed two different polymers, polyurethane and PVA, to create two different hydrogel bio-composites (Bhat et al., 2019).

Biosensors

A biosensor is a type of measuring device that uses an immobilized biological material (such as an enzyme, antibody, or whole cell) to interact with an analyte and generate chemical, electrical, or physical signals. It is often employed in the identification of analytical instruments, or biomolecules. These biosensors mix biological components like perspiration, saliva, or other vital body fluids with a physicochemical detector. These detectors could be wearable electronics or even massive pieces of equipment. The sophisticated sensors are receiving more and more attention due to their ease of testing, portability, fast diagnostic turnaround, lack of sample preparation requirements, etc. Materials based on nanocellulose have exceptional mechanical, thermal, chemical, and barrier qualities. By adding unique functional groups, the surface of nanocellulose augmented with hydroxyl groups can be modified in various ways. Esterification, silylation, oxidation, amidation, polymer grafting, sulfonation, etherification, and urbanization are the modification techniques used to create sensors (Langari et al., 2023).

2.4 Synthetic polymer

According to Sýkora and Švec (2014), synthetic polymers are those created by humans. Monomers, which are repeating structural units, make up polymers. Nylon and polyethylene are two well-known synthetic polymers that are commonly referred to as "plastics". Additional polymers, also known as chain-growth polymers, are polymers that are produced by connecting monomer units without changing the substance. All of these polymers are claimed to be synthetic.

So far, numerous synthetic polymers have been produced that can be used in everyday life. For biomedical applications, these types of polymers have chemical bonding that are oxidative and resistant to hydrolytic and other degradation mechanisms that have been extensively studied in biological environments. However, only a few have made it to medical devices or implants. "Poly" types such as olefins, urethanes, carbonates, siloxanes, amides, ethers, sulphones, and some forms of polyurethanes are the most common non-biodegradable synthetic polymers utilized in biomedical applications (esters) (Sýkora & Švec, 2014).

Table 2.4.1 showed the non-biodegradable synthetic polymers' general features and important biological uses. There are specific properties for each type of synthetic polymer and its major biomedical applications are shown in the Table 2.4.1 below:

Table 2.4.1: The non-biodegradable synthetic polymers' general features and important biological uses (Sýkora & Švec, 2014).

Polymer (Poly-)	Significant characteristics	Significant biomedical applications
Poly-ethylene	The linear thermoplastics that have either good chemical resistance or low and high-density grades are excellent.	Spine implant, total hip and knee
Poly-propylene	The line-form thermoplastics have flexible and tough characteristics. This type of polymer also has an excellent fatigue resistance.	Hernia mesh, nonabsorbable suture and medical pouches

Poly-methyl methacry-late	Even this polymer has poor resistance to chemicals but it has a good impact strength and also a lightweight transparent thermoplastic.	Slow release of peptide and protein drug, orthopaedic prosthesis and encapsulation device for cells
Poly-dimethyl siloxane	Excellent viscoelastic properties, nontoxic polymer, inert, non-flammable and also an elastic solid.	Prosthesis, breast implant and contact lens
Poly-ether ether ketone	Partially crystalline thermoplastics with good chemical resistance and also excellent mechanical properties.	Minimally invasive fusion surgery and motion preservation and dynamic stabilisation, interspinous spacers, pedicle-based rod systems for nun fusion, disc arthroplasty and spinal fusion
Poly-urethane	This type of polymer is the linear thermoplastics or cross-linked thermoset with biostable or biodegradable characteristics that can be manufactured as soft elastomers or stiff materials.	Catheters, pacemaker lead insulators and vascular grafts

The type of PVDFs, such as polyvinylidene fluoride or polyvinylidene difluoride, is used as an element polymer in biocompatible polymer materials for filters or protectors because it has a silkier anatomy, a homogeneous nanoparticle dimension and structure with a high binding ability. Other commercial polymer membranes such as spunbond polyethylene can be used. Polyethylene (PET), polyacrylonitrile (PAN), polypropylene (PP), polystyrene, polycarbonate, polyurethane (PU), or polyester are commonly used in facemasks. As shown in Figure 2.4.1 below, the surgical masks usually contain three polymeric layers which are a soft fiber inner layer, a middle layer with the main filter, and a hard plastic outer layer. Firstly, diverse layers of nonwoven textiles consisting of polymeric fibers are used to create facemasks. Fabrics are made in a number of ways, the most prevalent of which is meltblown and spunbond. Spunbonded nonwoven fabrics are nonwoven fabrics that have been spun together that are created in two different steps. In the first steps, a polymer is extruded and then stretched to create endless or continuous threads which then are arranged to become a fabric. After that, it is either self-adhered, mechanically reinforced, thermally bonded, or chemically bonded (Armentano et al., 2021).

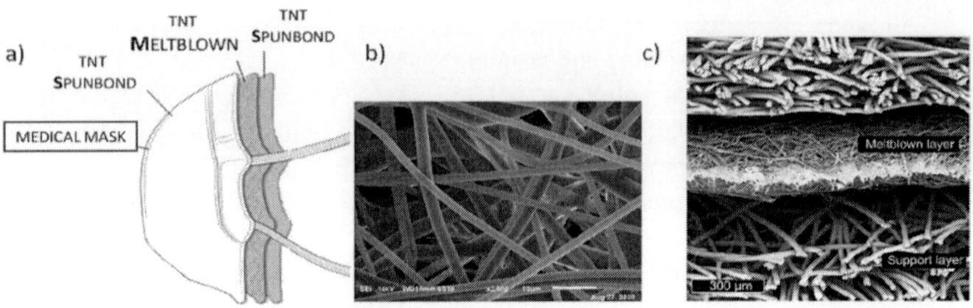

Figure 2.4.1: The figure shows the look of (a) three layers TNT of medical mask, (b) TNT meltblown layer and (c) SMS Web of the medical mask (Armentano et al., 2021).

Secondly, the manufacturing process of the surgical mask proceeds with electrospinning. Electrospinning is a versatile and successful process for creating polymeric porous membranes with morphology, controlled fiber dimension, and distribution for a number of applications, ranging from aligned to randomly oriented. According to the scientific community, electrospinning is the most necessary and feasible method for producing uniform nanofibers from polymer composites and nanocomposites, synthetic and natural polymer matrices based on ceramic nanoparticles or metal. The last process is electret technology. Coronavirus disease and viruses generally are not protected effectively by present filtration technologies. The coronavirus size was estimated at 60 until the $140\eta m$ range with a $100nm^4$ average which is comparable to the size of the SARS coronavirus and the common influenza virus in the laboratory in which a transmission electron microscope is used. Corona discharge, triboelectric charging, and thermal charging are all electret technologies. The charge storage capacity of a polymer is mostly determined by its properties. The storage capability and charge stability can also be affected by membrane morphology, additive type, and processing circumstances. Corona discharge is a technique for applying electrostatic charge to polymeric fabric materials that have been widely used (Armentano et al., 2021).

According to Armentano et al. (2021), for material selection or conceptual design, the chemistry of the polymer can be linked to the layer microstructure's distinctive absorbency qualities, both the void volume and capillary size are influenced by the geometry of the polymeric fiber. Face mask filters made of melted blown polypropylene are commonly used. Polypropylene is a semi-crystalline and nonpolar polyolefin thermoplastic polymer made from propylene polymerization. It is one of the most commonly made polymers and is used in a wide range of applications. To achieve a higher filtering efficiency, electrospinning was used as a processing technology. Cotton fibers have also been employed in natural materials. For example, due to the orientation of the cotton fibers, an elastic cotton-core membrane provides stretchability benefits as well as better wicking capabilities, making it perfect for facemasks. By combining a compact polymethyl methacrylate with a typical single-use filter, it may create long-lasting and clear face masks. This type of polymer has characteristics like impact resistant, viscoelastic, and lightweight. It is indeed durable and less expensive than polycarbonate, however it lacks the impact resistance of polycarbonate.

3. Biomedical and health applications of different types of modified polymers/ recycled polymers from 2012-2022

According to Hamad et al. (2013), polymer recycling is a way of addressing environmental problems caused by the accumulation of polymeric waste from ordinary polymer applications like packaging and construction. Plastics can be moulded into any shape or size, and chemical and functional properties can be added readily. Plastics are being researched extensively in medical disposable production due to the clear rise in illnesses connected to non-disposable materials. When employing plastics in biomedical applications, pay close attention to the material's chemical or thermal properties, sterilizing potential, and durability (Joseph et al., 2013).

The following are some of the polymers utilized in biomedical application which are acrylonitrile butadiene styrene, polyethylene terephthalate, polycarbonate, polyamide, polyethylene, polyurethane, polystyrene, polypropylene, and polyvinyl chloride. For its biocompatibility and mechanical qualities, polycarbonate is frequently used in biomedical applications. It is used to make stents, syringes, tubes, and other medical devices. Polyvinyl chloride (PVC) is the most widely used polymeric material in medical devices in terms of total volume consumed, and it is also one of the most recycled polymers. Polyvinyl chloride (PVC) offers a variety of qualities, including fluid inertness, high energy irradiation, transparency, steam sterilizability, gloss, flexibility, and chemical resistance. It is a reasonably inexpensive material. Face masks, drug solutions, nutrition, breathing and fluid suction, tubing for infusion, and injection are all made of highly plasticized and flexible polyvinyl chloride (PVC). Ultrahigh molecular weight polyethylene is a semicrystalline linear polymer that is utilized in hip replacement, orthopaedic implants, and primarily the knee. This type of polymer has high impact strength, low friction coefficient, good biocompatibility, excellent chemical resistance, great wear resistance, and is also inexpensive (Oral et al., 2017).

The process of recovering and converting plastic trash into valuable things is known as plastic recycling. There has been a growing awareness of the need to reduce our reliance on plastic. The first step in recycling is to identify recyclable items, which is followed by sorting them. Shape, color, or form and material type are among the variables used to sort plastics. Automated sorting strategies have sparked a lot of interest recently. Density separation, surfactant-based foam extraction, electrostatic sorting, and air sorting are among the separation processes employed. Following the collection, sorting, and cleaning of plastic trash, there are typically four major plastic recycling routes or processes, as shown in Figure 3.1, which show the different steps of primary recycling, secondary recycling, tertiary recycling, and quaternary recycling.

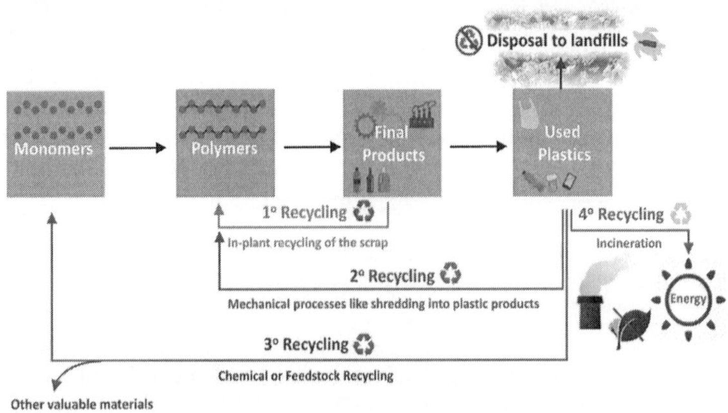

Figure 3.1: Recycling pathways or processes (Okan et al., 2019)

The first stage, which is primary recycling, recycles plastic scraps into items that have similar properties to the original. PET can be recovered from discarded bottles, for example, and utilized to make new bottles that are comparable. In order to achieve the needed product quality, recovered waste plastics or scrap can be blended with virgin material (Al-Sabagh et al., 2016). This method is well-established and is also known as the closed-loop method. Single-type plastics that are relatively clean or uncontaminated are used in this procedure. Closed loop recycling has the advantage of allowing rejected wastes to be promptly integrated back into the manufacturing cycle.

The next stage is secondary recycling. Mechanical recycling is the process of recovering plastic wastes using mechanical methods. Mechanical recycling includes separation, sorting and collection, as well as washing to remove organic or other impurities and grinding of the material. Mechanical characteristics are difficult to retain following mechanical recycling. Mechanical characteristics can be reduced due to a decrease in molecular weight following recycling and mixing with other polymers. Furthermore, as a result of treatment, recycled plastics are prone to thermomechanical degradation (Ragaert et al., 2017). Secondary recycling includes a variety of techniques such as blow moulding, screw extrusion, and injection moulding (Singh et al., 2017).

The next stage is tertiary recycling; chemical recycling is another name for it. In this stage, plastic waste is broken down into smaller molecules, primarily liquids or gases, which are subsequently used as feedstock in chemical and fuel manufacturing processes. Recycling process uses a chemical technique to recover the petrochemical components in polymers. Pyrolysis is a process that involves heating plastics to high temperatures in the presence of a catalyst. Cracking, gasification, and chemolysis are examples of other recycling techniques. Pyrolysis occurs in the absence of oxygen. Gasification refers to a procedure that takes place in a regulated atmosphere. The catalyst comes into close contact with the molten polymers in fluid catalytic cracking. In vapor phase catalytic cracking, the cracking vapor collides with the catalyst.

The last stage, which is quaternary recycling or incineration, minimizes waste volume while also recovering energy from wastes or discarded plastic through burning. When wastes are very polluted and cannot be recycled using traditional procedures, this is the option of choice. Quaternary recycling is demonstrated via tire-derived fuel. This method is also known as incineration, as it produces waste residues and ultimately, pollution. As a result, this is an excellent method of medical waste recycling.

4. Thermal and mechanical properties of polymeric materials for biomedical sectors

Every substance designed for biomedical use should be safe for the human body and functional. In terms of the body, the implantable substance must not contaminate the body and not cause an allergic reaction, not injure living cells, not cause cancer, not cause epitopes action, not cause protein and enzyme ruination and putrefaction, not disrupt electrolyte stability, and not cause irregularities in the metabolic systems (Babker et al., 2018). Adhesion, self-healing, and super hydrophobicity are three basic features present in nature that are required of bio composite materials in order for them to be suitable for biomedical applications (Ibrahim et al., 2017).

First, adhesion. Plants and animals both show this behavior, which is necessary for their survival. This enables creatures to adhere themselves to their hosts, whether temporarily or permanently, and vice versa. Bassas-Galia et al. (2016) show how microbes, animals, and plants employ their adherence skills in a variety of ways. According to Bassas-Galia et al. (2016), polymeric materials with this feature have been produced and employed in biomedical applications. Next, self-healing. The body of a human is designed in a manner that it heals and/or replaces injured tissue when a wound occurs; this is referred to as self-healing. If the wound isn't too severe, this is a viable option. When a damage or wound is too severe to recover on its own, an implant must be used. Implants are vulnerable to stress, ageing, and wear, which can trigger damage and necessitate replacement. As a result, scientists have been working to develop substances that could self-heal. The very first phase of self-healing materials consists of composites which are unalterably healed without recovering the injured matrix, whereas the second phase recovers the injured matrix in a reversible manner. Because of the virtues and capability to manage the limitations involved with the polymer matrix, several polymers and their composites have indeed been widely utilized. These properties should be present in self-healing substances. Firstly, the potential to cure ruined parts of substances multiple times and to repair the substances automatically, besides lower operating expenses. Finally, it should perform as well as or better than conventional substances. Lastly, super hydrophobicity. Wet resistance is well recognized for materials with extremely hydrophobic surfaces. Many insects and plants display this characteristic. Biomaterials with a super hydrophobic characteristic are advantageous in biomedical applications since they lower the risk of blood clotting due to poor platelet adherence

(Ibrahim et al., 2017). This characteristic has been demonstrated in a number of biomaterials.

Nanoparticles are used to enhance the physical and mechanical characteristics of polymeric-based self-healing materials. In comparison to metal and ceramic, polymers possess lesser strength and modulus, which requires development.

Thermal Properties

The glass transition temperature (T_g), melting temperature, and crystallization temperature of biomedical polymers are all determined through thermal characterization. These thermal characteristics are unavoidable in determining the processing conditions to transform a natural or synthetic biomedical polymer into the final implant or device even if they are unrelated to the polymers' biocompatibility. The technique used to calculate the temperature parameters described before is differential scanning calorimetry. Changes in thermal characteristics can also be caused by the polymer processing procedure itself. Electrospinning is a method of creating viscous solutions and melting for polymer filaments. The spinning technique is said to lower the crystallization temperatures of polyglycolic acid and polylactic acid from 68 to 49 °C and 108 to 77 °C, respectively. When performing differential scanning calorimetry on hygroscopic polymer fibers, extra caution should be exercised because the glass transition peak may not be seen during the first heating cycle. The biodegradability of a polymer is determined by its glass transition temperature. If a biomedical polymer's glass transition temperature is similar to that of the human body, the implanted polymer may be more flexible in the host environment than *in vitro*, thereby speeding up biodegradation *in vivo* (Pino-ramos et al., 2018). In Latin, *vitro* is within the glass: when something is performed in *vitro*, it happens outside of a living organism; while *vivo* is within the living thing, which refers to work that is performed in a whole, living organism.

Furthermore, the presence of water and other ions in the host environment may diminish the glass transition temperature of implanted polymers such as polyesters and hasten their biodegradation. Thermal compression, in which a polymer fabric or biotextile is subjected to varied stresses at varying temperatures, is another valuable thermal characterization approach. To generate appropriate scaffolds for tissue engineering, distribution, pore size, and thickness can be assessed at each condition. Thermal compression and simultaneous characterization were used to prepare nonwoven fiber scaffolds made of polyethylene terephthalate for tissue engineering.

Pressure applied at the polymer's glass transition temperature (70 °C) resulted in improved pore size distribution control and smaller pore diameters, resulting in faster and wider propagation of trophoblast cells on the scaffold (Pino-ramos et al., 2018).

Mechanical Properties

Before applying a material in any sector, it is critical to understand the basic mechanical characteristics, including how far it could be extended, how far it could be folded, how soft or hard it is, how it reacts under continuous pressure, and many more (Argawal et al., 2015).

a. Strength: Simply put, strength is the amount of tension needed to fracture the specimen. Impact (hammering), torsional (twisting of the polymer), flexural (bending of the polymer), compressional (compressing the polymer), tensile (stretching of the polymer), and other types of strength exist. The polymers are ranked in increasing strength order: Linear < Branched < Cross-linked < Network.

Variables Influencing Polymers' Strength

1. Molecular Weight: The tensile strength of the polymer increases with increasing molar mass and approaches saturation at certain molar mass value (Figure 4.1). The following equation relates tensile strength to molar mass.

$$\sigma = \sigma_{\infty} - \frac{A}{M}$$

σ_{∞} is the tensile strength of a polymer with an infinite molar mass. M is the molar mass, and A is a variable. At low molar mass, the polymer chains are weakly connected by weak van der Waals interactions and can shift readily, resulting in poor strength despite the presence of crystallinity. In the case of a polymer with a high molar mass, the chains grow long and intertwined, providing the polymer strength.

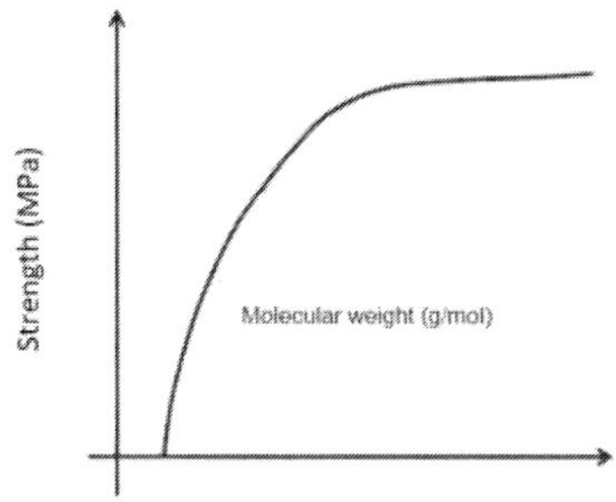

Figure 4.1: Tensile strength varies with polymer's molecular weight (Argawal et al., 2015).

2. Cross-linking: Cross-linking inhibits the chains' movement and boosts the polymer's strength.

3. Crystallinity: The polymer's crystallinity boosts its strength since intermolecular interaction is stronger in the crystalline phase. Thus, the deformation of polymer may lead to greater strength, resulting in orientated chains.

b. Percent Elongation to Break (Ultimate Elongation): This is the strain in the material upon breakage, as seen in Figure 4.2. It calculates the percentage change in material length before fracture. It is a measure of ductility. Ceramics have a very low elongation to break (1%), metals have a moderate elongation (1–50%), thermoplastics very high (>100%), and thermosets low (5%).

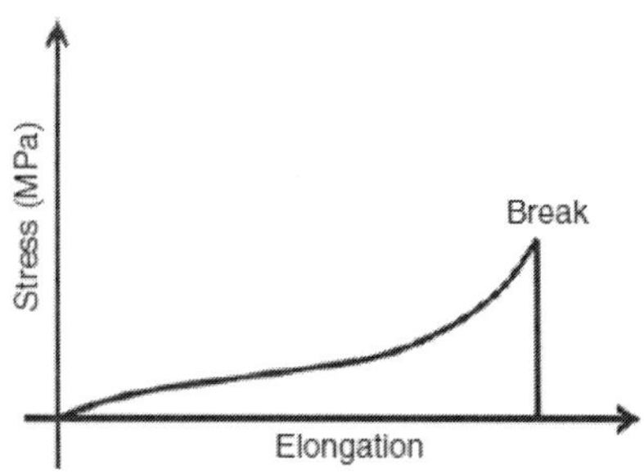

Figure 4.2: Polymer’s elongation to break (Argawal et al., 2015).

c. Modulus of Elasticity (Young's Modulus): The ratio of stress to strain in the linearly elastic zone is described as Young's Modulus (Figure 4.3). Modulus of elasticity is a measurement for a material's rigidity.

$$E = \frac{Tensile\ Stress\ (\sigma)}{Tensile\ Strain\ (\epsilon)}$$

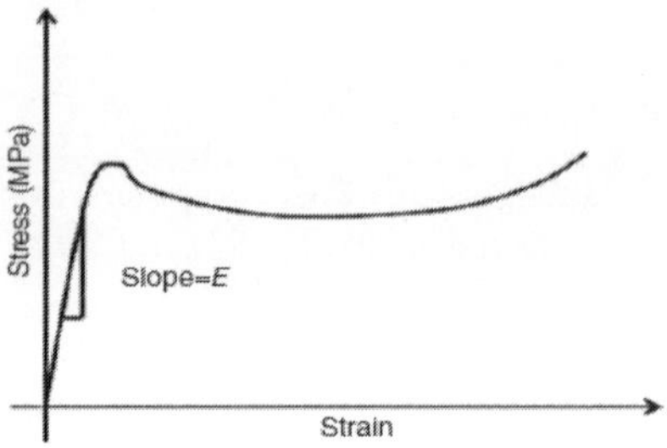

Figure 4.3: Polymer's Young's Modulus (Argawal et al., 2015).

d. Toughness: The area beneath a stress–strain curve represents a material's toughness. (Figure 4.4).

Toughness = $\int_{o}^{0} \sigma \, d\epsilon$

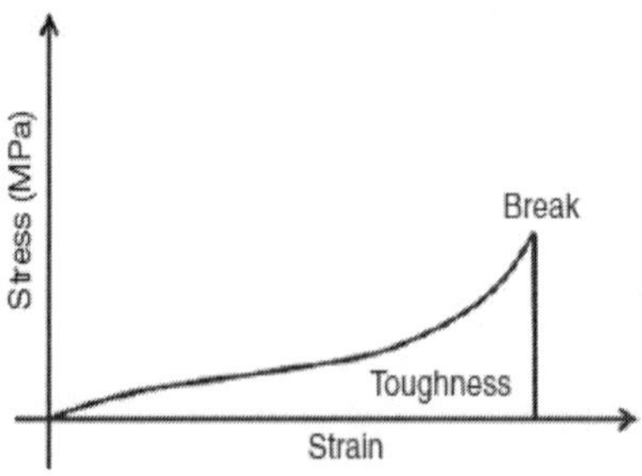

Figure 4.4: Polymer's Toughness (Argawal et al., 2015).

Toughness is a measure of how much energy a material absorbs before breaking.

Figure 4.5 depicts the basic stress–strain graph that analyzes the stress–strain behavior of several materials. Rigid materials (like brittle polymers) have great Young's modulus, whereas ductile polymers have the same modulus of elasticity but with greater break toughness. Elastomers, on the other hand, have a small modulus of elasticity and are rubbery in character.

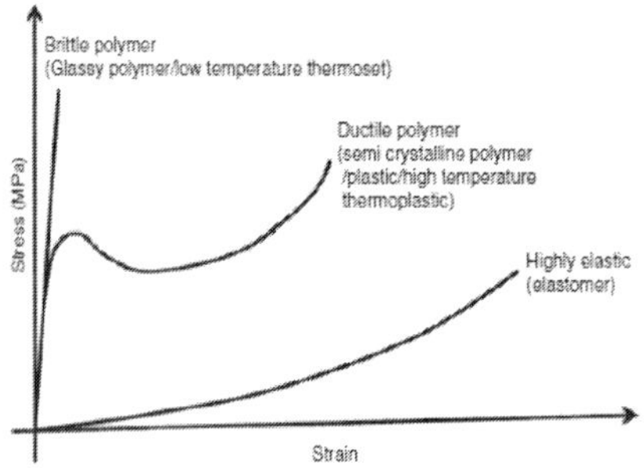

Figure 4.5: The stress–strain behavior of various materials (Argawal et al., 2015).

The plastic polymer's yield strength is the stress that corresponds to the end of the elastic range (Figure 4.6). Tensile strength is the tension associated with polymer's fracture. Tensile strength can be greater or less than yield strength (Figure 4.6).

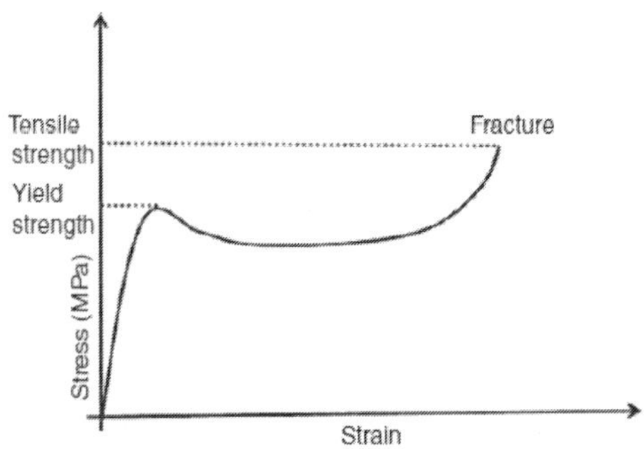

Figure 4.6: Polymer's yield strength and tensile strength (Argawal et al., 2015).

Temperature has a significant impact on the polymer's mechanical characteristics. Figure 4.7 depicts the basic stress-strain plot. The plot clearly shows that as temperature rises, the modulus of elasticity and tensile strength decrease, but ductility increases.

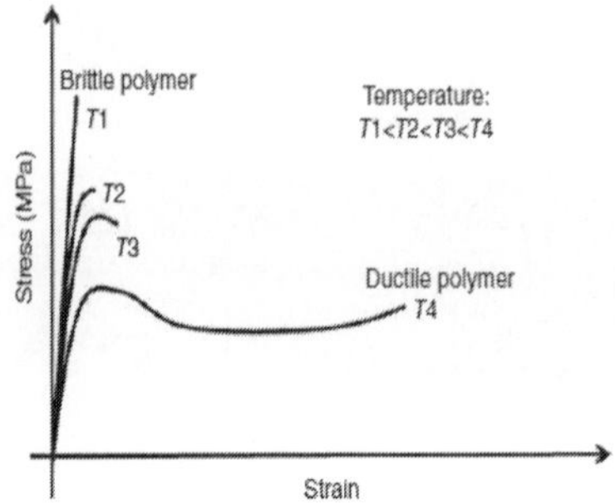

Figure 4.7: The effect of temperature on a polymer's mechanical characteristics (Argawal et al., 2015).

5. Challenges within fabrication and application of polymers, biopolymers, and recycled polymers in various applications

Trend of using polymer composite materials in additive manufacturing

According to Wong & Hernandez (2012), the additive manufacturing (AM) method involves using computer–aided design (CAD) and stereolithography (STL), where STL files are created by processing information from CAD. In this approach, the triangle mesh will be used for the drawing that has been created within CAD software and divided into sections comprising the details of each layer that will be printed. This type of manufacturing is used to produce lighter structures which help the aerospace sector lessen the weight. As for the profession of medicine, it is also evolving due to additive manufacturing, along with simplifying tasks for architects. More than three key technologies have emerged since 2004 when the Society of Manufacturing Engineers classified the numerous technologies. According to Regassa et al. (2019), ceramics for additive fabrication and concrete-based composite are one of the examples of additive manufacturing.

Ceramics for additive fabrication

Complex ceramics lattices may now be easily made for a variety of purposes due to additively built ceramics, which have sparked a trend toward custom-designing materials with high strength to weight ratio. One of the primary uses is in tissue engineering by developing robust and adaptable ceramic scaffolds with complicated forms. The biomedical industry has improved tremendously as a result of the ability to incorporate biosensors into human organs or medical devices. The microstructure and composition of the part can be controlled by using the ceramic additive method. Thus, there are still prospects for analysis and advancement in the refinement of additive manufacturing methods and increasing the material selection for ceramic production. Table 5.1 shows the application of ceramics and polymer composites as well as its advantages and disadvantages.

Table 5.1: The application, advantages, and disadvantages of ceramics and polymer composites (Regassa et al., 2019).

Material	Application	Advantages	Disadvantages
Ceramics	Chemical industries, biomedical, aerospace, and automotive	Shortened fabrication time, control of lattice porosity and printing complex structures and scaffolds for human organs, and many more	Inaccuracy if not post-processed, limited number accessible of printed ceramics, imprecision and bad surface finish
Polymer composites	Sports, toys, medical, architecture, and aerospace	Enable quick prototyping, reasonably priced, adaptable to complicated structure and personalized product	Restricted accessibility with reinforcing, anisotropic, mechanically weak

Concrete-based composite

The most traditional and often used composite material, especially in the building and infrastructure industries, is concrete. The first attempt in manufacturing a cement-based additive is to adhere sand layers using

Portland cement paste by utilizing the deposition process concept from relatively small size techniques such as 3D printing and finite deposition modelling technologies. The employment of cementitious composite in this process can be seen as development of 3D printing for extensive use in the fields of construction, architecture, and design. The extrusion process is one of the primary methods used to create concrete additively even though the powder-bed approach has also been extensively researched. However, there are numerous elements affecting the applicability of the concrete type; shape stability, modesty in extruding process; and dimensional certainty after fabrication.

Trend of Biodegradable Polymers

In accordance to Doppalapudi et al. (2014), biodegradable polymers are often categorized into two categories: natural and synthetic. In comparison to natural polymers, synthetic origin polymers are more adaptable, have a wider range of applications, can be customized in terms of their mechanical characteristic, and change their rate of degradation to suit specific requirements. However, despite the fact that natural polymers have good biocompatibility, they have not been thoroughly studied due to their unfavorable characteristics which are of the batch-to-batch variety and antigenicity. The instability of polymers, which lead to biodegradation, has proven to be extremely essential in many medicinal applications. The biodegradable polymer will have tremendous opportunities to grow with these numerous innovations: tissue engineering, medication delivery, regenerative medicine, medication delivery, coatings on implant, and gene therapy.

Agricultural Application

According to Luckachan & Pillai (2011), in agriculture fields, biodegradable polymers are usually used as agricultural mulches, planting containers, and are also contained in the controlled release of agricultural pesticide. Agricultural mulch film is produced by using starch-based biodegradable film along with the poly (vinyl alcohol), poly (vinyl chloride), and poly (ethylene-co-acrylic acid). There are a few natural polymers in utilizing the controlled release system which is cellulose, lignin, starch, chitin, and alginic acid. As for agricultural planting containers, the natural polymer used is polycaprolactone.

Medical application

In medical application, biodegradable is widely employed in surgery (orthopaedic and vascular) as implanted matrices for the regulated continual release of medication inside the body, as plastic staples and surgical sutures that can be absorbed, ligature clips and vascular grafts, numerous types of pins and screws, bone fixation devices, and artificial skin. Table 5.2 illustrates a few examples of the polylactide (PLA)-based biodegradable polymer that is regularly used in the medical field.

Table 5.2: The PLA-based biodegradable polymer used in medical field (Luckachan & Pillai, 2011).

Application	**Trade Name**	**Composition**	**Manufacturer**
Interference screw	Endofix	LPLA or PGA-TMC	Acufex
	Sysorb	DLPLA	Synos
	Biologically quiet	PGA-DLPLA	Instrument Makar
	Phantom	LPLA	DuPuy
Dental	Drilac	DLPLA	THM Biomedical AHP
Sutures	Polysorb	PGA-LPLA	U.S. surgical
	Vicryl	PLLA(8%)-co-PGA(92%)	Ethicon
Suture anchor	SmartAnchor-D	SR-LPLA L	Bionx implants
	Bio-statak	LPLA	Zimmer
	BioROC EZ 2.8 mm	LPLA	Innovasive devices
	Panalok	82/18 LPLG	Surgical dynamics
Screw	SmartScrew	LPLA	Lorenz
Angioplastic plug	Angioseal	PGA-DLPLA	Bionx
Plates, mesh, screws	LactoSorb	PGA-LPLA	
Tack	SmartTack	LPLA	Lorenz

Guided tissue	Resolut	PGA-DLPLA	W.L. Gore
	Guidor	DLPLA	Procordia
	Antisorb	DLPLA	Atrix
Ligating, clips, bone pins and rods	Lactasorb	LPLA	Davis and Geek
	Resor-Pin	LPLA-DLPLA	Geistlich
	Biofix	LPLA or PGA	Bionx
Drug delivery products	Zoladex	PLA/PLGA	AstraZencca
	Lupron Depot	PLA/PLGA	TAP
	Decapeptyl	PLA/PLGA	Ferring

Other than that, Luckachan & Pillai (2011) said that there are quite a few products that, based on stereoregular isotactic poly(L-lactide) PLLA, have emerged: Phantom Suture Anchor (DePuy), Phantom Soft Thread Soft Tissue Fixation Screws, BioScrew, Meniscal Stingers (Linvatec), Clearfix Meniscal Darts (Innovative Devices), and Bio-Anchors. The development of PLLA fiber was made in 1971 after receiving approval from the Food and Drug Administration (FDA) in order to create a better suture than the first biodegradable synthetic suture, called DEXONs, that are made from synthetic poly(-glycolic acid). The exceptional strength of PLLA fibers has led to research into using them as a scaffolding material for constructing augmentation devices or ligament replacement to substitute the non-degradable fibers.

Challenges within application of polymers, biopolymers, and recycled polymers

Despite the various applications or trends of polymers, biopolymers, and recycled polymers, there are challenges in the application itself. According to Regassa et al. (2019), Fused Deposition Modelling (FDM) is one of the machines used for additive manufacturing processes. FDM technique has serious drawbacks like inconsistent accuracy in dimension and surface appearance because of the polymers' low mechanical characteristics. There is no dependable method to manage the precision since many manufacturing aspects, including contour width, air gap, raster angle, and others, affect the accuracy and surface quality. Figure 5.1 shows the variables for FDM process technology.

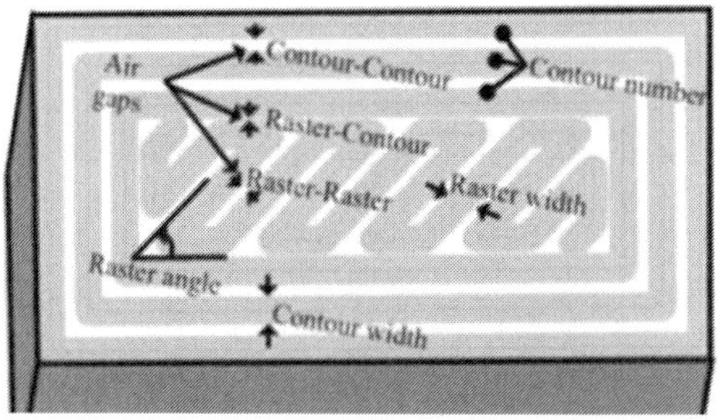

Figure 5.1: Variables for FDM process technology (Regassa et al., 2019).

The polymers in liquid form are by far the most popular materials in the additive manufacturing (AM) industry due to the relatively low melting temperature, lightweight, low price, and operating ease, although complicated geometry, lack of mechanical performance, and usefulness characteristics constantly repeated as the primary difficulties are restricting the variety of uses for additively produced polymer goods. From these few existing limitations, there has subsequently been a lot of interest in the creation of composite materials which are suitable with the accessible additive manufacturing techniques. For example, ceramics are used for additive fabrication as elaborated. However, this creation still records challenges in the usage for additive fabrication. The fundamental issue is the lack of readily available ceramic materials for additive manufacturing utilizing the current techniques, as well as the generally subpar dimensional tolerances and reliability. Despite the objection, as ceramic additive manufacturing offers greater influence over the part's microstructure and makeup, there are still chances for research and development, as in improvement of AM systems including the development of selection of materials for ceramics production (Regassa et al., 2019). Table 5.1 shows some other limitations of ceramics and polymer composites. Next, concrete-based composites are also used in additive manufacturing. Similar to ceramics, Regassa et al. (2019) mentioned that this composite material also faced challenges in the application. Because it might not maintain its geometry, the formwork-free production might not be suitable for self-compacting concrete. The key issues that need to be solved include other issues including uneven mechanical characteristics and poor cross-binding. Other than that, there also hasn't been much success in this industry's implementation. Nonetheless, before becoming popularized in the manufacturing sector, AM still has to go through a huge amount of analysis and research since some production materials are unaccustomed with this technology.

In addition, according to Ramesh Kumar et al. (2020), in the latest industry for bioplastics, academics and business are paying more attention to polylactic acid (PLA) because of its technical advancements in performance and usefulness. However, polyhydroxy alkanoates (PHA) consistently rank next to polybutylene adipate-co-terephthalate (PBAT) and polybutylene succinate (PBS) in terms of global production capacity. According to a recent research firm, the amount of PHA produced globally right now is around 25,320 tonnes, or 1.2%, compared to the respective shares held by PBAT and PBS of 13.4% and 4.3%, indicating that degradation rate is not always reliant on the point of origin or the polymer's constituent parts; polycaprolactone (PCL) and PBAT are fossil-based polymers that have a propensity to decompose. Nevertheless, nowadays, plastic waste has been emphasized as an international concern at every level, from plastic production to disposal and combustion. According to Ramesh Kumar et al. (2020), several bio-based polymers are not considered to be biodegradable. It is true that well-known bioplastics families like Polyhydroxybutyrate (PHB), PCL, and starch and their mixtures have been shown to biodegrade both in controlled and particular uncontrolled environments; regrettably, if their disposal is not managed, unmanaged biodegradation will increase the amount of plastic pollution already in the ecosystem. Since some bioplastics are made to only biodegrade under carefully controlled settings, disposing of them in less-than-ideal ecosystems like fresh water, soil, and the ocean has significant negative consequences which makes this one of the challenges in using polymers in daily life. This issue can be reduced with recycling. However, post-consumer bioplastic's waste variability, poor market sizes, different origins, and strong probability for infection are considered to be the main problems in recycling it. These can be related to the agricultural and medical applications from the previous elaborated trend that used biodegradable polymers.

6. Conclusion

In conclusion, there are different types of polymers that have been used in biomedical application; thermoplastic, natural polymer, elastomer, and synthetic polymer. Thermoplastic has the most fundamental molecular structure, which is composed of chemically unreactive macromolecules, and the application of thermoplastic can be through nanofluid devices. Natural polymer, also called a biopolymer, is an organic substance produced by the life processes of living things, where it can be categorized into polysaccharides, polypeptides, and polynucleotides. Cancer treatment and respiratory injury treatment is one of the biomedical applications for natural polymers. Polymers that are flexibly cross-linked can be referred to as elastomers, and they have three different kinds of elastomer; commercial elastomer, clinical grades of elastomer, and a low number of elastomers. All these types are used for different sectors. Synthetic polymer is made up of monomers, which are repeating structural elements produced by humans; there are two types, that have been referred to as plastics (nylon and polyethylene). The suitable types of polymers for biomedical application are chemical bonding, that is oxidative, resistant to hydrolytic, and other degradation mechanisms.

The modified or recycled polymer involved in biomedical application are polyethylene terephthalate, polyethylene, polystyrene, polyurethane, polypropylene, polyvinyl chloride, and acrylonitrile butadiene styrene, polycarbonate, and polyamide. There are four steps in the process of recovering and converting plastic trash which are primary recycling, secondary recycling (mechanical recycling), tertiary recycling (chemical recycling), and quaternary recycling (incineration). In order for bio composite materials to be suitable for biomedical applications, they must possess three fundamental characteristics found in nature: adhesion, self-healing, and super hydrophobicity. The thermal characterization involved in biomedical polymers are glass transition temperature (T_g), melting temperature, and crystallization temperature; which is necessary in defining the processing conditions to turn a natural or synthetic biomedical polymer into the finished device or implant. As for mechanical properties, it is important to know the strength that influences the polymer (molecular weight, cross-linking, and crystallinity), percent elongation to break (Ultimate Elongation), modulus of elasticity (Young's Modulus), and toughness.

There are a few trends in using polymers in various applications such as polymer composite material in additive manufacturing (ceramics for additive fabrication and concrete-based composite) and the trend of biodegradable polymers (agricultural application and medical application). Additive manufacturing is related to the files that are created by processing information using CAD and STL and ceramics for additive fabrication and concrete-based composite are examples of AM. One of the significant uses of additive ceramics is in tissue engineering, where strong and adaptable ceramic scaffolds with intricate designs are created. It also can control the microstructure and composition of the part. However, the challenges occur due to availability of ceramic material, and need to undergo development and improvement in terms of selection of materials for ceramic production. Concrete-based composite is related to concrete that is traditionally used in building and infrastructure. The first manufacturing of a cement-based additive is adhering sand layer using Portland cement paste by applying the deposition process concept. Nonetheless, it is not suitable for self-compacting concrete, especially for formwork-free, and has issues with the uneven mechanical characteristic and poor cross-binding, which are the reasons it is rarely used in the industry. Biodegradable polymers are categorized into two groups; natural and synthetic. Agricultural application used biodegradable polymers as agricultural mulches, planting containers, and also contained in the controlled release of agricultural pesticide. As for medical application, biodegradables are frequently used in orthopaedic and vascular surgery, mostly PLA-based biodegradable polymer and PLLA. Anyhow, the degradation rate is not always reliant on the point of origin or the polymer's constituent part, and there are several bio-based polymers that are not considered to be biodegradable.

Acknowledgements

First and foremost, we give gratitude and praise to God, the Almighty, for His numerous blessings while we worked on this project. We would like to express our deep and sincere gratitude to our lecturer, Professor Ir. Dr. Mohd Sapuan Salit, for giving us the opportunity to do research and providing helpful advice throughout this project. Other than that, we would like to thank Dr. Mohd Omar Syafiq and Dr Nazrin Nur Arief Mardi for their help and guidance during the preparation of this paper throughout the semester. As our advisers, their support, advice, valuable comments, and suggestions helped us in completing our paper successfully. Moreover, we want to thank our parents for their continuous support and understanding as it provideed us with strength and moral support throughout this semester. Lastly, we also want to give a special thanks and appreciation to our colleague and anyone who willingly helped us out with their abilities.

References

Agarwal, A., Balani, K., Narayan, R., & Verma, V. (2015). *Physical, thermal, and mechanical properties of polymers*. 329–344.

Al-Sabagh, A. M., Yehia, F. Z., Eshaq, G., Rabie, A. M., &ElMetwally, A. E. (2016). Greener routes for recycling of polyethylene terephthalate. *Egyptian Journal of Petroleum*, *25*(1), 53–64. https://doi.org/10.1016/j.ejpe.2015.03.001

Armentano, I., Barbanera, M., Carota, E., Crognale, S., Marconi, M., Rossi, S., Rubino, G., Scungio, M., Taborri, J., &Calabrò, G. (2021). Polymer Materials for Respiratory Protection: Processing, End Use, and Testing Methods. *ACS Applied Polymer Materials*, *3*(2), 531–548. https://doi.org/10.1021/acsapm.0c01151

Babker, A. M., Sotnik, S., & Lyashenko, V. (2018). *Scholars Journal of Applied Medical Sciences (SJAMS) Polymeric Materials in Medicine. February*. https://doi.org/10.21276/sjams.2018.6.1.33

Biron, M. (2012). *Thermoplastics and Thermoplastic Composites* [E-book]. William Andrew Publishing.

Biron, M. (2016). *Thermoplastics Material Selection. InMaterial Selection for Thermoplastic Parts*. https://doi.org/10.1016/b978-0-7020-6284-1.00001-5

Doppalapudi, S., Jain, A., Khan, W., & Domb, A. J. (2014). Biodegradable polymers-an overview. *Polymers for Advanced Technologies*, *25*(5), 427–435. https://doi.org/10.1002/pat.3305

Ferreira, P., Coimbra, U. De, & Gil, H. (2012). *Biomedical Polyurethane-Based Materials. May 2014*.

Hamad, K., Kaseem, M., & Deri, F. (2013). Recycling of waste from polymer materials: An overview of the recent works. *Polymer Degradation and Stability*, *98*(12), 2801–2812. https://doi.org/10.1016/j.polymdegradstab.2013.09.025

H. Iván Meléndez-Ortiz, Gustavo H.C. Varca, Edgar Zavala-Lagunes, E. B. (2016). State of the art of smart polymers: from fundamentals to final applications. *Polymer Science: Research Advances, Practical Applications and Educational Aspects, July*, 476–487. https://www.researchgate.net/publication/304950935_State_of_the_art_of_smart_polymers_from_fundamentals_to_final_applications

Ibrahim I., Sadiku E., Jamiru T., Hamam A., & Kupolati W. (2017). Applications of Polymers in the Biomedical Field. *Current Trends in Biomedical Engineering & Biosciences*, *4*(5), 102–104. https://doi.org/10.19080/ctbeb.2017.04.555650

John, M. J., & Thomas, S. (Eds.). (2012). *Natural polymers: composites* (Vol. 1). Royal society of chemistry.

Joseph, B., James, J., Kalarikkal, N., & Thomas, S. (2021). Recycling of medical plastics. *Advanced Industrial and Engineering Polymer Research*, *4*(3), 199–208. https://doi.org/10.1016/j.aiepr.2021.06.003

Kaushik, K., Sharma, R. B., & Agarwal, S. (2016). Natural polymers and their applications. *International Journal of Pharmaceutical Sciences Review and Research*, *37*(2), 30–36.

Khan, M. J., Svedberg, A., Singh, A. A., Ansari, M. S., & Karim, Z. (2019). Use of Nanostructured Polymer in the Delivery of Drugs for Cancer Therapy. In *Nanostructured Polymer Composites for Biomedical Applications*. Elsevier Inc. https://doi.org/10.1016/b978-0-12-816771-7.00013-2

Luckachan, G. E., & Pillai, C. K. S. (2011). Biodegradable Polymers- A Review on Recent Trends and Emerging Perspectives. *Journal of Polymers and the Environment*, *19*(3), 637–676. https://doi.org/10.1007/s10924-011-0317-1

Okan, M., Aydin, H. M., &Barsbay, M. (2019). Current approaches to waste polymer utilization and minimization: a review. *Journal of Chemical Technology and Biotechnology*, *94*(1), 8–21. https://doi.org/10.1002/jctb.5778

Oral, E., Kurtz, S. M., &Muratoglu, O. K. (2017). 7.3 Ultrahigh molecular weight polyethylene total joint implants. In *Comprehensive Biomaterials II* (Vol. 7, Issue November 2016). Elsevier Ltd. https://doi.org/10.1016/B978-0-12-803581-8.10219-X

Pino-ramos, V. H., Bucio, E., & Grafting, B. (2018). *Biomedical Polymer Radiation Grafting of Biopolymers and Synthetic Polymers Improving the hemocompatibility of biomedical polymers Polyacetals*.

Ragaert, K., Delva, L., & Van Geem, K. (2017). Mechanical and chemical recycling of solid plastic waste. *Waste Management*, *69*, 24–58. https://doi.org/10.1016/j.wasman.2017.07.044

Ramesh Kumar, S., Shaiju, P., O'Connor, K. E., & Babu, P. R. (2020). Bio-based and biodegradable polymers - State-of-the-art, challenges and emerging trends. *Current Opinion in Green and Sustainable Chemistry*, *21*, 75–81. https://doi.org/10.1016/j.cogsc.2019.12.005

Rana, M. M. (2021). Polymer-based nano-therapies to combat COVID-19 related respiratory injury: progress, prospects, and challenges. *Journal of Biomaterials Science, Polymer Edition*, *32*(9), 1219–1249. https://doi.org/10.1080/09205063.2021.1909412

Rajeswari, S. (2017). Natural Polymers: A Recent Review. *World Journal of Pharmacy and Pharmaceutical Sciences*, *June*, 472–494. https://doi.org/10.20959/wjpps20178-9762

Regassa, Y., Lemu, H. G., & Sirabizuh, B. (2019). Trends of using polymer composite materials in additive manufacturing. *IOP Conference Series: Materials Science and Engineering*, *659*(1), 012021. https://doi.org/10.1088/1757-899x/659/1/012021

Sastri, V. R. (2010). Materials Used in Medical Devices. *Plastics in Medical Devices*, 21–32. https://doi.org/10.1016/b978-0-8155-2027-6.10003-0

Shrivastava, A. (2018). Introduction to Plastics Engineering. *Introduction to Plastics Engineering*, 1–16. https://doi.org/10.1016/b978-0-323-39500-7.00001-0

Singh, N., Hui, D., Singh, R., Ahuja, I. P. S., Feo, L., &Fraternali, F. (2017). Recycling of plastic solid waste: A state of art review and future applications. Composites Part B: Engineering, 115, 409–422. https://doi.org/10.1016/j.compositesb.2016.09.013

Sýkora, D., & Švec, F. (2014). Synthetic polymer. *Chemical and Engineering News*, *79*(31), 16. https://doi.org/10.1016/b978-0-12-801889-7.00028-5

Ter Horst, B., Moiemen, N. S., & Grover, L. M. (2019). Natural polymers. In *Biomaterials for Skin Repair and Regeneration*. Elsevier Ltd. https://doi.org/10.1016/B978-0-08-102546-8.00006-6

Weerakoon-Ratnayake, K. M., O'Neil, C. E., Uba, F. I., & Soper, S. A. (2017). Thermoplastic nanofluidic devices for biomedical applications. *Lab on a Chip*, *17*(3), 362–381. https://doi.org/10.1039/c6lc01173j

Wong, K. V., & Hernandez, A. (2012). A Review of Additive Manufacturing. *ISRN Mechanical Engineering*, *2012*, 1–10. https://doi.org/10.5402/2012/208760

Yoda, R. (1998). Journal of Biomaterials Science, Elastomers for biomedical applications. *Journal of Biomaterials Science, Polymer Edition*, *9*(6), 561–626.

Aditya T, Allain JP, Jaramillo C, Restrepo AM (2022) Surface Modification of Bacterial Cellulose for Biomedical Applications. Int J Mol Sci 23:. https://doi.org/10.3390/ijms23020610

Bhat AH, Khan I, Usmani MA, et al (2019) Cellulose an ageless renewable green nanomaterial for medical applications: An overview of ionic liquids in extraction, separation and dissolution of cellulose. Int J Biol Macromol 129:750–777. https://doi.org/https://doi.org/10.1016/j.ijbiomac.2018.12.190

Czaja WK, Young DJ, Kawecki M, Brown RMJ (2007) The future prospects of microbial cellulose in biomedical applications. Biomacromolecules 8:1–12. https://doi.org/10.1021/bm060620d

Huang W, Wang Y, Huang Z, et al (2018) On-Demand Dissolvable Self-Healing Hydrogel Based on Carboxymethyl Chitosan and Cellulose Nanocrystal for Deep Partial Thickness Burn Wound Healing. ACS Appl Mater Interfaces 10:41076–41088. https://doi.org/10.1021/acsami.8b14526

Lam NT, Chollakup R, Smitthipong W, et al (2017) Utilizing cellulose from sugarcane bagasse mixed with poly(vinyl alcohol) for tissue engineering scaffold fabrication. Ind Crops Prod 100:183–197.

https://doi.org/https://doi.org/10.1016/j.indcrop.2017.02.031

Langari MM, Nikzad M, Labidi J (2023) Nanocellulose-based sensors in medical/clinical applications: The state-of-the-art review. Carbohydr Polym 304:120509. https://doi.org/10.1016/j.carbpol.2022.120509

Migliaresi C, Nicolais L (1980) Composite Materials for Biomedical Applications. Int J Artif Organs 3:114–118. https://doi.org/10.1177/039139888000300213

Potter KA, Jorfi M, Householder KT, et al (2014) Curcumin-releasing mechanically adaptive intracortical implants improve the proximal neuronal density and blood–brain barrier stability. Acta Biomater 10:2209–2222. https://doi.org/https://doi.org/10.1016/j.actbio.2014.01.018

Seifu DG, Meghezi S, Unsworth L, et al (2018) Viscoelastic properties of multi-layered cellularized vascular tissues fabricated from collagen gel. J Mech Behav Biomed Mater 80:155–163. https://doi.org/10.1016/j.jmbbm.2018.01.021

Shi Y, Jiao H, Sun J, et al (2022) Functionalization of nanocellulose applied with biological molecules for biomedical application: A review. Carbohydr Polym 285:119208. https://doi.org/https://doi.org/10.1016/j.carbpol.2022.119208

Tavşanlı B, Okay O (2018) Mechanically robust and stretchable silk/hyaluronic acid hydrogels. Carbohydr Polym 208:. https://doi.org/10.1016/j.carbpol.2018.12.088

CHAPTER XI

BIOCOMPOSITES FOR BIOMEDICAL APPLICATION

ABIR KHAN, MOHD SAPUAN SALIT

Abstract

Biocomposites present a distinctive combination of properties that make them exceptionally well-suited for a wide range of biomedical applications. These materials are typically composed of a polymer matrix—either derived from natural sources or synthesized—reinforced with natural fibers or nanoparticles. This composition significantly improves the mechanical, thermal, and biological characteristics. The adaptability of biocomposites allows them to closely replicate the mechanical properties of natural tissues, making them particularly effective in fields such as tissue engineering, drug delivery, wound care, and the development of medical devices. Moreover, the ability to precisely control the degradation rate and biocompatibility of biocomposites enhances their applicability in temporary implants and various therapeutic procedures. This chapter offers a comprehensive examination of the different categories of biocomposites, including polymer-based composites, nanocomposite hydrogels, and bioactive composites, among others. It highlights their applications, methods of fabrication, and the challenges they pose. Additionally, it explores the future trajectory of biocomposite research, stressing the importance of standardized safety assessments, the development of multifunctional materials, and breakthroughs in biosensing and 4D printing technologies.

1. Introduction

Biocomposites have gained prominence as an important material in the biomedical field, providing a distinctive set of properties that render them exceptionally well-suited for a wide array of medical applications (Ramakrishna, 2004). These materials are composed of two or more distinct phases, typically involving a natural or synthetic polymer matrix reinforced with natural fibers or particles. This blend of components enables biocomposites to showcase superior mechanical, thermal, and biological characteristics, making them highly suitable for use in tissue repair, tissue replacement, and a variety of therapeutic treatments and interventions (Benjamin et al., 2024). One of the key advantages of biocomposites is their ability to closely mimic the mechanical properties of natural tissues, which is crucial for successful integration into the human body (Ullah & Chen, 2020). Traditional biomaterials like metals and ceramics often pose challenges due to their high stiffness or brittleness, which can lead to complications such as stress shielding or implant failure. In contrast, biocomposites can be engineered with high precision to match the mechanical properties of both hard and soft tissues, thereby reducing the risk of complications and improving patient outcomes. The versatility of biocomposites extends beyond their mechanical properties. These materials can be designed to degrade within the body at a controlled rate, making them particularly useful for temporary implants that support tissue regeneration (Basha et al., 2015). Additionally, biocomposites are often biocompatible and eco-friendly, especially when natural fibers are used as the reinforcing phase. This biocompatibility is essential for ensuring that the materials do not cause adverse reactions when implanted in the body, further enhancing their suitability for medical applications.

Biocomposites are extensively utilized across numerous biomedical disciplines, including the development of dental implants, advancements in regenerative medicine, innovations in tissue engineering, and the production of artificial body parts (Aisyah et al., 2020). The ability to customize these materials for particular applications by adjusting variables such as fiber dimensions, shape, alignment, and the type of matrix offers considerable flexibility in design. This adaptability allows for precise tailoring to meet the specific demands of various biomedical uses. This customization potential allows biocomposites to meet the specific demands of different medical applications, making them a cornerstone of modern medical material science. In conclusion, biocomposites mark a substantial leap forward in the field of materials science, providing a highly adaptable and customizable option for a diverse array of biomedical applications. Their versatility and innovative potential position them as a key development in the advancement of medical materials. Their ability to provide a balance of strength, flexibility, and biocompatibility, along with their potential for controlled degradation and eco-friendliness, makes them an increasingly important material in the development of medical devices and therapeutic interventions. As research in this field progresses, biocomposites are expected to become increasingly vital in enhancing patient care and driving advancements in medical technology. Their growing importance suggests they will play a pivotal role in the future of healthcare innovations.

2. Overview on biocomposites for biomedical applications

Biocomposite materials are gaining prominence in biomedical applications due to their ability to combine the advantageous properties of different components, such as polymers, nanoparticles, and natural materials like cellulose and chitosan. These composites provide a flexible foundation for a wide range of applications, encompassing tissue engineering, drug delivery systems, wound healing processes, and the creation of medical devices, as outlined in **Table 1**. Polymer-based composites provide structural support and controlled degradation, making them ideal for scaffolds in tissue regeneration. Nanocomposite hydrogels enhance mechanical properties and responsiveness to stimuli, while bioactive composites promote cell adhesion and tissue growth, particularly in bone regeneration. Chitosan and bacterial cellulose composites are valued for their biocompatibility and are explored in biosensors, drug delivery, and wound care. Additionally, electrically conductive and shape-memory polymer composites bring advanced functionalities, such as electrical stimulation and minimally invasive deployment, expanding the possibilities for medical innovations.

Table 1: Overview of biomedical applications of various biocomposite materials

Type of Biocomposite	Biomedical Applications
Polymer-Based Composites	-Tissue engineering (scaffolds for cell growth) -Drug delivery systems -Medical device development -Advanced wound dressings (offering protection, antimicrobial properties, and therapeutic delivery)
Nanocomposite Hydrogels (NCHs)	-Tissue engineering -Regenerative medicine -Controlled drug delivery

Bioactive Composites	-Bone tissue engineering (scaffolds for bone regeneration) -Bone implants (enhancing osteoconductivity and osteoinductivity)
Chitosan Composites	-Drug delivery systems -Tissue engineering (scaffolds for bone and cartilage repair) -Wound healing -Biosensors (e.g., glucose sensing, DNA hybridization detection) -Imaging and targeted drug delivery
Bacterial Cellulose (BC) Composites	-Wound healing (promoting tissue regeneration and providing a moist environment) -Tissue engineering (scaffolds mimicking extracellular matrix) -Drug delivery systems (controlled release of therapeutic agents)
Electrically Conductive Polymeric Composites	-Biosensors (enhanced sensitivity and selectivity) -Drug delivery systems (controlling release of therapeutic agents in response to external stimuli) -Biomedical implants -Tissue engineering (scaffolds providing electrical stimulation for enhanced tissue regeneration)
Shape Memory Polymer (SMP) Composites	-Minimally invasive medical devices (e.g., stents, sutures) -Tissue engineering scaffolds (responsive to external stimuli like temperature and light) -Smart medical devices (adaptable to the body's environment for enhanced precision in medical interventions)”

2.1 Polymer-based composites

Polymer composite materials are becoming more prevalent in biomedical applications, owing to their remarkable versatility, robust mechanical properties, and compatibility with biological systems. Their significance is especially noted in tissue engineering, drug delivery systems, and the development of medical devices, where they are highly effective in replicating the characteristics of natural tissues. These composites are utilized in the fabrication of scaffolds for tissue engineering, offering essential structural support while ensuring controlled degradation that aligns with the natural tissue regeneration process. This allows the scaffolds to effectively integrate and support the development of new tissue (Biswal et al., 2020). Advanced fabrication techniques like electrospinning enable precise scaffold design, promoting cell adhesion and growth. Additionally, polymer composites are utilized in advanced wound dressings, offering protection, antimicrobial properties, and therapeutic delivery to enhance healing. However, challenges such as ensuring long-term stability, biocompatibility, and controlled degradation rates remain, necessitating ongoing research (Zagho et al., 2018). Overall, polymer composites hold great potential for improving patient outcomes and advancing medical technology.

2.2 Nanocomposite hydrogels

Nanocomposite hydrogels (NCHs) are cutting-edge materials that combine the water-retaining properties of polymeric hydrogels with the enhanced functionality of embedded nanoparticles or nanostructures (Biondi et al., 2015). These hydrogels create a three-dimensional, cross-linked framework that closely replicates the physical, chemical, and biological characteristics of natural tissues. This makes them exceptionally well-suited for biomedical applications such as tissue engineering, regenerative medicine, and the precise delivery of drugs.

Integrating different nanomaterials, such as carbon-based, polymeric, ceramic, and metallic nanoparticles, into the hydrogel matrix significantly improves the mechanical strength, biocompatibility, and sensitivity of nanocomposite hydrogels (NCHs) to external stimuli like temperature, magnetic fields, and electrical signals. This adaptability allows NCHs to provide more tailored and effective solutions in medical applications; particularly, mimicking the extracellular matrix and supporting tissue growth is crucial. Future research is focused on understanding the interactions between the hydrogel matrix and the embedded nanomaterials at different scales.

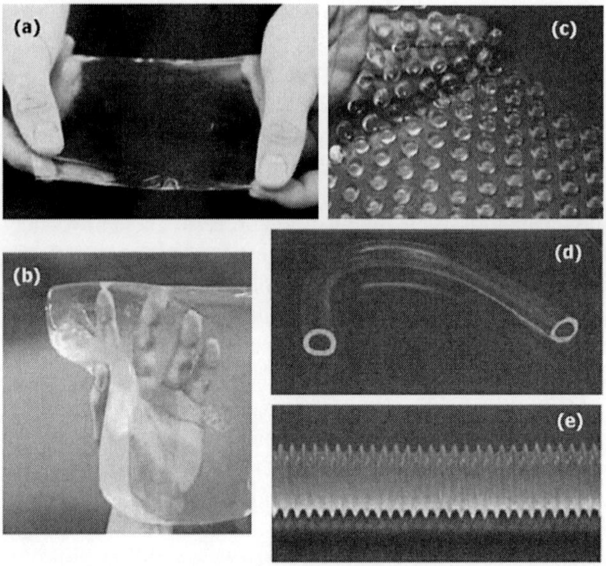

Figure 1 : Various shapes of nanocomposite hydrogels (a) thin film, (b) sheet, (c) uneven sheet, (d) hollow tube, and (e) bellows (Vashist et al., 2018)

This understanding will enable the custom design of NCHs for specific applications, potentially leading to the development of more sophisticated, stimuli-responsive hydrogels that closely replicate the properties of native tissues, further advancing the field of biomedicine. **Figure 1** shows different forms of nanocomposite hydrogels utilized in biomedical applications.

2.3 Bioactive composites

Bioactive composites are increasingly central to biomedical applications, especially in bone tissue engineering, where their capacity to replicate natural tissue properties and facilitate healing is essential. These composites merge bioactive substances, including bioactive glass, calcium phosphate ceramics, and bioactive polymers, with various polymer matrices to form scaffolds that encourage cell adhesion, growth, and differentiation (Guo et al., 2021). A key benefit of bioactive composites lies in their capacity to promote bone formation (osteoconductivity) and even stimulate the growth of new bone (osteoinductivity). For example, composites that combine polymers like poly(lactic acid) (PLA) or polycaprolactone (PCL) with ceramics such as hydroxyapatite (HA) or β-tricalcium phosphate (β-TCP) have demonstrated significant potential in bone regeneration. These composites not only provide mechanical strength but also enhance bioactivity, aiding in the seamless integration of implants with natural bone tissue. Advances in fabrication techniques, including electrospinning, additive manufacturing, and 3D printing, have made it possible to produce scaffolds that are highly porous and structurally intricate. These methods allow for precise control over the scaffold's architecture, which is essential for replicating the extracellular matrix and creating the optimal environment for tissue regeneration. Overall, bioactive composites represent a significant advancement in biomaterials, offering tailored solutions that combine the best properties of their individual components, thus playing a critical role in improving patient outcomes in various medical applications.

2.4 Chitosan composites

Chitosan composites have gained recognition as a highly adaptable material in biomedical applications, thanks to their natural biocompatibility, biodegradability, and bioresorbability (as shown in Figure 2). Sourced primarily from the shells of crustaceans, chitosan is a cationic polysaccharide renowned for its ability to form stable dispersions in water-based environments, making it an excellent choice for drug delivery, tissue engineering, and wound healing. In the field of tissue engineering, chitosan composites are frequently combined with inorganic substances such as calcium phosphate or hydroxyapatite to boost mechanical strength and osteoconductivity, making them particularly effective for repairing bone and cartilage (Hein et al., 2008). These composites can be fabricated into scaffolds that provide structural support for cell growth and tissue regeneration, with some studies exploring their use in controlled drug release within these scaffolds.

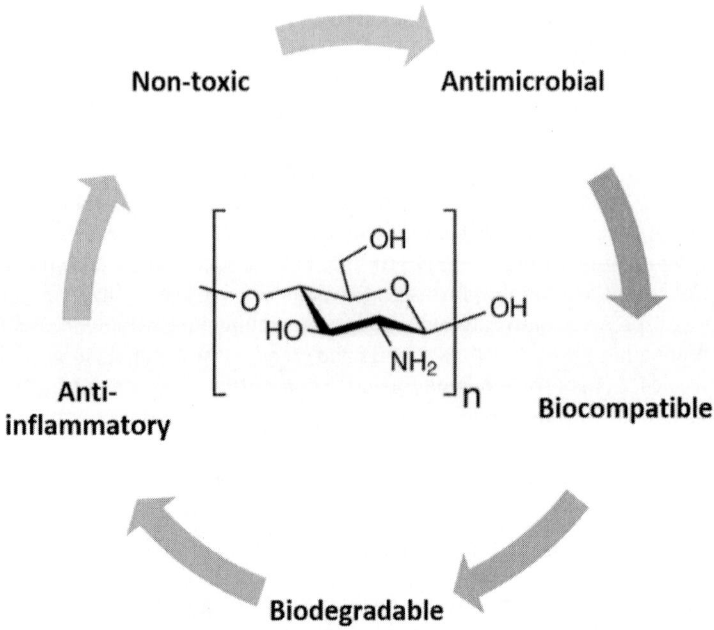

Figure 2 : Characteristics of chitosan composites for biomedical applications (Kravanja et al., 2019)

Chitosan composites are also gaining attention in biosensor development. When combined with electroconducting nanomaterials like gold nanoparticles or carbon nanotubes, chitosan enhances the sensitivity, selectivity, and stability of biosensors. These composite biosensors are being explored for applications ranging from glucose sensing to DNA hybridization detection, offering high precision and reliability in medical diagnostics. Moreover, chitosan's ability to form stable complexes with nanoparticles has opened avenues for its use in imaging and drug delivery systems. For instance, chitosan-coated gold nanoparticles have been investigated for targeted drug delivery and imaging applications, while chitosan-quantum dot composites are being studied for both cellular imaging and gene delivery. Overall, chitosan composites present significant potential in advancing biomedical technologies, though challenges related to mechanical properties and long-term stability remain areas of active research.

2.5 Bacterial cellulose composites

Bacterial cellulose (BC) has attracted considerable interest in the biomedical sector due to its exceptional characteristics, including non-toxicity, high purity, excellent biocompatibility, and robust mechanical strength. Produced by certain bacteria, BC features a distinct three-dimensional (3D) network structure with high porosity and the ability to absorb large amounts of water, making it highly suitable for a variety of medical applications. Recent innovations have centered on creating BC-based composite scaffolds by integrating nanoparticles and polymers, further enhancing the material's capabilities for biomedical purposes (Liu et al., 2020). BC's high mechanical strength and biocompatibility make it particularly suitable for wound healing, where it can provide a moist environment and promote tissue regeneration. Furthermore, BC-based scaffolds are gaining prominence in tissue engineering, functioning as structural frameworks that closely resemble the extracellular matrix, thereby supporting cell adhesion, growth, and differentiation. Additionally, BC has demonstrated significant potential in drug delivery systems, where its capacity to form stable hydrogels enables the controlled release of therapeutic substances. BC-derived nanocrystals (BCNCs) and nanofibrils (BCNFs) have also been investigated as reinforcing components in polymeric scaffolds, expanding the range of BC-based composites in various biomedical applications.

2.6 Electrically conductive polymeric composites

Electrically conductive polymeric composites have become highly promising materials in the biomedical field, thanks to their distinctive blend of electrical conductivity, biocompatibility, and mechanical strength. This unique combination makes them well-suited for a variety of medical applications. Traditional conductive polymers like polypyrrole and poly(3,4-ethylenedioxythiophene) (PEDOT) offer good electrical properties but often lack the necessary mechanical strength, biocompatibility, and processability required for biomedical use (Kaur et al., 2015). In order to overcome these challenges, researchers have engineered conductive polymeric composites by integrating conductive fillers like graphene, carbon nanotubes, and metallic nanoparticles into biocompatible polymer matrices. These advanced composites are being investigated for a variety of applications, including biosensors, drug delivery systems, biomedical implants, and tissue engineering. For instance, in tissue engineering, these materials can be used to create scaffolds that not only support cell growth but also provide electrical stimulation to enhance tissue regeneration. In drug delivery, the electrical properties of these composites can be leveraged to control the release of therapeutic agents in response to external stimuli. The development of these composites requires fine-tuning the balance between electrical conductivity and other essential properties like mechanical strength and biocompatibility. Progress in this area is ushering in a new era of "smart" biomaterials capable of interacting with biological systems in more advanced ways, ultimately leading to enhanced medical devices and therapies.

2.7 Shape memory polymer composites

Shape Memory Polymer (SMP) composites are a major breakthrough in the biomedical field, distinguished by their remarkable capacity to react to external stimuli and return to their original form after being deformed (Zhao et al., 2019). SMP composites have emerged as a groundbreaking development in the biomedical field, combining unique properties such as the shape memory effect, biocompatibility, and suitability for minimally invasive procedures. These materials can be activated by various stimuli, including temperature, light, or magnetic fields, making them versatile for use in a wide range of medical devices like stents, sutures, and tissue engineering scaffolds. A key advantage of SMPs is their ability to be implanted in a compact form and then expand or unfold into their functional shape at the target location within the body, making them ideal for minimally invasive surgeries. This feature not only reduces patient recovery times but also enhances the precision of medical interventions. Additionally, by incorporating fillers such as carbon nanotubes or metallic nanoparticles, the mechanical properties, biocompatibility, and activation methods of SMPs can be customized to meet specific biomedical needs. Recent advancements have also explored the development of multi-shape memory polymers, capable of undergoing multiple shape changes, further broadening their potential in complex medical applications. The integration of SMP composites into medical technology is poised to revolutionize the field, offering smarter, more adaptable materials for improved patient care. **Figure 3** shows different biomedical applications of shape memory polymers.

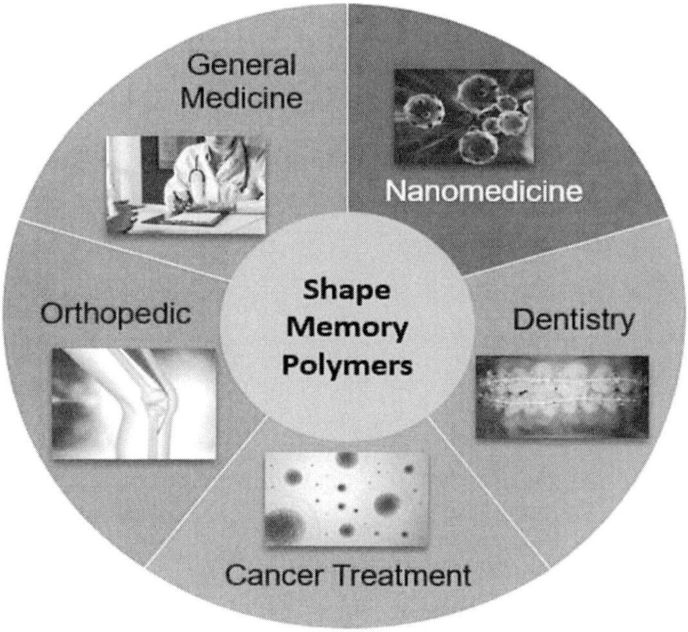

Figure 3 : Biomedical applications of shape memory polymers (Rokaya et al., 2023)

3. Fabrication Methods of Biocomposites for Biomedical Applications

Biocomposites for biomedical application are developed through various fabrication methods, each offering unique advantages for specific applications. These techniques include electrospinning, melt-extrusion, solution mixing, latex technology, in situ polymerization, particle leaching, additive manufacturing, and powder metallurgy.

3.1 Electrospinning

Electrospinning is a highly versatile and widely adopted technique for creating nanofibrous scaffolds in biomedical applications (Stocco et al., 2018). This process involves the application of a high-voltage electric field to a polymer solution, which is then forced through a needle to produce fine fibers with diameters ranging from nanometers to micrometers **(Figure 4)**.

The resulting fibers are collected on a grounded surface, forming a non-woven mat characterized by a high surface area and porosity. These features make electrospunfibers particularly well-suited for tissue engineering, as they closely mimic the extracellular matrix, creating an environment that supports cell adhesion, growth, and differentiation.

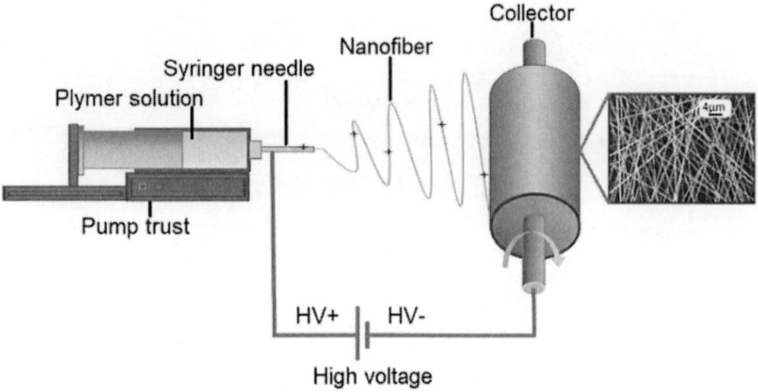

Figure 4 : Electrospinning process (Flores-Rojas et al., 2023)

The electrospinning process enables the integration of various functional materials, including nanoparticles, carbon nanotubes, or bioactive molecules, directly into the fibers. This capability allows for the development of composite scaffolds that possess improved mechanical strength, electrical conductivity, or biological functionality. For example, incorporating carbon nanotubes into electrospunfibers can improve the electrical conductivity of the scaffold, which is beneficial for applications like nerve regeneration or muscle tissue engineering (Suh et al., 2022). Additionally, electrospun scaffolds can be engineered to deliver drugs or growth factors in a controlled manner, further expanding their utility in regenerative medicine. However, electrospinning also presents challenges, particularly in scaling up for clinical applications. The technique is highly dependent on various parameters, such as solution viscosity, voltage, and environmental conditions, which can affect fiber uniformity and reproducibility. Moreover, creating complex, load-bearing structures using electrospinning remains difficult, limiting its use in applications requiring high mechanical strength. Despite these challenges, electrospinning continues to be a promising technique for developing advanced biomedical scaffolds.

3.2 Melt-Extrusion

Melt-extrusion is a commonly employed method for fabricating polymer composites, particularly in biomedical applications where mechanical strength and thermal stability are crucial. The process involves melting a polymer and mixing it with various fillers or additives using a twin-screw extruder. The melted mixture is then shaped into the desired form using compression molding, where it is cooled and solidified. This method allows precise control over the composite's properties by adjusting factors such as the filler content, screw speed, and extrusion temperature. Melt-extrusion is especially useful for creating composites that require uniform dispersion of fillers, such as ceramic particles or carbon nanotubes, within the polymer matrix. For instance, in the production of bioresorbable composites for bone implants, materials like polylactic acid (PLA) can be mixed with ceramic powders to enhance the mechanical properties and biocompatibility of the final product. The extrusion process ensures that the filler is evenly distributed throughout the polymer, resulting in consistent performance characteristics. One of the main advantages of melt-extrusion is its scalability and efficiency, making it suitable for large-scale production of biomedical devices and implants (Zagho et al., 2018). The process is also compatible with a wide range of polymers, including biodegradable and bioresorbable materials, which are essential for temporary implants that degrade within the body over time. However, the high temperatures required for extrusion can limit the use of heat-sensitive materials, and achieving optimal filler dispersion can be challenging, particularly with nanomaterials that tend to agglomerate.

3.3 Solution Mixing

Solution mixing is a straightforward and effective technique for producing polymer composites, particularly when working with polymers that dissolve in specific solvents (Campbell Jr, 2003). The process begins by dissolving the polymer in a solvent, creating a solution to which fillers, such as nanoparticles or fibers, are added. The mixture is stirred continuously to ensure uniform dispersion of the fillers within the polymer solution. Once fully mixed, the solution is cast into a mold and dried, either at room temperature or under vacuum, to remove the solvent and solidify the composite. This method is particularly advantageous for incorporating fillers that are difficult to disperse in a molten polymer, as the solution phase allows for better mixing and distribution. Solution mixing is commonly used for creating thin films, coatings, and membranes with specific properties tailored for biomedical applications. For instance, poly(vinyl alcohol) (PVA) can be mixed with clay nanoparticles to produce nanocomposite films with enhanced mechanical strength and barrier properties, suitable for wound dressings or drug delivery systems.

One of the key benefits of solution mixing is its versatility; it can be applied to a wide range of polymers and fillers. Additionally, the process can be conducted at relatively low temperatures, making it suitable for materials that are sensitive to heat. However, the choice of solvent is critical, as it must be capable of dissolving the polymer without adversely affecting the filler material. Moreover, the removal of the solvent must be carefully controlled to prevent defects such as voids or cracks in the final composite.

3.4 Latex Technology

Latex technology is a unique method used to incorporate conductive fillers into polymer matrices, particularly for creating electrically conductive composites. This process involves dispersing the fillers, such as carbon nanotubes or graphene, in a polymer latex, which is a colloidal suspension of polymer particles in water or another solvent (Cai & Song, 2008). The fillers are mixed into the latex, ensuring they are evenly distributed throughout the polymer matrix. The mixture is then dried, typically by lyophilization (freeze-drying), to remove the solvent and solidify the composite.

The key advantage of latex technology is its ability to produce composites with a highly uniform distribution of fillers, which is essential for achieving consistent electrical properties. This method is particularly useful for fabricating conductive composites used in biosensors, electrodes, or other biomedical devices that require precise electrical conductivity. Additionally, the latex process is scalable and can be easily adapted for large-scale production, making it suitable for industrial applications. However, latex technology also has its limitations. The process requires careful control of the dispersion and drying conditions to prevent aggregation of the fillers or defects in the composite structure. Moreover, the choice of polymer and filler must be compatible with the latex process, as some materials may not form stable colloidal suspensions or may be difficult to mix uniformly. Despite these challenges, latex technology remains a valuable tool for developing advanced conductive composites in the biomedical field.

3.5 In-situ Polymerization

In-situ polymerization is a fabrication technique where the polymerization of the monomer occurs within a system already containing the filler material. The process begins by dispersing fillers such as graphene or modified graphene in a liquid monomer (Mao & Wang, 2020). Once the fillers are adequately swollen by the monomer, an initiator is added to trigger the polymerization process, often using heat or radiation. This results in the formation of a polymer composite with a high content of well-dispersed fillers, leading to enhanced properties such as electrical conductivity or mechanical strength. This method is particularly advantageous for creating composites with a high loading of nanomaterials, as it allows for uniform distribution and strong interfacial bonding between the polymer matrix and the fillers. In situ polymerization is commonly used in the fabrication of advanced materials for biomedical applications, such as drug delivery systems, where the polymer matrix can be designed to degrade at a controlled rate, releasing therapeutic agents in a sustained manner. However, in-situ polymerization requires precise control over the polymerization conditions to avoid premature gelation or incomplete polymerization, which can result in heterogeneous materials. The process is also sensitive to the compatibility between the monomer and the filler, as poor interaction can lead to phase separation or agglomeration of the fillers. Despite these challenges, in-situ polymerization remains a powerful technique for producing high-performance composites with tailored properties for specific biomedical applications.

3.6 Particle Leaching

Particle leaching is a fabrication method used to create porous scaffolds, particularly in tissue engineering (Hokmabad et al., 2017). The process involves mixing a polymer solution with uniformly distributed leachable particles, such as salt or sugar. The mixture is cast into a mold, and the solvent is removed, leaving behind a solid polymer matrix embedded with the particles. The composite is then immersed in a solvent that dissolves the leachable particles, leaving behind a porous structure with interconnected pores. This technique is valued for its simplicity and ability to control the porosity and pore size of the scaffold by adjusting the amount and size of the leachable particles. Porous scaffolds created by particle leaching are ideal for applications that require high surface area and permeability, such as bone regeneration or vascular grafts. The method can also be combined with other techniques, such as extrusion-based 3D printing, to create more complex and precise structures.

The main limitation of particle leaching is the difficulty in achieving uniform porosity throughout the scaffold, especially in thicker or more complex structures. Additionally, the leaching process must be carefully controlled to ensure complete removal of the particles without compromising the integrity of the scaffold. Despite these challenges, particle leaching remains a widely used and effective method for creating porous materials in biomedical applications.

3.7 Additive Manufacturing

Additive manufacturing (AM), often referred to as 3D printing, is a groundbreaking fabrication method that enables the accurate creation of complex structures through a layer-by-layer approach. In the biomedical field, AM is employed to produce patient-specific implants, tissue engineering scaffolds, and other medical devices with intricate designs that are challenging or even impossible to realize using conventional manufacturing techniques (Singh et al., 2017). AM techniques include stereolithography (SLA), fused deposition modeling (FDM), selective laser sintering (SLS), and extrusion-based printing, each with its advantages depending on the material and application. One of the key benefits of AM is its ability to produce highly customizable structures tailored to the specific needs of the patient, such as implants that perfectly fit a bone defect or scaffolds with precisely controlled pore sizes and shapes for tissue engineering. AM also allows for the integration of multiple materials within a single structure, enabling the creation of composites with enhanced mechanical properties, bioactivity, or drug delivery capabilities. However, AM is not without challenges. The process is limited by the range of materials that can be 3D printed, particularly in terms of biocompatibility and biodegradability. Additionally, the mechanical properties of 3D printed structures may not match those of traditionally manufactured materials, and optimizing the design and printing parameters to achieve the desired performance can be complex. Nevertheless, AM continues to revolutionize biomedical engineering, offering new possibilities for patient-specific treatments and advanced medical devices as displayed in **Figure 5**.

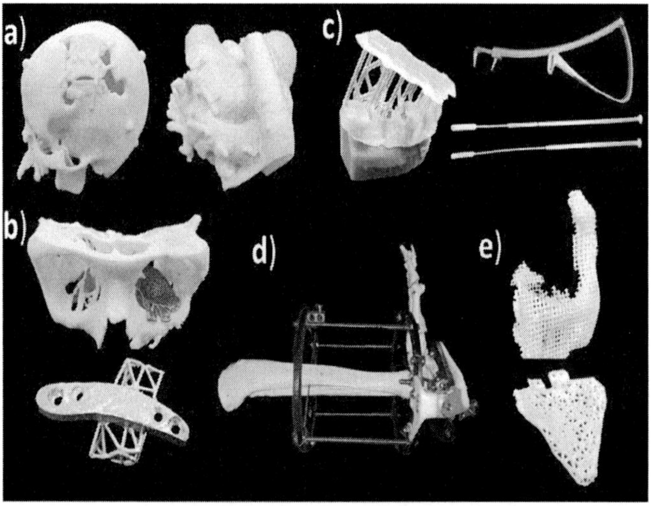

Figure 5 : (a) Medical models; (b) implants; (c) tools, instruments, and parts for medical devices; (d) medical aids, supportive guides, splints, and prostheses; (e) biomanufacturing developed using additive manufacturing (Salmi, 2021)

3.8 Powder Metallurgy

Powder metallurgy (PM) is an advanced fabrication technique widely employed for creating metal matrix composites (MMCs), particularly those utilized in biomedical applications such as implants and prosthetics. This method involves three primary stages: mixing metal powders with reinforcing particles, compacting the mixture into a desired shape, and sintering the compact at elevated temperatures to bond the materials together and solidify the composite (Kumar et al., 2021). In the first stage, metal powders, such as magnesium, titanium, or stainless steel, are thoroughly mixed with reinforcing agents like ceramics, carbon nanotubes, or bioactive glass. The mixing process is often enhanced by mechanical alloying, where high-energy ball milling ensures uniform distribution of the reinforcing particles within the metal matrix. This uniform dispersion is crucial for achieving consistent mechanical properties throughout the composite. Once mixed, the powder is compacted into a mold under high pressure to form what is known as a "green compact." This compacted shape is then subjected to sintering, a heat treatment process carried out at temperatures below the metal's melting point. Sintering helps to bond the particles together, enhancing the material's density, strength, and integrity. In some cases, additional steps like hot pressing or extrusion are used to further densify the material and improve its mechanical properties.

Powder metallurgy is particularly advantageous for producing complex, net-shaped components that require minimal machining (Munir et al., 2020). It also allows for the incorporation of bioactive materials, which can promote bone integration or other desirable biological responses when used in implants. Furthermore, the method is energy-efficient and generates minimal waste, making it a sustainable choice for biomedical manufacturing. A major application of powder metallurgy in the biomedical field is the production of biodegradable magnesium-based composites for temporary implants. These implants naturally dissolve within the body over time, thereby removing the need for a second surgery to extract them. By incorporating ceramic reinforcements or other bioactive agents through powder metallurgy, the mechanical strength and corrosion resistance of these composites are significantly improved, making them ideal for load-bearing orthopedic applications. In summary, powder metallurgy is a highly flexible and efficient technique for creating biocompatible composites with customized properties, providing substantial benefits in the design and manufacturing of advanced biomedical devices.

4. Applications of Biocomposites in Biomedical Fields

4.1 Tissue engineering

Tissue engineering involves regenerating damaged tissues using biocompatible scaffolds that support new cell growth. These scaffolds mimic the natural extracellular matrix (ECM) and are crucial for tissue-specific integration. Polymer composites, due to their unique morphology, facilitate cell adhesion and growth, making them ideal for tissue reconstruction.

4.1.1 Bone Tissue Engineering

Bone, a composite material consisting mainly of collagen fibers and hydroxyapatite (HA), remodels itself in response to mechanical stress, a principle known as Wolff's law (Ficai et al., 2011). Fracture healing requires effective implants that provide biomechanical stability and minimize tissue trauma. Researchers are focusing on biodegradable bone graft substitutes, using both natural and synthetic polymers such as chitosan, poly(propylene fumarate), and poly(lactide-co-glycolide) (PLAGA) (Rossi et al., 2015). For instance, PLAGA/calcium phosphate composite scaffolds have shown promising results in bone regeneration (Zagho et al., 2018). Polylactic acid (PLA) has been reinforced with materials like glass and ceramics to improve its mechanical properties for weight-bearing applications. Additionally, PLA composites with phosphate glass fibers have demonstrated good biological and mechanical characteristics. Innovations such as plasma-treated bioglassfibers and PLA composites have enhanced the mechanical strength and bone integration properties, essential for successful bone healing.

4.1.2 Blood Vessel Engineering

The repair of blood vessels is challenging due to the inadequacies of artificial scaffolds, leading to prolonged recovery times. Tissue-engineered vascular grafts, which combine synthetic materials with patient cells, offer a promising solution. Techniques like electrospinning are widely used to create 2D and 3D fibers that mimic the ECM, facilitating vascular graft development (Z. Wang et al., 2019). Bacterial nanocellulose (BNC) blended with polyvinyl alcohol (PVA) has been explored for artificial blood vessels, showing improved mechanical properties (Sharip et al., 2020). Additionally, biopolymers like chitosan and gelatin have been used to fabricate bilayered composites that mimic blood vessel characteristics. These materials provide a porous structure for cell proliferation and adhesion while offering the necessary mechanical support.

4.1.3 Skin Tissue Engineering

The skin, as the body's first defence against pathogens, requires effective regeneration strategies when damaged. Polymeric materials, particularly in 3D nanofibrous structures, are advantageous due to their high surface area and mechanical strength. Hydrogels, both natural (e.g., alginate, collagen) and synthetic (e.g., PVA), are commonly used in skin regeneration (Qi et al., 2022). Fibrous scaffolds containing collagen and drug molecules are commercially available for skin grafting. Recent advances include electrospun scaffolds combining modified hyaluronic acid and chondroitin sulfate, as well as alginate-based thin films for burn treatments.

4.1.4 Oral Tissue Engineering

Oral tissues are prone to damage from various factors, including dental caries and periodontitis. Tissue engineering in this area has made significant strides, such as the regeneration of tooth structures using biodegradable scaffolds seeded with porcine tooth buds (Monteiro & Yelick, 2017). These approaches have led to the successful development of mature tooth tissues. Additionally, biodegradable polymer scaffolds have been used to create 3D salivary glandular tissue and temporomandibular joint constructs, demonstrating the potential for regenerating complex oral structures.

4.1.5 Peripheral Nerve Injury Repair

Traditional methods for repairing peripheral nerve injuries, such as nerve autografts or artificial nerve elongation, often result in secondary damage to the injured tissue. To overcome these limitations, researchers like Hu et al. have explored the use of poly sebacic acid esters of glycerol as nerve scaffolds (Hu et al., 2019). By implanting Schwann cells onto the surface of these scaffolds, they observed promising results in nerve repair. The study demonstrated that shape memory polymer (SMP) nerve conduits, which exhibit self-healing behavior in bodily fluids, could safely and efficiently promote the orientated growth of tissue and cells, leading to successful nerve regeneration. This approach also reduces the risk of secondary damage, making it a significant advancement in nerve repair techniques.

4.1.6 Cell Culture and Mechanobiology

The mechanical properties of substrates used in cell culture, such as topography and elastic modulus, play a crucial role in cell behavior, including growth, migration, and differentiation. Traditional technologies have been limited to passively controlling these substrate properties. However, recent advancements in using shape memory polymers (SMP) for cell culture systems have shown that the surface morphology of these substrates can be actively changed during culturing (Chen et al., 2022). This ability to remodel the cytoskeleton and control cell behavior through topographic changes, without compromising cell viability, represents a significant breakthrough in mechanobiology research. For example, cells seeded on grooved SMP substrates adhered and spread, with their morphology being stably maintained even after the substrate's shape was altered by external stimuli.

4.1.7 Modulatable Hydrogels for Cellular Growth

The rigidity of extracellular substrates significantly impacts cell morphology, growth, and differentiation. To address the need for dynamic control of substrate mechanical properties, researchers have developed modulatable hydrogels that can change their rigidity in response to UV light (Wang, 2018). This innovation allows for the spatial and temporal manipulation of mechanical properties, thereby providing a more versatile platform for studying cellular behavior and enhancing tissue engineering applications.

4.2 Dental applications

Biocomposite materials play a crucial role in dental applications, particularly in the development of dental implants and resin-based restorative systems. In dental resin composites, biocomposites are widely used for aesthetic restorations due to their versatility and ability to mimic the natural appearance of teeth (Cho et al., 2022). However, challenges such as polymerization shrinkage, stress, and the presence of unreacted monomers can affect their performance and biocompatibility. Researchers have explored various resin formulation variables, including catalyst type, diluent content, and curing methods, to improve the mechanical properties and conversion efficiency of these composites. Moreover, the potential cytotoxicity and estrogenic effects of materials like bisphenol A (BPA) in these composites have been areas of concern, prompting further research into safer alternatives.

In dental implants, biocomposites offer advantages over traditional materials like ceramics and metals. Composite materials combining silicon carbide (SiC) and carbon fibers, or titanium (Ti) with hydroxyapatite (HA), provide a favorable balance of strength and biocompatibility (Beyth et al., 2014). These materials mimic the elastic modulus of natural teeth, reducing stress on surrounding tissues and improving long-term success. Additionally, graded composites with varying compositions of Ti and HA along the implant's length have shown promising results in promoting bone integration while maintaining mechanical stability. Despite these advancements, ongoing research and clinical studies are essential to fully realize the potential of biocomposites in dental applications.

4.3 Wound healing and dressing

Biocomposites play a pivotal role in the field of wound healing and dressing applications due to their ability to provide essential properties such as maintaining a moist environment, preventing microbial infections, and promoting tissue regeneration. Various biocomposite materials, such as chitosan, collagen, and nanocellulose, have been extensively studied for their potential in accelerating the wound healing process. These materials are often combined with other components to enhance their mechanical strength, biocompatibility, and functional properties. For instance, chitosan-based composites are known for their excellent biocompatibility and antibacterial properties, making them suitable for wound dressing applications (Khan & Mujahid, 2019). These composites can be further functionalized with additives like titanium oxide (TiO_2) or silver nanoparticles (AgNPs) to improve their antimicrobial efficacy. Such enhancements are crucial for preventing infections and promoting faster wound healing, especially in chronic wounds where bacterial invasion is a significant concern. Figure 6 shows some requirements for wound dressing materials. Collagen-based biocomposites have been developed to mimic the natural extracellular matrix, providing structural support and promoting the proliferation of fibroblasts and keratinocytes, which are essential for tissue regeneration. Additionally, composites incorporating nanocellulose have shown promise due to their high surface area, water-holding capacity, and mechanical strength, which are vital for maintaining a conducive environment for wound healing.

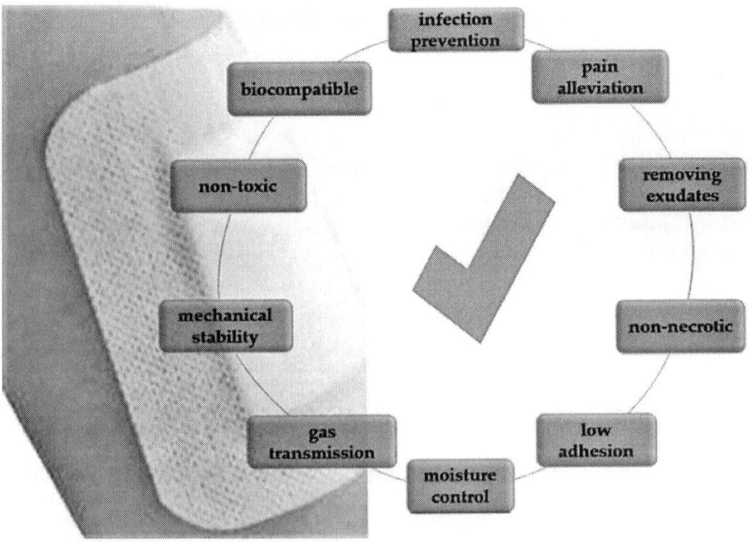

Figure 6 : Important aspects of wound dressing materials (Madej-Kiełbik et al., 2022)

Recent advancements in biocomposites also include the development of multifunctional wound dressings that combine drug delivery systems with wound healing properties. For example, alginate-based composites have been enhanced with bioactive glass nanoparticles to promote angiogenesis and accelerate tissue regeneration, while also providing sustained drug release to prevent infections. Overall, biocomposites represent a versatile and effective solution for wound healing and dressing applications, offering a combination of structural support, antimicrobial activity, and enhanced healing properties. These materials continue to be a focus of research, with ongoing efforts to optimize their performance and expand their clinical applications.

4.4 Drug and gene delivery

Biocomposites have emerged as promising materials for drug and gene delivery applications, offering tailored solutions to the challenges associated with conventional therapies. These composites combine the benefits of biocompatible polymers with nanoscale materials, enabling targeted and controlled delivery of therapeutic agents, thereby minimizing side effects and improving treatment efficacy. One of the key advantages of biocomposites in drug delivery is their ability to respond to specific stimuli, such as pH, temperature, or light, which can be exploited for targeted release in cancer therapy. For instance, MXene-based composites have demonstrated the ability to target tumor cells with low pH and release chemotherapeutic drugs like doxorubicin upon activation by near-infrared (NIR) light (George & Kandasubramanian, 2020). The inherent photothermal properties of MXenes, combined with their high drug-loading capacity, allow for efficient tumor ablation with minimal impact on healthy cells. These composites can be further functionalized with biocompatible coatings, such as hyaluronic acid, to enhance selectivity towards tumor cells and reduce cytotoxicity during circulation.

Similarly, carbon dot (CD)/polymer composites have shown potential in both drug and gene delivery (Adam et al., 2022). Their unique optoelectronic properties, such as photoluminescence and photoresponsiveness, make them ideal candidates for controlled drug release triggered by external stimuli like light. Moreover, these composites can be engineered to carry therapeutic genes, providing a dual function of diagnostics and treatment. The biocompatibility and tunability of CD/polymer composites make them suitable for a wide range of biomedical applications, including photothermal and photodynamic therapies for cancer treatment, as well as bioimaging and biosensing. Furthermore, biocomposites like hydrogels reinforced with nanoparticles and cellulose nanofibers (CNFs) demonstrate enhanced mechanical properties and controlled drug release profiles, making them suitable for various drug delivery systems. These smart materials are particularly valuable for creating responsive drug delivery platforms that can adapt to the physiological conditions of the body, ensuring that the therapeutic agents are released at the optimal time and location. In conclusion, biocomposites offer a versatile and effective platform for advanced drug and gene delivery systems, with the potential to significantly enhance the precision and efficacy of treatments in various medical fields.

4.5 Biosensors

Biocomposites play a crucial role in enhancing the performance of biosensors by integrating advanced nanomaterials, leading to improved sensitivity, selectivity, and rapid detection. For instance, graphene, with its exceptional electrical conductivity and high surface area, significantly enhances signal transduction in biosensors, making it an ideal material for sensor electrodes (Bahadır & Sezgintürk, 2016). Graphene-based biosensors have shown detection limits as low as 0.6 nM for fructose and 0.8 nM for glucose, demonstrating their high sensitivity. Molecularly imprinted polymers (MIPs) incorporated into biocomposites provide high specificity by forming nano-cavities that precisely match the target molecules, enabling accurate detection even at concentrations as low as 10^{-4} ng/mL for specific biomolecules. Additionally, magnetic nanoparticles (NPs) within biocomposites facilitate the separation and concentration of target analytes, further enhancing sensitivity. For example, biosensors using magnetic NPs have achieved detection limits of 0.006 μM for antithyroid drugs.

These advancements have led to the development of biosensors capable of detecting a wide range of analytes, including toxins, pathogens, and biomarkers, making them invaluable in medical diagnostics, environmental monitoring, and food safety. **Figure 7** shows the application of biosensors for controlling the various parameters of wound treatment. The combination of biocomposites with microfluidic systems has further increased their efficiency, enabling fast detection times of around 15 minutes and achieving detection limits as low as 1 ng/L for tumor markers like prostate-specific antigen (PSA). Thus, biocomposites are pivotal in advancing biosensor technology, offering efficient and versatile solutions for modern applications.

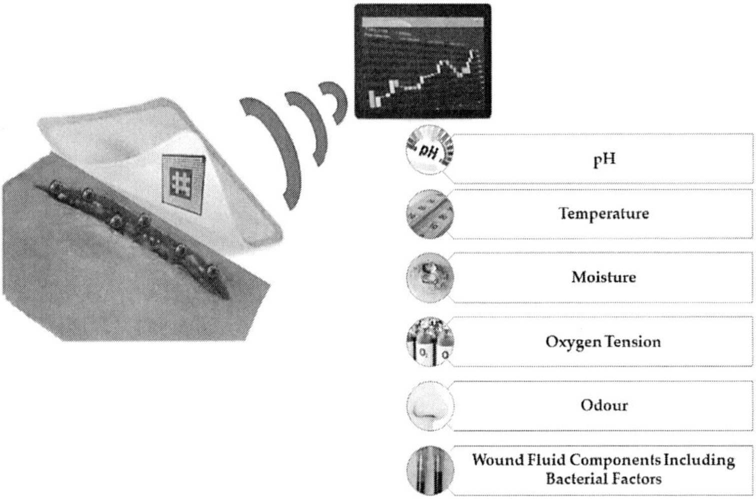

Figure 7: Application of biosensors for controlling the various parameters of wounds (Madej-Kiełbik et al., 2022)

5. Challenges and Future Research Directions of Biocomposites in Biomedical Applications

Biocomposite materials, including advanced nanoparticles and nanodiamonds (ND), exhibit substantial promise in the biomedical field due to their versatile and multifunctional properties. However, to fully harness their potential in clinical applications, several significant challenges need to be addressed.

i. Standardization and Safety Challenges

A major hurdle in the medical application of advanced nanomaterials is the absence of standardized protocols for safety testing (Saifi et al., 2018). Current research often involves diverse cellular models, varying doses, and different animal studies, resulting in inconsistent and non-comparable findings. Moreover, there is a critical need to conduct toxicity evaluations that encompass both healthy and diseased biological models. This issue is particularly relevant for materials like ND, where controlling degradation by-products and ensuring compatibility with biological systems are vital concerns.

ii. Development of Multifunctional Nanomaterials

The creation of multifunctional nanomaterials, such as those that combine various functional components within a polymer matrix, offers significant potential. These composites can be engineered for a range of applications, including diagnostics, therapy, and theranostics. For example, nano diamond composites with biopolymers are receiving increased attention, with future research likely to focus on developing devices that can simultaneously target specific cells, deliver therapeutic agents, and facilitate imaging (Mochalin & Gogotsi, 2015).

iii. Advances in Medical Diagnostics and Treatments

Nanomaterials present exciting opportunities in the field of medical diagnostics, particularly through the customization of their surfaces with imaging agents or therapeutic molecules. Advanced nanoparticles can be designed to treat deep-seated tumors or viral infections by exploiting their sensitivity to temperature changes (Bajpai et al., 2021). Future research is expected to explore the hybridization of these nanoparticles with other elements (such as silver, bismuth, or gold) to enhance their capabilities in complex medical procedures.

iv. Innovations in Biosensing and Programmable Sensors

The application of nanomaterials in biosensing technologies is another promising area, especially in the development of lab-on-a-chip devices and wearable sensors (Di Pietrantonio et al., 2019). By modifying nanomaterials with various elements, their ability to detect diseases can be significantly improved. The development of programmable electrochemical sensors offers a promising avenue for the real-time detection and monitoring of a wide range of diseases.

v. Stimuli-Responsive Materials and 4D Printing

Polymers, known for their adjustable properties, are ideal for developing materials that can react to environmental changes like temperature, pH, or light. The innovative technology of 4D printing, which incorporates the element of time into the conventional 3D printing process, allows for the creation of complex, multifunctional structures (Agarwal et al., 2021). Polymer-based hybrid materials are anticipated to be key players in 4D printing, leading to groundbreaking advancements in tissue engineering and regenerative medicine.

vi. Overcoming Production and Surface Modification Challenges

For nanomaterials to move from the laboratory to real-world medical applications, challenges such as large-scale production, cost-efficiency, and surface modification must be addressed. Current techniques like mechanical exfoliation for producing 2D nanomaterials are effective but require further improvement (Gao et al., 2018). Additionally, preventing the re-aggregation of nanoparticles and achieving better control over surface chemistry are essential to improving the performance and safety of these materials.

vii. Potential in Stem Cell Research and Tissue Engineering

The impact of nanomaterials on stem cell growth through thermoregulation and electronic signalling is an emerging area of study (Singla et al., 2019). Gaining a deeper understanding of these effects could lead to the development of novel 3D cell culture systems and advanced tissue engineering techniques, offering new possibilities for regenerative medicine. In summary, while nanomaterials like ND hold immense potential for biomedical applications, addressing the challenges of standardization, safety, production, and functionality will be critical for their successful adoption in clinical practice. Future research is expected to focus on enhancing the multifunctionality of these materials, advancing biosensing technologies, and developing cutting-edge fabrication techniques like 4D printing to expand the horizons of biomedical science.

6. Conclusions

The chapter concludes that biocomposites represent a pivotal advancement in biomedical applications due to their ability to combine the strengths of natural and synthetic materials. These materials offer the versatility to mimic the mechanical properties of various tissues, making them suitable for diverse applications such as tissue engineering, drug delivery, and wound healing. However, the successful implementation of biocomposites in clinical practice faces several challenges. Key among these is the need for standardized safety testing protocols, overcoming fabrication difficulties, and improving the scalability of production methods. The chapter emphasizes the importance of continued research into multifunctional materials, such as those enabled by 4D printing and nanotechnology, to address these challenges. As biocomposites evolve, their potential to revolutionize patient care through more effective, personalized medical solutions becomes increasingly clear. The insights provided suggest that addressing these challenges will be crucial for biocomposites to achieve their full potential in advancing biomedical technology.

References

Adam, G. O., Sharker, S. M., & Ryu, J. H. (2022). Emerging biomedical applications of carbon dot and polymer composite materials. *Applied Sciences*, *12*(20), 10565.

Agarwal, T., Hann, S. Y., Chiesa, I., Cui, H., Celikkin, N., Micalizzi, S., Barbetta, A., Costantini, M., Esworthy, T., & Zhang, L. G. (2021). 4D printing in biomedical applications: emerging trends and technologies. *Journal of Materials Chemistry B*, *9*(37), 7608–7632.

Aisyah, H. A., Paridah, M. T., Sapuan, S. M., Khalina, A., Ilyas, R. A., & Nurazzi, N. M. (2020). A review of biocomposites in biomedical application. *Composites in Biomedical Applications*, 31–48.

Bahadır, E. B., &Sezgintürk, M. K. (2016). Applications of graphene in electrochemical sensing and biosensing. *TrAC Trends in Analytical Chemistry*, *76*, 1–14.

Bajpai, S., Tiwary, S. K., Sonker, M., Joshi, A., Gupta, V., Kumar, Y., Shreyash, N., & Biswas, S. (2021). Recent advances in nanoparticle-based cancer treatment: a review. *ACS Applied Nano Materials*, *4*(7), 6441–6470.

Basha, R. Y., TS, S. K., & Doble, M. (2015). Design of biocomposite materials for bone tissue regeneration. *Materials Science and Engineering: C*, *57*, 452–463.

Benjamin, S. R., & Júnior, E. J. M. R. (2024). Biocomposites for biomedical applications. In *Advances in Biocomposites and their Applications* (pp. 227–255). Elsevier.

Beyth, N., Farah, S., Domb, A. J., & Weiss, E. I. (2014). Antibacterial dental resin composites. *Reactive and Functional Polymers*, *75*, 81–88.

Biondi, M., Borzacchiello, A., Mayol, L., & Ambrosio, L. (2015). Nanoparticle-integrated hydrogels as multifunctional composite materials for biomedical applications. *Gels*, *1*(2), 162–178.

Biswal, T., BadJena, S. K., & Pradhan, D. (2020). Synthesis of polymer composite materials and their biomedical applications. *Materials Today: Proceedings*, *30*, 305–315. https://doi.org/https://doi.org/10.1016/j.matpr.2020.01.567

Cai, D., & Song, M. (2008). Latex technology as a simple route to improve the thermal conductivity of a carbon nanotube/polymer composite. *Carbon*, *46*(15), 2107–2112.

Campbell Jr, F. C. (2003). *Manufacturing processes for advanced composites*. Elsevier.

Chen, J., Hamilton, L. E., Mather, P. T., & Henderson, J. H. (2022). Cell-responsive shape memory polymers. *ACS Biomaterials Science & Engineering*, *8*(7), 2960–2969.

Cho, K., Rajan, G., Farrar, P., Prentice, L., & Prusty, B. G. (2022). Dental resin composites: A review on materials to product realizations. *Composites Part B: Engineering*, *230*, 109495.

Di Pietrantonio, F., Cannatà, D., & Benetti, M. (2019). Biosensor technologies based on nanomaterials. In *Functional nanostructured interfaces for environmental and biomedical applications* (pp. 181–242). Elsevier.

Ficai, A., Andronescu, E., Voicu, G., & Ficai, D. (2011). *Advances in collagen/hydroxyapatite composite materials*. InTech Rijeka, Croatia.

Flores-Rojas, G. G., Gómez-Lazaro, B., López-Saucedo, F., Vera-Graziano, R., Bucio, E., & Mendizábal, E. (2023). Electrospun Scaffolds for Tissue Engineering: A Review. *Macromol*, *3*(3), 524–553. https://doi.org/10.3390/macromol3030031

Gao, E., Lin, S.-Z., Qin, Z., Buehler, M. J., Feng, X.-Q., & Xu, Z. (2018). Mechanical exfoliation of two-dimensional materials. *Journal of the Mechanics and Physics of Solids*, *115*, 248–262.

George, S. M., & Kandasubramanian, B. (2020). Advancements in MXene-Polymer composites for various biomedical applications. *Ceramics International*, *46*(7), 8522–8535.

Guo, Z., Poot, A. A., &Grijpma, D. W. (2021). Advanced polymer-based composites and structures for biomedical applications. *European Polymer Journal*, *149*, 110388. https://doi.org/https://doi.org/10.1016/j.eurpolymj.2021.110388

Hein, S., Wang, K., Stevens, W. F., & Kjems, J. (2008). Chitosan composites for biomedical applications: status, challenges and perspectives. *Materials Science and Technology*, *24*(9), 1053–1061.

Hu, T., Wu, Y., Zhao, X., Wang, L., Bi, L., Ma, P. X., & Guo, B. (2019). Micropatterned, electroactive, and biodegradable poly (glycerol sebacate)-aniline trimer elastomer for cardiac tissue engineering. *Chemical Engineering Journal*, *366*, 208–222.

Kaur, G., Adhikari, R., Cass, P., Bown, M., & Gunatillake, P. (2015). Electrically conductive polymers and composites for biomedical applications. *Rsc Advances*, *5*(47), 37553–37567.

Khan, M. A., & Mujahid, M. (2019). A review on recent advances in chitosan based composite for hemostatic dressings. *International Journal of Biological Macromolecules*, *124*, 138–147.

Kravanja, G., Primožič, M., Knez, Ž., & Leitgeb, M. (2019). Chitosan-Based (Nano)Materials for Novel Biomedical Applications. *Molecules*, *24*(10). https://doi.org/10.3390/molecules24101960

Kumar, A., Vichare, O., Debnath, K., & Paswan, M. (2021). Fabrication methods of metal matrix composites (MMCs). *Materials Today: Proceedings*, *46*, 6840–6846.

Liu, W., Du, H., Zhang, M., Liu, K., Liu, H., Xie, H., Zhang, X., & Si, C. (2020). Bacterial cellulose-based composite scaffolds for biomedical applications: a review. *ACS Sustainable Chemistry & Engineering*, *8*(20), 7536–7562.

Madej-Kiełbik, L., Gzyra-Jagieła, K., Jóźwik-Pruska, J., Dziuba, R., & Bednarowicz, A. (2022). Biopolymer Composites with Sensors for Environmental and Medical Applications. *Materials*, *15*(21). https://doi.org/10.3390/ma15217493

Mao, H., & Wang, X. (2020). Use of in-situ polymerization in the preparation of graphene/polymer nanocomposites. *New Carbon Materials*, *35*(4), 336–343.

Mochalin, V. N., &Gogotsi, Y. (2015). Nanodiamond–polymer composites. *Diamond and Related Materials*, *58*, 161–171.

Monteiro, N., &Yelick, P. C. (2017). Advances and perspectives in tooth tissue engineering. *Journal of Tissue Engineering and Regenerative Medicine*, *11*(9), 2443–2461.

Munir, K., Biesiekierski, A., Wen, C., & Li, Y. (2020). Powder metallurgy in manufacturing of medical devices. In *Metallic Biomaterials Processing and Medical Device Manufacturing* (pp. 159–190). Elsevier.

Qi, L., Zhang, C., Wang, B., Yin, J., & Yan, S. (2022). Progress in hydrogels for skin wound repair. *Macromolecular Bioscience*, *22*(7), 2100475.

RaeisdastehHokmabad, V., Davaran, S., Ramazani, A., & Salehi, R. (2017). Design and fabrication of porous biodegradable scaffolds: a strategy for tissue engineering. *Journal of Biomaterials Science, Polymer Edition*, *28*(16), 1797–1825.

Ramakrishna, S. (2004). *An introduction to biocomposites*. Imperial College Press.

Rokaya, D., Skallevold, H. E., Srimaneepong, V., Marya, A., Shah, P. K., Khurshid, Z., Zafar, M. S., & Sapkota, J. (2023). Shape Memory Polymeric Materials for Biomedical Applications: An Update. *Journal of Composites Science*, *7*(1). https://doi.org/10.3390/jcs7010024

Rossi, F., Santoro, M., & Perale, G. (2015). Polymeric scaffolds as stem cell carriers in bone repair. *Journal of Tissue Engineering and Regenerative Medicine*, *9*(10), 1093–1119.

Saifi, M. A., Khan, W., &Godugu, C. (2018). Cytotoxicity of nanomaterials: Using nanotoxicology to address the safety concerns of nanoparticles. *Pharmaceutical Nanotechnology*, *6*(1), 3–16.

Salmi, M. (2021). Additive Manufacturing Processes in Medical Applications. *Materials*, *14*(1). https://doi.org/10.3390/ma14010191

Sharip, N. S., Yasim-Anuar, T. A. T., Norrahim, M. N. F., Shazleen, S. S., Nurazzi, N. M., Sapuan, S. M., & Ilyas, R. A. (2020). A review on nanocellulose composites in biomedical application. *Composites in Biomedical Applications*, 161–190.

Singh, S., & Ramakrishna, S. (2017). Biomedical applications of additive manufacturing: Present and future. *Current Opinion in Biomedical Engineering*, *2*, 105–115.

Singla, R., Abidi, S. M. S., Dar, A. I., & Acharya, A. (2019). Nanomaterials as potential and versatile platform for next generation tissue engineering applications. *Journal of Biomedical Materials Research Part B: Applied Biomaterials*, *107*(7), 2433–2449.

Stocco, T. D., Bassous, N. J., Zhao, S., Granato, A. E. C., Webster, T. J., & Lobo, A. O. (2018). Nanofibrous scaffolds for biomedical applications. *Nanoscale*, *10*(26), 12228–12255.

Suh, T. C., Twiddy, J., Mahmood, N., Ali, K. M., Lubna, M. M., Bradford, P. D., Daniele, M. A., & Gluck, J. M. (2022). Electrospun carbon nanotube-based scaffolds exhibit high conductivity and cytocompatibility for tissue engineering applications. *ACS Omega*, *7*(23), 20006–20019.

Ullah, S., & Chen, X. (2020). Fabrication, applications and challenges of natural biomaterials in tissue engineering. *Applied Materials Today*, *20*, 100656.

Vashist, A., Kaushik, A., Ghosal, A., Bala, J., Nikkhah-Moshaie, R., A. Wani, W., Manickam, P., & Nair, M. (2018). Nanocomposite Hydrogels: Advances in Nanofillers Used for Nanomedicine. *Gels*, *4*(3). https://doi.org/10.3390/gels4030075

Wang, Y. (2018). Programmable hydrogels. *Biomaterials*, *178*, 663–680.

Wang, Z., Mithieux, S. M., & Weiss, A. S. (2019). Fabrication techniques for vascular and vascularized tissue engineering. *Advanced Healthcare Materials*, *8*(19), 1900742.

Zagho, M. M., Hussein, E. A., &Elzatahry, A. A. (2018). Recent overviews in functional polymer composites for biomedical applications. *Polymers*, *10*(7), 739.

Zhao, W., Liu, L., Zhang, F., Leng, J., & Liu, Y. (2019). Shape memory polymers and their composites in biomedical applications. *Materials Science and Engineering: C*, *97*, 864–883. https://doi.org/https://doi.org/10.1016/j.msec.2018.12.054

CHAPTER XII

APPLICATION OF FIBER MATERIAL AS SUSTAINABLE FILLING MATERIAL

AIDA SAFITRI, WASLIYATUL HASANAH, TEUKU RIHAYAT, SURYANI

Abstract

The growing demand for sustainable materials has prompted the exploration of fiber-based materials as viable alternatives to conventional fillers. Fiber materials, derived from both natural and synthetic sources, offer a wide range of environmental and performance benefits, including biodegradability, recyclability, and improved mechanical properties. This book reviews and discusses the applications of various fiber materials, such as natural fibers (e.g., jute, sisal, hemp) and synthetic fibers (e.g., polypropylene, polyester), as sustainable filling materials in various industries, including construction, automotive, and packaging. This book focuses on discussing the mechanical properties as well as the environmental impact of these fiber materials compared to traditional non-sustainable options. Research shows that fiber materials not only meet the required performance standards but also contribute to the reduction of carbon footprint and waste, making them a promising solution for sustainable development.

1 Introduction

1.1 Definition and Use of Composites

Fundamentally, materials consisting of at least two different materials are called composites. The combination of two or more materials either engineered or natural aims to obtain better physical and mechanical properties of the materials. Composites are formed from two elements, namely matrix and reinforcement (filler), and generally composites consist of a weak matrix and rigid and strong fibers. Composites are widely applied in various aspects of human life such as in home furnishings, motor vehicles, ships, and airplanes. Some of the advantages of composites include light weight, durability, low abrasion, and affordable procurement costs [1]. Bricks, wood, man-made, concrete, and bone composites, for example, like fiber-reinforced plastics (FRP), fall under this definition of composites [2].

Based on the type of reinforcement, there are three divisions of composites, namely fiber composites, particulate composites, and structural composites. While based on the nature of the reinforcement, composites are divided into isotropic and anisotropic composites. In its application, isotropic composites provide the same effect both in terms of stress or strain due to getting the same reinforcement for all directions (longitudinal and transverse). Thus, anistotropic composites have the opposite way of working, namely the strength is not the same value of either the effect of stress or strain because the reinforcement is not able to provide the same reinforcement for various directions both transversal and longitudinal [3].

The polymer matrix which is then reinforced with fibers is the main material that forms the composite. Generally, fibers such as carbon, glass, and aramid are widely used, although this does not rule out the use of other fibers such as wood, asbestos, and paper, which are applied on several occasions as needed. The polymers used are usually thermosetting plastics such as epoxy, vinylester, or polyester, and phenolformaldehyde resins are still commonly used. The polymers used are usually thermosetting plastics such as epoxy, vinylester, or polyester, and phenolformaldehyde resins are still frequently used [4].

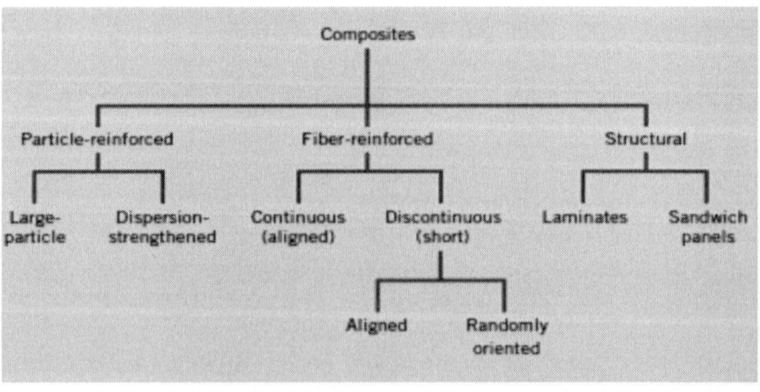

Figure 1. Classification of Composites (Rosa et al., 2016)

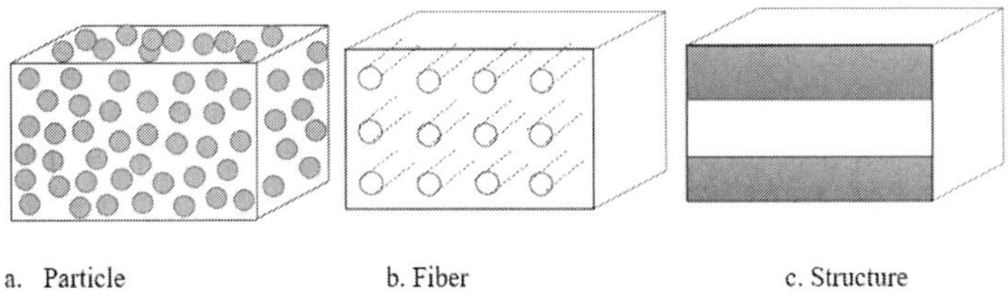

Figure 2. Illustration of Composites Based on Reinforcement

1.2 Fiber-Reinforced Polymers (FRP)

Fiber-reinforced polymers (FRPs) are composites applied in various types of advanced engineering structures, encompassing use in helicopters, aircraft and spacecraft, as well as ships, and offshore platforms. In addition, FRPs are also used in vehicles, chemical processing equipment, and civil infrastructure such as bridges and buildings [4].

With their high strength and stiffness, FRPs are often applied to create various structures. Moreover, they have a combination of low weight as a property of epoxy resin and synthetic fiber. Therefore, they are considered as high-performance composite materials. Among the techniques considered popular for various applications is the addition of natural fibers (NF), especially in view of sustainable materials. Many benefits are offered by NF in addition to its potential to replace synthetic fibers, namely producing bio-composites such as NFRPC (Natural Fiber-Reinforced Polymers Composite) [5].

Fiber and matrix are two elements contained in composite materials, including FRP composites. However, the difference is that there are additional compositions in the form of fillers and additives. Fiber is a material with a high modulus of elasticity. This certainly greatly affects the mechanical properties possessed by FRP. Not only that, the resin, which plays an important role in distributing pressure and preventing it from damage caused by the environment, also works together. In addition to fiber and matrix, fillers as additional elements also have an important role, especially in reducing procurement or overall production cost requirements. Other additional elements, namely additives, are also known to contribute to increasing the strength of FRP, both mechanical properties and physical properties, so as to increase its capabilities and performance.

In physics, the abbreviated name of a composite often reflects the type of reinforcing fiber used; for example, carbon fiber-reinforced polymers are known as CFRP. On the other hand, the difference between fiber types is in their tensile strain capability and stiffness values. Figure 3 presents a summarized classification of FRP composites [6].

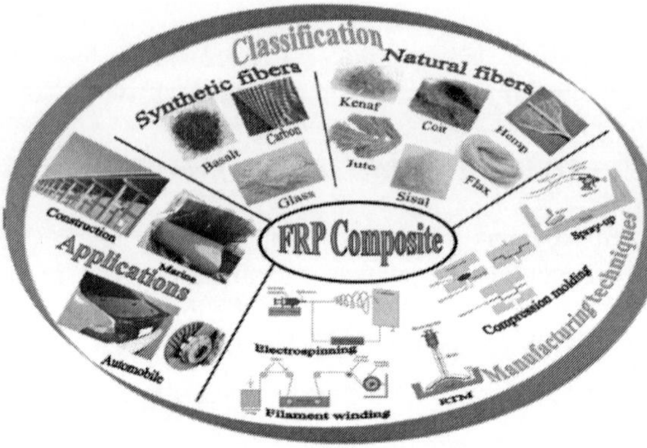

Figure 3. Classification of FRP Composite (Rajak et al., 2019)

A major factor driving the increased use of composites in recent years has been advances in the development of more advanced types of FRP materials. These include innovations in high-performance resin systems as well as the use of new reinforcement styles, such as carbon nanotubes and nanoparticles. This book offers an up-to-date explanation of the mechanical properties, resistance to impact tolerance, and applications of 3D FRP composites.

1.3 Natural Fibers as Reinforcement in Composites (NFRPC)

There are various advantages possessed by Natural Fiber-Reinforced Polymer composites (NFRPC), including being cheaper, easy to find, having a light weight, being easily decomposed, but still having good mechanical properties. With these advantages, NFRPC is a highly recommended material for various applications, starting from applications to transportation such as motor vehicles, trains, to aircraft. In the construction field, NFRPC also takes part in making foundations that are more affordable but still have good strength and quality. NFRPCs can also be referred to as bio-composites which are categorized into full or partial green composites [7].

The weaknesses of structural concrete elements such as easy deterioration due to weathering and aging, design defects, etc., have led scientists to conduct research and come up with various alternatives to correct these problems. This has encouraged the research and development of various reinforcement methods. One of the alternatives used is the addition of a cement matrix to increase the strength of the component. Steel fibers and conventional fibers are conventionally more commonly used to increase the ductility of composites. However, with the need to use and apply more environmentally friendly materials, the demand for ecological materials has led to the exploration of plant-based fibers as an alternative to conventional fibers. Thus, not only contributing to improved strength, toughness, and durability at a better cost, the use of materials from plant-based fibers is also more environmentally friendly as they are recyclable and biodegradable. Therefore, the review contained in this book aims to present the current trends in the use of plant-based fibers as reinforcement in cement matrix composites. The literature review and analysis will facilitate the identification of: (1) the most widely used vegetable fibers in cement-based composites, (2) methods to improve the mechanical properties of reinforced composites, (3) the effect of fiber length (short or long) on the mechanical properties of these composites, and (4) the main applications in structural and nonstructural elements to evaluate the applicability of these fibers in the construction industry.

This book is based on a review of experimental research conducted in the field of vegetable fiber-reinforced cement matrix composites using databases and search engines, such as Science Direct, Scopus, EBSCO, Springer, ASCE Journals, SAGE Journals, and Google Scholar. In addition, doctoral dissertations and books were also reviewed. The keywords used for the literature review search were vegetable fibers, cement matrix composites, reinforcement, treatment, structural elements, and mechanical properties. The initial analysis identified experimental studies where the effects of these fibers on the mechanical properties of various cementitious composites were investigated, showing a significant increase in technological advances from 2010 onwards (Figure 4).

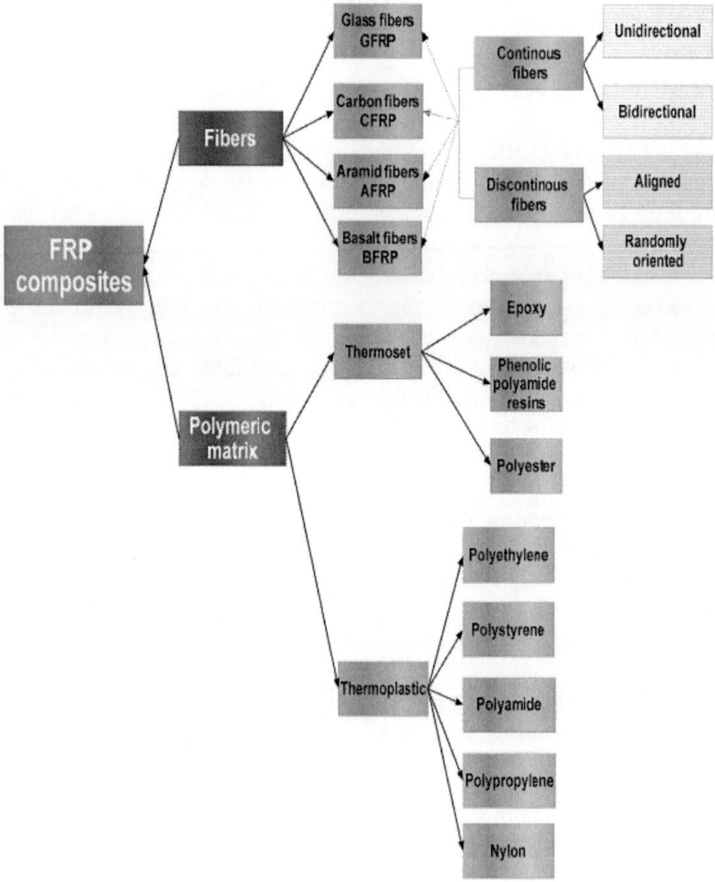

Figure 4. Classification of Fiber-reinforced Composites (FRPs) (Dinita et al., 2024)

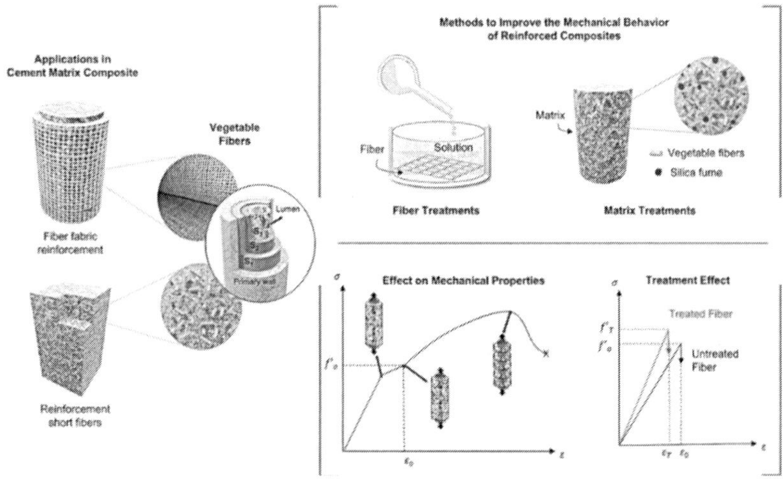

Figure 5. Effects of Fibers on Mechanical Properties of Cementitious Composites (Valeria et al., 2022)

On the other hand, the most widely used vegetable fibers in cement matrices were also identified at this stage, as seen in Table 1.

Table 1 . Chemical Composition of Common Natural Plant Fiber Used in Cement Composites (Labib WA, 2022)

Fibre	Cellulose %	Hemicellulose %	Lignin %
Abaca	65–63	15–17	7–9
Bamboo	26–43	15–26	21–31
Banana	60–65	6–8	5 –10
Coir	32–45	27–29	40–45
Date palm	39–44	22–26	11–15
Flax	72–75	5–14	2–4
Hemp	57–77	14–17	9–13
Hibiscus	28–38	25–27	17–23
Jute	41–48	18–22	21–24
Kenaf	44–57	22–23	15–19
Pineapple	70–82	18–20	5–12
Ramie	87–91	5–8	0.8
Sisal	47–62	21–24	12–20

Similarly, studies related to the application of conventional treatments, such as alkalinization, silanization, hornification, accelerated carbonation, and pozzolan addition, to the cement matrix and vegetable fibers were selected to evaluate their effects on the mechanical performance (bending, tensile, compressive, and toughness) of the cement matrix reinforced with various vegetable fibers (both short fibers and continuous fabrics). Finally, a review of the applications of these cementitious composites is summarized and analyzed to determine their maximum level of technological development in structural and nonstructural elements.

1.4 Enviromental Perspective

Compared to conventional aluminum, composite materials' ecological benefits on aircraft have been demonstrated and researched through various life cycle assessment (LCA)-based studies. The global advancement of the aircraft industry with composite designs is estimated to contribute to a 15%-20% reduction in CO_2 content by 2050. NFRPC is more environmentally friendly compared to traditional composites due to the content of cellulose, lignin, and also hemicellulose in its stucture [15]. Lightweight composites offer advantages in renewable energy sectors like wind, solar (photovoltaics), and hydrogen, with each area having extensive needs for conservation, storage, transportation, and utilization [16].

Cellulose is an abundantly available molecule that biodegrades naturally with the molecular shape shown in Figure 6 below [17]. In addition, after serving its valuable life, NFRPC can decompose along with various polymers. For example, the combustion-related energy for Chinese reed fibers is estimated at 14 MJ/kg addition [8]. This combustion results in no CO_2 emissions to the atmosphere. Thus, the combustion of natural fiber composites results in positive carbon credits and a lower hazard to atmospheric impact.

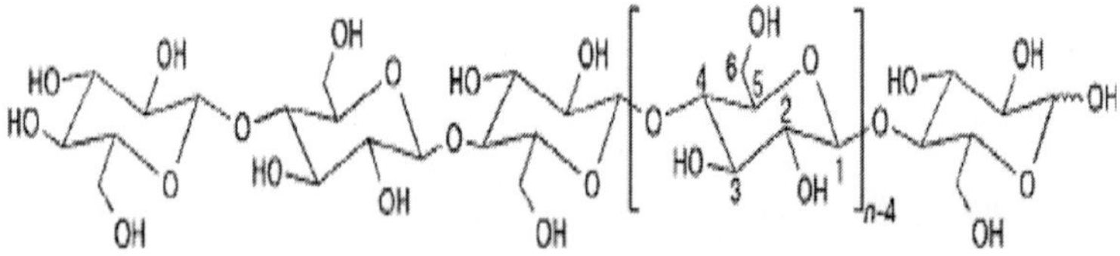

Figure 6. Molecular Structure of Cellulose (N = DP, Degree of Polymerization) (Praveen et al., 2019)

1.5 Chemical Treatment

To improve the mechanical properties of composites, it is certainly necessary to apply various chemical treatments, especially to metal matrix composites and fiber-reinforced composites, due to the fundamental problems associated with natural fibers (NF), which is their hydrophobic nature [9]. Exhibiting less than optimal mechanical properties even when hybridized with synthetic fibers, natural fibers tend to absorb moisture more easily and have lower strength than synthetic fibers [9]. Researchers aim to develop a pre-treatment process to increase the compatibility of natural fibers with the polymer matrix and to improve the mechanical characteristics of composites. Researchers aim to develop a pre-treatment process to improve the compatibility of NF with the polymer matrix and to improve the mechanical characteristics of composites [10].

The alkali treatment approach is one of the most simple, economical, and robust techniques applied to improve the attachment of NF to epoxy resins as show in Figure 7. The impact of NaOH in alkali treatment has been studied by various researchers. The results showed that NaOH concentrations (1, 5, and 10%) with soaking times for (0, 24, and 48 hours) had different consequences on the surface morphology of the fiber [11] [55].

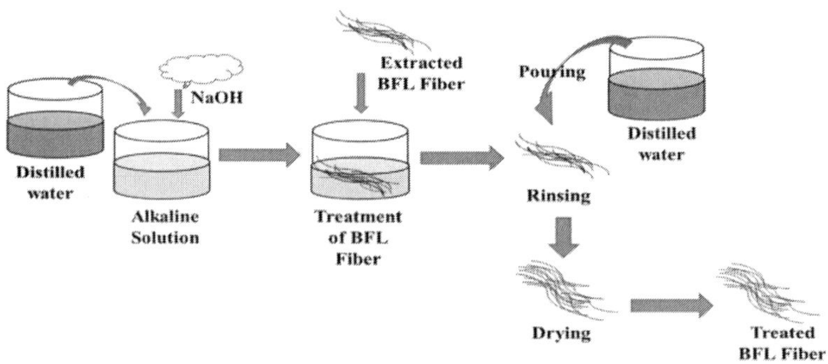

Figure 7. Alkali Treatment (Mustapha et al., 2023)

Sahril et al., in their research, found that FF fibers treated by adding 9 wt.% alkali produced young modulus values and mechanical strength that were qualified because of the high cellulose fraction after the non-cellulose material was removed. In addition, the alkali- and sodium bicarbonate-treated fibers have improved thermal stability, crystallinity, and surface roughness compared to hemp fibers that did not undergo processing [12].

Silane consists of a bifunctional structure and belongs to a material that can form a chemical platform between the matrix and the fiber as a chemical enhancer. Silane is also an adhesion promoter in various applications and is used as a substrate primer. With the main task of producing a practical assembly and collaboration of maleic anhydride, it is expected that the maleic bridging agent is expected to reinforce composites containing various NFs from its treatment [13].

Modification of natural fibers through chemical processes, such as acetylation, has been recognized for enhancing their structural performance in composite applications. This technique involves replacing some of the hydroxyl (–OH) groups on the fiber surface with acetyl (CH_3CO) groups, which leads to a decrease in the fiber's affinity for water. As a result, the dimensional stability and durability of the fiber improve significantly under varying environmental conditions. Natural fibers, due to their biodegradability and availability, are increasingly considered suitable alternatives to synthetic reinforcements in polymer composites [9]. Their performance can be further optimized by integrating functional agents during processing. One such approach involves incorporating graphene oxide (GO) into piassava fibers before embedding them into an epoxy matrix, allowing for better interfacial bonding and improved composite behavior compared to untreated fibers [18]. Hemicellulose and lignin degradation occurred at relatively higher temperatures for GO-coated piassava fibers. Pure piassava fiber, based on the results of thermogravimetric analysis, showed a maximum thermal stability of 200 °C [10]. In pure piassava fibers, no GO was found to coat the fibers within the time frame of the study conducted at an ambient temperature of 125°C for pure piassava fibers which was not found in GO-coated fibers within the maximum time frame studied [11]. Dynamic mechanical analysis (DMA) testing showed significant changes in the viscous stiffness and damping capacity of epoxy composites due to GO coating on piassava fibers [18].

1.6 Natural Fiber-reinforced Hybrid Composite Material

Relatively inexpensive hybrid composites are formed from one or more types of materials in a matrix. It is possible that fiber-reinforced polymer composites have:

1. Two types of phases, consisting of one matrix and reinforcement for each phase.
2. Two-phase type, consisting of one reinforcing phase and yet more than one matrix phase.

3. Consisting of many matrix phases and more than one lasing phase [9].

Earlier in many studies, it has been observed that hybridization with any synthetic fiber improves many structural properties of NFRPC. By adding glass volume fraction, the Young's modulus of tensile strength increases significantly, this happens due to better compatibility between the matrix and glass fibers. Therefore, Venkateshwaran et al. concluded that to improve the mechanical properties of hybridization, carbon fibers and glass fibers are largely responsible for this [22][56]. Compared to hemp reinforced with epoxime

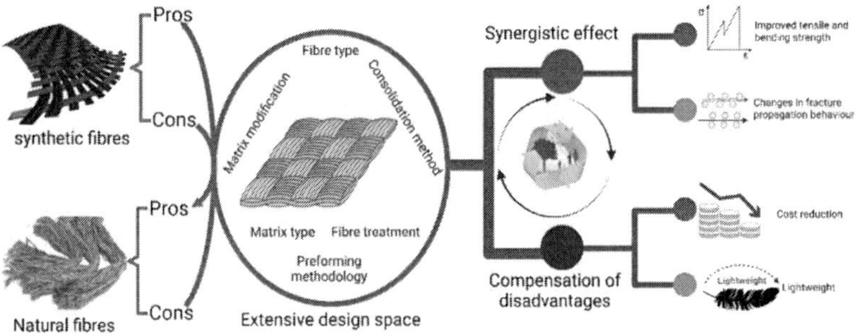

Figure 8 . The General Concept of Natural FRP Hybrid Composites (Sikiru et al., 2022)

composites, studies show that polyester composites reinforced with hemp fibers are inferior in properties. In another study, Sanjay et al. observed that by hybrid reinforcement of jute fibers and glass fibers, the properties of composites can be improved compared to composites reinforced with jute fibers or glass fibers individually [23]. A tensile strength of about 59 MPa was obtained for 3 cm long and 51% weight banana fiber-reinforced with polyester [9]. Hybridization is usually categorized into categories. Interlamination, or basically overlay, is obtained by placing layers made of various fibers. Whereas, in intralamine, two filaments are trapped in a single layer of [24].

The hybridization effect of different layers of laminates composed of different layers of hemp/cloth/E-Glass woven fabrics made using the vacuum bagging method has excellent hybridization results, especially in mechanical properties. In addition, the increasing consequences of hybridization to improve structural, mechanical, and thermal properties as a result of hybridization to NFRPC have been reported in many studies [25].

Hybridization by adding synthetic fiber laminates can give NFRPC better moisture absorption properties as well as nobler mechanical properties [9]. For example, glass filament is resistant to moisture absorption and has relatively higher mechanical strength compared to NF [26]. Thus, hybridization through synthetic fibers achieves an agreement between superior mechanical properties and composite cost. For hybrid fiber composites, the Rule of Hybrid Mixture (RoHM) is a widely used technique that can predict the Young's modulus of these hybrid NFRPCs. However, the impact of hybridization on mechanical degradation caused by moisture remains unclear [26].

Weapons-related threats have greatly affected the lives of people, especially public servants and soldiers. Multi-layered Armor Systems (MAS) provide effective protection against these threats. Aluminum and Kevlar (Dyneema) fibers, as synthetic materials, are conventionally used in the manufacture of MAS. Therefore, we can say that NFRPC is now considered as one of the most developed materials for various engineering application. NFRPC is now considered as the second MAS layer [9].

The comprehensive study that has been conducted using NFRPC as well as conventional materials (Dyneema plates) in addition to hard armor is certainly not without reason. The PC is produced by utilizing pineapple leaves (PALF) which serve as an extra hard armor system inside a regular level IIIA protective vest made with Kevlar. By combining NFRPC with a ceramic front panel, the result is level III ballistic protection. After research, it was found that this level of protection has a positive and effective effect on the user, especially for 7.62mm rifle-type ammunition for the National Institute of Justice (NIJ) global standard, which is equivalent to the Dyneema plate typically used in protective vests [9].

Khalid et al. carried out a comparable study on polyester composites made using fique fibers for the aforementioned tests. After evaluating different volume fractions, the fique fibers were tested, and the findings indicated that the composite with 30% fiber exhibited the best performance volume fraction of the polyester composite having 30% fiber volume fraction was the best formulation for MAS. In addition, the composite enriched with fique fiber was also able to replace Kevlar thanks to its high energy-absorbing ability and integrity after impact [9].

1.7 Cement Matric Composites

Cement matrix composites (CMC) are materials formed from a mixture of cement as the main binding matrix with the addition of aggregates such as sand and gravel and reinforcing materials such as fibers. In this case, the cement matrix is able to unify the additional materials. As a result, when compared to conventional concrete, its physical and mechanical properties are significantly superior [18]. The significant improvement in both the mechanical and dynamic aspects of reinforced concrete is an advantage of adding composite fibers. One of its important effects is the ability to control tensile and compressive strength in composite materials [10].

Overcoming brittleness and poor crack resistance, which are weaknesses of conventional concrete, is a key advantage of CMC [11]. As a result, its tensile strength becomes superior to that of ordinary concrete [12]. In addition, CMC showed a significant improvement in resistance to micro-cracking [13]. Various types of fibers have been used in the development of CMCs, including steel fibers, glass fibers, polymer fibers, and natural fibers. Each type of fiber imparts unique characteristics to the composite. For example, research by Arturs, et al. (2018). Concrete reinforced with composite fibers showed flexural behavior compared to straight and corrugated steel fibers at the same dosage up to a strength of 125 MPa. Meanwhile, CMC with steel fibers showed more stable strain hardening [14].

One interesting aspect of CMC is its potential to support sustainable development. The use of natural fibers such as coconut fiber or bamboo fiber in CMC can also improve the mechanical properties of the material, while it is also biodegradable so that its application in industry can match metals, especially with stronger corrosion resistance and better flexibility. In the case of coconut fiber, it is the most ductile and energy-absorbing material [15-16].

CMC applications in the construction industry are vast, ranging from large-scale infrastructure to complex architectural components. In the infrastructure sector, CMC has been used in the construction of earthquake-resistant bridges, with much better energy-absorbing capabilities than conventional reinforced concrete [17]. In architecture, CMC allows the creation of complex and thin shapes that were previously difficult to achieve with traditional materials.

A major challenge in the development of CMCs is the optimization of the interaction between the cement matrix and reinforcing fibers. A recent study by Suman et al. (2020) that focused on surface modification and interfacial engineering on fiber and matrix bonding shows that the increase in bond strength depends on the polymer matrix that is directly related to the composite [8].

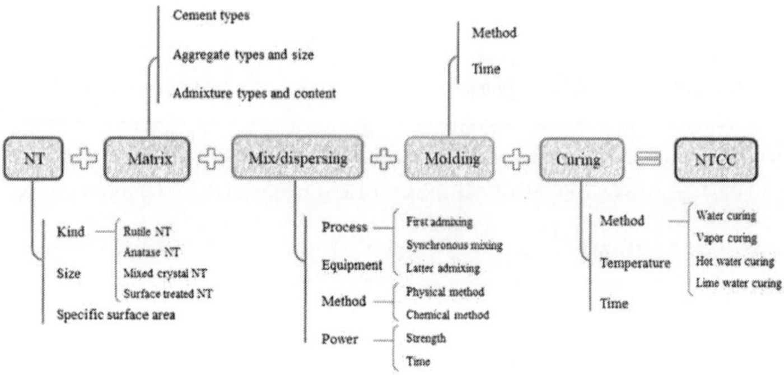

Figure 9 . Schematic Presentation of Nano Titanium Dioxide/ Cementitious Composites (NTCC) Production Methods (Li et al., 2018)

The development of nanotechnology is also a new opportunity for cement matrix composites, especially in their design. In addition, for the addition of strength as well as electrical and thermal conductivity of CMCs, the addition of nanoparticles such as nanosilica is highly relevant [19]. This is certainly a new breakthrough that allows for smart concrete that is capable of automonitoring and even generating its own energy. With continuous advances in the field of production technology, the future of CMC is very promising. Market projections show significant growth, with the estimated global market value reaching 150.28 billion by 2028, a sharp increase from the previous value of 86.56 in 2021. Of course, this progress reflects the great potential of CMC in shaping the future of the construction and materials industry [20].

2. Natural Fibers as Cement Matrix Composite Reinforcement

2.1 Characteristic of Natural Fibers-Reinforced

The characteristics of Natural Fiber-reinforced Cementitious Composites (NFRCC) have been the focus of research in recent years. The presence of these compounds provides many advantages, including low cost, easy to find, as they are derived from the availability of renewable natural resources, and biodegradability [21]. In this case, NFRCC combines a cement matrix with natural fibers derived from plants to offer an alternative to commonly used synthetic fibers. In general, natural fibers often used in NFRCC are sisal, jute, kenaf, and coconut fibers, although there are several other sources as shown in Figure 10 below.

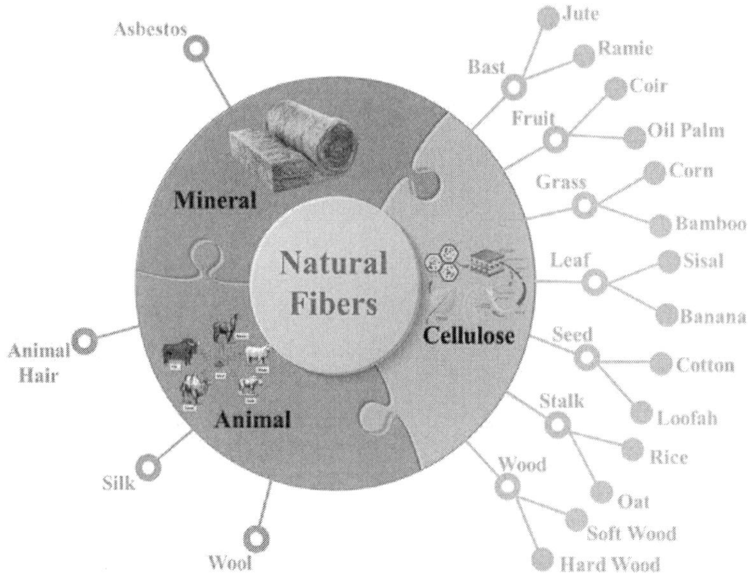

Figure 10 . Classification of Plant Fibers According to the Origin (Issam et al., 2023)

Research by Marzena et al. (2022) showed that synthetic fibers can cause a decrease in compressive strength up to 4%. In contrast to this, the use of jute fiber can increase the compressive strength up to 27% [27]. Meanwhile, a study by Mahyuddin et al. (2013) in the "Construction and Building Materials" journal mentioned that coconut fiber can increase the crack resistance and ductility of cementitious composites up to 13% and 9% of the initial value, respectively [28]. One of the main characteristics of NFRCC is its superior mechanical properties. Research conducted by Jawad et al. (2023) in the "Journal of Materials Research and Technology" showed that highlighting 1 to 1.5% by weight of cement is the optimal proportion of bamboo fiber addition [29]. In addition, NFRCC exhibits better impact resistance than conventional concrete [30].

Although NFRCC exhibits many beneficial characteristics, challenges remain in terms of standardization and quality control. Variability in the properties of natural fibers between production batches can affect the consistency of composite performance. However, with advances in technology and further research, NFRCC holds great potential to become a primary construction material in the future, combining high performance with environmental sustainability [32].

Cellulose is the primary component of plant fibers, whereas protein is a key element of animal fibers [33]. Plant-based fibers are primarily sourced from tropical regions and agricultural crops. Fibers like sisal, jute, ramie, cotton, banana, and hemp are widely used in the construction and automotive industries because of their low cost, lightweight nature, excellent mechanical properties, and abundant availability, as shown in Figure 11 [34].

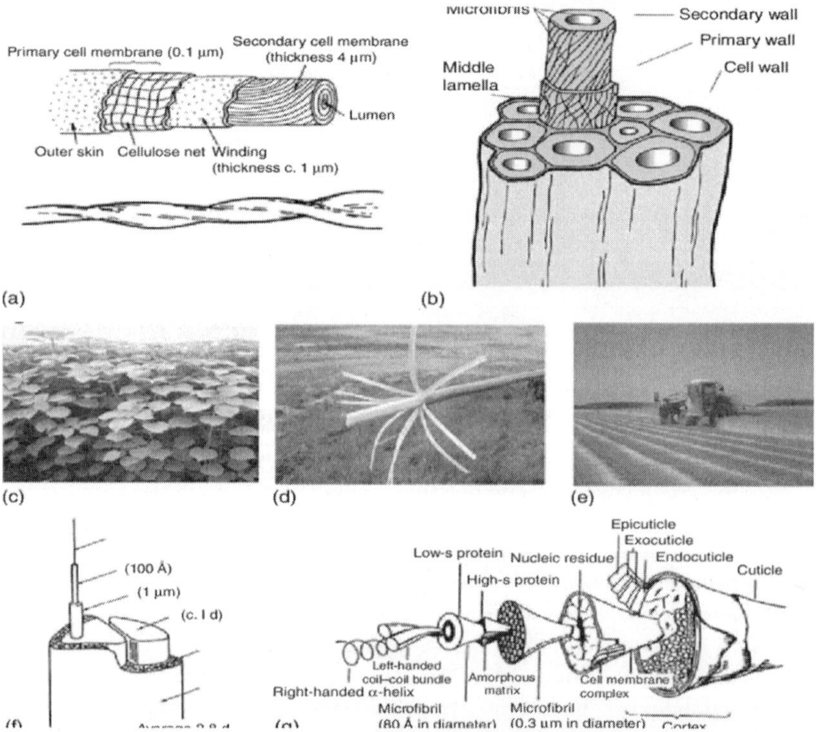

Figure 11 . Structural And Photographical Images of Natural and Animal Fibers. (A) Cotton. (B) Jute. (C) Ramie. (D) Hemp. (E) Flax. (F) Silk. (G) Wool (Hu, 2020)

Mineral-based fibers and animal-based fibers are rarely applied as reinforcing media in NFRPC compared to plant-based fibers due to their low availability and high price. Animal-based fibers consist of wool, feathers, silk, and hair [35]. In contrast, mineral fibers are derived from minerals, with asbestos and basalt being the most commonly found types [18][36].

These fibers are environmentally friendly and biodegradable, positively impacting the environment. Additionally, they offer greater stiffness and tensile strength than animal fibers, with properties comparable to synthetic glass fibers. Furthermore, animal fibers have lower hydrophobicity compared to plant fibers, are highly sensitive to certain alkalis, and are poor heat conductors. Reinforcing synthetic polymers with these fibers produces composites with excellent mechanical properties [4]. Several researchers have noted that plant-based fibers begin to degrade at an initial temperature of 240°C. Following this, cellulosic materials break down within the range of 260°C to 350°C. In contrast, hemicellulose decomposes between 200°C and 260°C, releasing a higher proportion of non-combustible gases. Meanwhile, the degradation of lignin, a component of plant fibers, occurs over a wider range, from 160°C to 400°C [37].

2.2 How to Prepare Natural Fiber-Reinforced

1. Natural Fiber Selection and Collection: The first step is to select suitable natural fibers. Sisal, jute, and coconut fibers are popular choices due to their strength and availability [21].
2. Fiber Treatment: Fiber treatment is a crucial step to improve compatibility with the cement matrix and long-term durability. Common methods include: a) Alkali Treatment: Research reports that immersion of fibers in alkaline solutions can make the fibers have better tensile strength compared to those without [38]. In bamboo fibers, 48 hours of soaking with NaOH solution gave high fiber strength [39]. b) Hornification: This process involves repeated wetting and drying cycles. Research found that fiber hornification due to its reduced hydrophilic properties [40]. c) Silaniation: Based on the research conducted, high interfacial shear strength was obtained after this treatment [41].
3. Preparation of Cement Matrix: The composition of the cement matrix should be optimized for NFRCC. Research recommends a water-cement ratio between 0.3-0.8 to ensure good workability and a strong bond with the fibers [42].
   - Mixing: The mixing process is key for even fiber distribution. The two-stage mixing method involves:

- Stage 1: Mixing of fibers with fine aggregates.
- Stage 2: Addition of cement and water. This method results in 20% more even fiber distribution compared to single-stage mixing [43].

4. Casting and Compaction: Special casting techniques may be required for NFRCC. Studies have found that the spray-up technique results in better fiber orientation, improving tensile strength better than conventional casting methods [44].
5. Care: Proper care is essential for NFRCC. Ardanuy et al. (2015) reported that 28 days of treatment in water resulted in higher compressive strength than air treatment [45].
6. Testing and Characterization: After preparation, the NFRCC must be tested to ensure its quality. Standard tests include:
   - Compressive and tensile strength (ASTM D695 and D8289)
   - Flexural strength (ASTM d790)
   - Ductility and crack resistance (ASTM E2472)

2.3 Effect of Fiber on Mechanical Properties of Various Cement Composites

The incorporation of fibers into cement-based matrices has been shown to significantly enhance the mechanical properties of the resulting composites. Fibers, which can be made from materials such as steel, glass, synthetic polymers, or natural materials, act as reinforcement within the cement matrix. They primarily contribute to improving the tensile strength, flexural strength, and toughness of the composite.

2.2.4 How to Improve the Mechanical Properties of Composites

The first thing that must be done to improve the mechanical properties of a composite is to give maximum effort from the fiber reinforcement side. Some of the things that determine the type and quality of a fiber are fiber volume fraction, type, and orientation. In addition, strength and stiffness determine the toughness of the composite. For example, carbon fibers are known for their high strength-to-weight ratio and stiffness, making them ideal for aerospace applications. Research shows that increasing the volume fraction of fiber applied in (CFRP) from 50% to 60% can result in a 20-30% increase in tensile strength and modulus [46]. Not only that, alignment of the fibers towards the load direction can also increase the strength of the composite in that area [47]. One strategy to improve the mechanical properties of composites is to focus on improving the interface between the fibers and the matrix. A strong interface plays a critical role in ensuring efficient load transfer from the matrix to the reinforcing fibers, which in turn greatly affects the overall performance of the composite. Treatments to the fiber surface, such as applying sizing to glass fibers or oxidizing carbon fibers, can significantly improve adhesion at the interface. In addition, modifications at the nanoscale, such as grafting nanoparticles onto the fiber surface or adding nanomaterials into the matrix, have also been shown to improve interfacial properties. For example, the addition of 0.5 wt% graphene oxide into a glass fiber-reinforced epoxy composite matrix has shown an increase in interfacial shear strength of up to 40% [48].

The choice and modification of the matrix material are crucial in defining the characteristics of a composite. Epoxy matrices, a popular thermoset option, are favored for their strength and effective fiber wetting, although they can be prone to brittleness. To address this, toughening agents like rubber particles or thermoplastic phases can be introduced to significantly boost fracture toughness. For example, adding 10 wt% of carboxyl-terminated butadiene acrylonitrile (CTBN) rubber to an epoxy matrix can double its fracture toughness while minimally impacting other mechanical properties. Additionally, advancements in high-performance thermoplastic matrices, such as polyetheretherketone (PEEK), have created new opportunities for composites with enhanced toughness and processability, offering significant advantages for aerospace and automotive applications.

2.4 Effect of Natural Fibers on Mechanical Properties of Composites

Natural fibers from plants like flax, hemp, jute, sisal, and kenaf have proven to be promising reinforcement materials for polymer composites. These fibers bring a range of advantages, including low density, high strength and stiffness, biodegradability, and cost-effectiveness. When these fibers are incorporated into polymer matrices, they can significantly boost the mechanical properties of the composites. For instance, research by Pickering et al. (2016) revealed that adding 30 wt% of aligned hemp fibers to a polypropylene matrix led to a 115% increase in tensile strength and a 165% rise in Young's modulus compared to the unreinforced polymer [50]. Similarly, Sanjay et al. (2019) found that epoxy composites reinforced with 40 wt% jute fibers exhibited a 78% improvement in flexural strength and a 328% increase in impact strength compared to pure epoxy [51].

The mechanical performance of natural fiber composites is influenced by several factors, including the type of fiber used, the amount of fiber present, its orientation, and the quality of the bond between the fiber and the matrix.

To boost the compatibility between hydrophilic natural fibers and hydrophobic polymer matrices, various fiber treatment techniques are employed. These treatments play a crucial role in enhancing the overall mechanical properties of the composites.

For example, alkali treatment is commonly applied to eliminate hemicellulose and lignin from fiber surfaces, thereby increasing the cellulose content and improving the fibers' bond with the matrix. Additionally, the use of coupling agents has been shown to enhance composite performance. Ferreira et al. (2019) demonstrated that adding 3% maleic anhydride grafted polypropylene (MAPP) to curaua fiber/polypropylene composites led to a 62% boost in tensile strength and a 51% increase in flexural strength compared to composites that were not treated.

While natural fiber composites offer numerous advantages, they face several challenges, including moisture absorption, thermal instability, and inconsistencies in fiber properties. These factors can affect the durability and dependability of the composites over time. To address these issues, researchers are exploring various approaches, such as combining different types of fibers and developing new surface treatments.

For example, Jawaid et al. (2013) investigated hybrid composites made from jute and oil palm fibers within an epoxy matrix, discovering that these hybrid composites exhibited superior mechanical properties compared to single-fiber-reinforced composites, thanks to a synergistic effect [53]. Additionally, recent advancements in nanotechnology have led to the creation of nanocellulose-reinforced composites, which demonstrate remarkable mechanical properties. Lee et al. (2014) reported that adding just 5 wt% of cellulose nanocrystals to a poly(vinyl alcohol) matrix resulted in a 200% increase in tensile strength and a 400% increase in Young's modulus [54].

References

[1] S. Choudhary, M. Kumar, V. Kumar, P. Saraswat, and M. Kumar, “Materials Today : Proceedings Advantages and applications of sisal fiber reinforced hybrid polymer composites in automobiles : A literature review,” *Mater. Today Proc.*, no. xxxx, 2023, https://doi.org/10.1016/j.matpr.2023.02.072

[2] Rodiawan, Rodiawan & Suhdi, Suhdi & Rosa, Firlya. (2017). Analisa Sifat-Sifat Serat Alam Sebagai Penguat Komposit Ditinjau Dari Kekuatan Mekanik. Turbo: Jurnal Program Studi Teknik Mesin. 5.

https://doi.org/10.24127/trb.v5i1.117

[3] Swaroop G Nagaraja, Martin Pletz, Clara Schuecker, Constitutive modeling of anisotropic plasticity with application to fiber-reinforced composites, International Journal of Solids and Structures,Volumes 180–181, 2019, Pages 84-96, ISSN 0020-7683, https://doi.org/10.1016/j.ijsolstr.2019.07.002

[4] Rajak, D.K.; Pagar, D.D.; Menezes, P.L.; Linul, E. Fiber-Reinforced Polymer Composites: Manufacturing, Properties, and Applications. *Polymers* **2019**, *11*, 1667. https://doi.org/10.3390/polym11101667

[5] M. Abdul, M. Zaki, A. Farouk, and M. Azab, "Results in Engineering An assessment of the processing parameters and application of fiber reinforced polymers (FRPs) in the petroleum and natural gas industries : A review," *Results Eng.*, vol. 18, no. April, p. 101091, 2023, https://doi.org/10.1016/j.rineng.2023.101091

[6] Jin Zhang, Gang Lin, Uday Vaidya, Hao Wang, Past, present and future prospective of global carbon fibre composite developments and applications, Composites Part B: Engineering, Volume 250, 2023, 110463, ISSN 1359-8368, https://doi.org/10.1016/j.compositesb.2022.110463

[7] Diniță, A.; Ripeanu, R.G.; Ilincă, C.N.; Cursaru, D.; Matei, D.; Naim, R.I.; Tănase, M.; Portoacă, A.I. Advancements in Fiber-Reinforced Polymer Composites: A Comprehensive Analysis. *Polymers* 2024, *16*, 2. https://doi.org/10.3390/polym16010002

[8] Suman Chhetri, Habiba Bougherara, A comprehensive review on surface modification of UHMWPE fiber and interfacial properties, Composites Part A: Applied Science and Manufacturing, Volume 140, 2021, 106146, ISSN 1359-835X, https://doi.org/10.1016/j.compositesa.2020.106146

[9] M. Yasir, A. Al, Z. Ullah, W. Ahmed, H. Arshad, and A. Ali, "Results in Engineering Natural fiber reinforced composites : Sustainable materials for emerging applications," *Results Eng.*, vol. 11, no. April, p. 100263, 2021, https://doi.org/10.1016/j.rineng.2021.100263

[10] Yang J, Ching YC, Chuah CH. Applications of Lignocellulosic Fibers and Lignin in Bioplastics: A Review. Polymers (Basel). 2019 Apr 28;11(5):751. doi: 10.3390/polym11050751. PMID: 31035331; PMCID: PMC6572173.]

[11] Jerachard Kaima, Itthichai Preechawuttipong, Robert Peyroux, Pawarut Jongchansitto, Tawanpon Kaima, Experimental investigation of alkaline treatment processes (NaOH, KOH and ash) on tensile

strength of the bamboo fiber bundle, Results in Engineering, Volume 18, 2023, 101186, ISSN 2590-1230, https://doi.org/10.1016/j.rineng.2023.101186

[12] M. Shahril, M.J.M. Ridzuan, M.S. Abdul Majid, A.M.N. Bariah, M.T.A. Rahman, P. Narayanasamy, Alkali treatment influence on cellulosic fiber from Furcraea foetida leaves as potential reinforcement of polymeric composites, Journal of Materials Research and Technology, Volume 19, 2022, Pages 2567-2583, ISSN 2238-7854, https://doi.org/10.1016/j.jmrt.2022.06.002

[13] Konakanchi, Anusha & Alla, Rama Krishna & Guduri, Vineeth. (2017). Silane coupling agents – Benevolent binders in composites. Trends in Biomaterials and Artificial Organs. 31. 108-113.

[15] Timmis, A.J., Hodzic, A., Koh, L. *et al.* Environmental impact assessment of aviation emission reduction through the implementation of composite materials. *Int J Life Cycle Assess* **20**, 233–243 (2015). https://doi.org/10.1007/s11367-014-0824-0\ https://doi.org/10.1016/j.compositesb.2022.110463

[16] Structures and M. B. Science, "Applications of Fibber Reinforced Polymer Composites (FRP) in Civil Engineering," vol. 03, no. 03, 2014.

[17] Gupta, Praveen Kumar & Raghunath, Shreeya & Venkatesh Prasanna, Deepali & Venkat, Priyadharsini & Shree, Vidhya & Chithananthan, Chandrananthi & Choudhary, Shreya & Surender, Krithika & Geetha, Keerthana. (2019). An Update on Overview of Cellulose, Its Structure and Applications. https://doi.org/10.5772/intechopen.84727

[18] Muhammad Yasir Khalid, Ans Al Rashid, Zia Ullah Arif, Waqas Ahmed, Hassan Arshad, Asad Ali Zaidi, Natural fibre reinforced composites: Sustainable materials for emerging applications, Results in Engineering, Volume 11, 2021, 100263, ISSN 2590-1230, https://doi.org/10.1016/j.rineng.2021.100263

[19] Ghazanlou SI, Ghazanlou SI, Ashraf W. Improvement in the physical and mechanical properties of the cement-based composite with the addition of nanostructured BN-Fe_3O_4 reinforcement. Sci Rep. 2021 Sep 29;11(1):19358. https://doi.org/10.1038/s41598-021-98800-4

[20] Duggal, Purva. (2022). Composites Market Size is Evaluated 86.56 Billion by 2021 and Estimated To Reach 150.28 billion by 2028.

[21] Issam Elfaleh, Fethi Abbassi, Mohamed Habibi, Furqan Ahmad, Mohamed Guedri, Mondher Nasri, Christian Garnier, A comprehensive review of natural fibers and their composites: An eco-friendly alternative to conventional materials, Results in Engineering, Volume 19, 2023, 101271, ISSN 2590-1230,

https://doi.org/10.1016/j.rineng.2023.101271
[22] Neha Uppal, Asokan Pappu, Vijaya Kumar Sorna Gowri, Vijay Kumar Thakur, Cellulosic fibres-based epoxy composites: From bioresources to a circular economy, Industrial Crops and Products, Volume 182, 2022, 114895, ISSN 0926-6690, https://doi.org/10.1016/j.indcrop.2022.114895
[23] Sabri, Habinata & Fajrin, Jauhar & Anshari, Buan. (2020). PERILAKU LENTUR KOMPOSIT POLIESTER YANG DIPERKUAT DENGAN LIMBAH SERAT RAMI DARI INDUSTRI TEKSTIL. 15. 3987-3992.
[24] G. Lubineau, A. Rahaman, A review of strategies for improving the degradation properties of laminated continuous-fiber/epoxy composites with carbon-based nanoreinforcements, Carbon, Volume 50, Issue 7, 2012, Pages 2377-2395, ISSN 0008-6223, https://doi.org/10.1016/j.carbon.2012.01.059
[25] Muhammad Yasir Khalid, Ans Al Rashid, Zia Ullah Arif, Muhammad Fahad Sheikh, Hassan Arshad, Muhammad Ali Nasir, Tensile strength evaluation of glass/jute fibers reinforced composites: An experimental and numerical approach, Results in Engineering, Volume 10, 2021, 100232, ISSN 2590-1230, https://doi.org/10.1016/j.rineng.2021.100232
[26] Kuan HTN, Tan MY, Shen Y, Yahya MohdY. Mechanical properties of particulate organic natural filler-reinforced polymer composite: A review. Composites and Advanced Materials. 2021;30. https://doi.org/10.1177/26349833211007502
[28] Ramli, Mahyuddin & Kwan, Wai Hoe & Abas, Noor. (2013). Strength and durability of coconut-fiber-reinforced concrete in aggressive environments. Construction and Building Materials. 38. 554–566. https://doi.org/10.1016/j.conbuildmat.2012.09.002
[29] Jawad Ahmad, Zhiguang Zhou, Ahmed Farouk Deifalla, Structural properties of concrete reinforced with bamboo fibers: a review, Journal of Materials Research and Technology, Volume 24, 2023, Pages 844-865, ISSN 2238-7854, https://doi.org/10.1016/j.jmrt.2023.03.038
[30] Chowdhury, I.R.; Pemberton, R.; Summerscales, J. Developments and Industrial Applications of Basalt Fibre reinforced Composite Materials. *J. Compos. Sci.* **2022**, *6*, 367. https://doi.org/10.3390/jcs612036
[31] Srinidhi, P & Gokulapriya, K & P., Parthiban. (2008). Experimental Study on Strength Enhancement of Concrete by using Magnetic Water and Coconut Fiber. https://doi.org/10.13140/RG.2.2.14164.76160

[32] Thyavihalli Girijappa Yashas Gowda , Mavinkere Rangappa Sanjay , Parameswaranpillai Jyotishkumar , Siengchin Suchart (2019) Natural Fibers as Sustainable and Renewable Resource for Development of Eco-Friendly Composites: A Comprehensive Review. 6. https://doi.org/10.3389/fmats.2019.00226

[33] A.V. Kiruthika, A review on physico-mechanical properties of bast fibre reinforced polymer composites, Journal of Building Engineering, Volume 9, 2017, Pages 91-99, ISSN 2352-7102, https://doi.org/10.1016/j.jobe.2016.12.003

[34] Hu, Jinlian & Jahid, Md Anwar & Kumar, Narayana & Venkatesan, Harun & Kumar, Bipin & Lu, Jing. (2020). Fundamentals of the Fibrous Materials. 1-36. https://doi.org/10.1002/9783527342587.ch1

[35] J. Robertson, C. Roux, Fibers: Overview, Editor(s): Jay A. Siegel, Pekka J. Saukko, Max M. Houck, Encyclopedia of Forensic Sciences (Second Edition), Academic Press, 2013, Pages 109-112, ISBN 9780123821669, https://doi.org/10.1016/B978-0-12-382165-2.00088-X

[36] Umran Dogan, Meral Dogan, Salih Emri, Erionite, Editor(s): Philip Wexler, Encyclopedia of Toxicology (Second Edition), Elsevier, 2005, Pages 237-241, ISBN 9780123694003, https://doi.org/10.1016/B0-12-369400-0/00380-X

[37] Jechan Lee, Ajit K. Sarmah, Eilhann E. Kwon, Chapter 1 - Production and Formation of Biochar, Editor(s): Yong Sik Ok, Daniel C.W. Tsang, Nanthi Bolan, J.M. Novak, Biochar from Biomass and Waste, Elsevier, 2019, Pages 3-18, ISBN 9780128117293, https://doi.org/10.1016/B978-0-12-811729-3.00001-7

[38] Jayabal, S. & Sathiyamurthy, S. & LOGANATHAN, K & Subramanian, Kalyana Sundaram. (2012). Effect of soaking time and concentration of NaOH solution on mechanical properties of coir–polyester composites. Bulletin of Materials Science. 35. https://doi.org/10.1007/s12034-012-0334-2

[39] Siew Choo Chin, Kong Fah Tee, Foo Sheng Tong, Huei Ruey Ong, Jolius Gimbun, Thermal and mechanical properties of bamboo fiber reinforced composites, Materials Today Communications, Volume 23, 2020, 100876, ISSN 2352-4928, https://doi.org/10.1016/j.mtcomm.2019.100876

[40] Valeria Laverde, Angie Marin, José M. Benjumea, Mauricio Rincón Ortiz, Use of vegetable fibers as reinforcements in cement-matrix composite materials: A review, Construction and Building Materials, Volume 340, 2022, 127729, ISSN 0950-0618, https://doi.org/10.1016/j.conbuildmat.2022.127729

[41] Adwani S, Elsubeihi E, Zebari A, Aljanahi M, Moharamzadeh K, Elbishari H. Effect of Different Silane Coupling Agents on the Bond Strength between Hydrogen Peroxide-Etched Epoxy-Based- Fiber-Reinforced Post and Composite Resin Core. Dent J (Basel). 2023 May 29;11(6):142. https://doi.org/10.3390/dj11060142
[42] Ayanlere, Samuel & Olalere, Ajamu & Samson O., Odeyemi & Ajayi, O.E. & Kareem, Mutiu. (2023). Effects of water-cement ratio on bond strength of concrete. Materials Today: Proceedings. 86. https://doi.org/10.1016/j.matpr.2023.04.686
[43] Li, Zhenxing & Wang, Xin & Zhu, Zhongguo & Wu, Zhishen. (2022). Effect of mixing methods on the dispersion of fibers in the gypsum matrix and performance improvement mechanism. Construction and Building Materials. 320. 126193. https://doi.org/10.1016/j.conbuildmat.2021.126193
[44] Kikuchi, Tetsuo & Fudauchi, Akira & Koshino, Tetsushi & Narita, Chieko & Endo, Atsushi & Takai, Yuka & Goto, Akihiko & Ohtani, Akio & Nakai, Asami & Hamada, Hiroyuki. (2013). Mechanical Property of CFRP by Carbon Spray Up Method. https://doi.org/10.1115/IMECE2013-64144
[45] Anwar, Ammar & Tariq, Hamza & Adil, Shaheryar & Iftikhar, Muhammad. (2022). Effect of curing techniques on compressive strength of concrete. World Journal of Advanced Research and Reviews. 16. 694-710. https://doi.org/10.30574/wjarr.2022.16.3.1379
[46] Fu, S.-Y & Lauke, B. & Mai, Y.-W. (2009). Science and Engineering of Short Fiber reinforced Polymer Composites. Science and Engineering of Short Fiber reinforced Polymer Composites. 1-338.
[47] Yentl Swolfs, Larissa Gorbatikh, Ignaas Verpoest, Fibre hybridisation in polymer composites: A review, omposites Part A: Applied Science and Manufacturing, Volume 67, 2014, Pages 181-200, ISSN 1359-835X, https://doi.org/10.1016/j.compositesa.2014.08.027
[48] Qin, W., Vautard, F., Drzal, L.T. and Yu, J. (2015) Mechanical and Electrical Properties of Carbon Fiber Composites with Incorporation of Graphene Nanoplatelets at the Fiber-Matrix Interphase. Composites Part B: Engineering, 69, 335-341. https://doi.org/10.1016/j.compositesb.2014.10.014
[49] Pickering, K.L., Efendy, M.G.A. and Le, T.M. (2016) A Review of Recent Developments in Natural Fibre Composites and Their Mechanical Performance. Composites Part A: Applied Science and Manufacturing, 83, 98-112. https://doi.org/10.1016/j.compositesa.2015.08.038

[50] Sanjay, M.R., Madhu, P., Jawaid, M., Senthamaraikannan, P., Senthil, S. and Pradeep, S. (2018) Characterization and Properties of Natural Fiber Polymer Composites: A Comprehensive Review. Journal of Cleaner Production, 172, 566-581. https://doi.org/10.1016/j.jclepro.2017.10.101

[51] M.M. Kabir, H. Wang, K.T. Lau, F. Cardona, Chemical treatments on plant-based natural fiber reinforced polymer composites: An overview, Composites Part B: Engineering, Volume 43, Issue 7, 2012, Pages 2883-2892, ISSN 1359-8368. https://doi.org/10.1016/j.compositesb.2012.04.053

[52] M. Jawaid, H.P.S. Abdul Khalil, Azman Hassan, Rudi Dungani, A. Hadiyane, Effect of jute fibre loading on tensile and dynamic mechanical properties of oil palm epoxy composites, Composites Part B: Engineering, Volume 45, Issue 1,2013, Pages 619-624, ISSN 1359-8368, https://doi.org/10.1016/j.compositesb.2012.04.068

[53] Koon-Yang Lee, Yvonne Aitomäki, Lars A. Berglund, Kristiina Oksman, Alexander Bismarck, On the use of nanocellulose as reinforcement in polymer matrix composites, Composites Science and Technology, Volume 105, 2014, Pages 15-27, ISSN 0266-3538, https://doi.org/10.1016/j.compscitech.2014.08.032

[54] Ferreira, F.V., Cividanes, L.S., Gouveia, R.F. and Lona, L.M.F. (2019), An overview on properties and applications of poly(butylene adipate-*co*-terephthalate)–PBAT based composites. Polym Eng Sci, 59: E7-E15. https://doi.org/10.1002/pen.24770

[55] Abubakar, M.N., Bello, T.K., Isa, M.T. *et al.* Effect of alkaline treatment on the physical and mechanical properties of *borassus flabellifer* leaf fiber. *Polym. Bull.* 80, 12577–12590 (2023). https://doi.org/10.1007/s00289-022-04666-5

[56] Sikiru Oluwarotimi Ismail, Emmanuel Akpan, Hom N. Dhakal, Review on natural plant fibres and their hybrid composites for structural applications: Recent trends and future perspectives, Composites Part C: Open Access, Volume 9, 2022, 100322, ISSN 2666-6820, https://doi.org/10.1016/j.jcomc.2022.100322